An Introduction to Medical Statistics

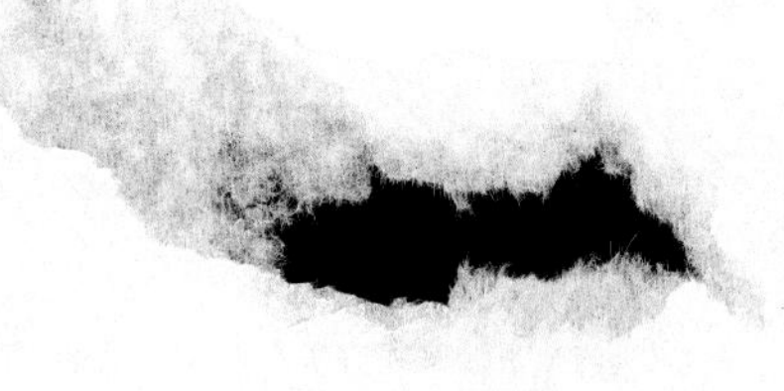

An Introduction to Medical Statistics

Second Edition

Martin Bland

Reader in Medical Statistics,
St George's Hospital Medical School, London

OXFORD
UNIVERSITY PRESS
Great Clarendon Street, Oxford OX2 6DP
Oxford University Press is a department of the University of Oxford.
It furthers the University's objective of excellence in research, scholarship,
and education by publishing worldwide in
Oxford New York
Athens Auckland Bangkok Bogotá Buenos Aires Calcutta
Cape Town Chennai Dar es Salaam Delhi Florence Hong Kong Istanbul
Karachi Kuala Lumpur Madrid Melbourne Mexico City Mumbai
Nairobi Paris São Paulo Singapore Taipei Tokyo Toronto Warsaw
with associated companies in Berlin Ibadan
Oxford is a registered trade mark of Oxford University Press
in the UK and in certain other countries
Published in the United States
by Oxford University Press Inc., New York

First published 1995
Reprinted 1996, 1999

A catalogue record for this book is available from the British Library
Library of Congress Cataloging in Publication Data
Data available
ISBN 0 19 262428 8
Printed in Great Britain
on acid-free paper by
Biddles Ltd, Guildford and King's Lynn

To Ernest and Phyllis Bland

Preface to the Second Edition

This is a textbook of statistics for medical students, doctors, medical researchers, and others concerned with medical data. The fundamental concepts of study design, data collection and data analysis are explained by illustration and example. For those who wish to go a little further in their understanding, some of the mathematical background to the techniques described is also given, largely as appendices to the chapters rather than in the main text.

The material covered includes all the statistical work that would be required for a course in medicine and for the examinations of most of the royal colleges. It includes the design of clinical trials and epidemiological studies, data collection, summarizing and presenting data, probability, the Binomial, Normal, Poisson, t and Chi-squared distributions, standard errors, confidence intervals, tests of significance, large sample and small sample comparisons of means, the use of transformations, regression and correlation, methods based on ranks, contingency tables, measurement error, reference ranges, mortality data, vital statistics, and the choice of the statistical method.

Since I wrote the first edition, powerful microcomputers have enabled medical researchers to carry out analyses easily which were previously very difficult; as a result the journals are filled with multiple logistic regression and similar techniques. I have added a chapter on these multifactorial methods, which includes multiple regression, logistic regression and Cox regression, and into which meta-analysis seemed to fit nicely. I have also added sections on other topics: serial data, analysis of variance and multiple comparisons, correlation using repeated observations, odds ratios, relative risks, goodness of fit tests, and the log-rank test. I have collected the material on sample size determination into a single chapter. Some of the old material has been deleted to make way for the new, and some exercises have been replaced.

The book is firmly grounded in medical data, particularly in medical research, and the interpretation of the results of calculations in their medical context is emphasized. Except for a few obviously invented numbers used to illustrate the mechanics of calculations, all the data in the examples and exercises are real, from my own research and statistical consultation or from the medical literature.

There are two kinds of exercise. Each chapter has a set of multiple choice questions of the 'true or false' type. Multiple choice questions can cover a large amount of material in a short time, so are a useful tool for revision. As MCQs are widely used in postgraduate examinations, these exercises should also be useful to those preparing for memberships. All the MCQs have solutions, with reference to an appropriate part of the text or a detailed explanation for most of the answers. Each chapter also has one long exercise. Although these usually involve calculation, I have tried to avoid merely slotting figures into formulae. These exercises include not only the application of statistical techniques, but also the interpretation of the results in the light of the source of the data.

I wish to thank many people who have contributed to the writing of this book. First, there are the many medical students, doctors, research workers, and nurses whom it has been my pleasure to teach, and from whom I have learned so much. Second, the book contains many examples drawn from research carried out with other statisticians, epidemiologists, and social scientists, particularly Douglas Altman, Ross Anderson, Mike Banks, Beulah Bewley and Walter Holland. These studies could not have been done without the assistance of Patsy Bailey, Bob Harris, Rebecca McNair, Janet Peacock, Swatee Patel and Virginia Pollard. Third, the clinicians and scientist with whom I have collaborated or who have come to me for statistical advice not only taught me about medical data but many of them have left me with data which are used here, including Naib Al-Saady, Thomas Bewley, Nigel Brown, Peter Fish, Caroline Flint, Nick Hall, Tessi Hanid, Michael Hutt, Riahd Jasrawi, Ian Johnston, Moses Kipembwa, Pam Luthra, Hugh Mather, Daram Maugdal, Douglas Maxwell, Charles Mutoka, Tim Northfield, Paul Richardson and Alberto Smith. I am particularly indebted to John Morgan, as Chapter 16 is partly based on his work. The manuscript was typed by Sue Nash, Sue Fisher, Susan Harding, Sheilah Skipp, and myself. An earlier draft of the book was read by David Jones, Douglas Altman, Robin Prescott, Klim McPherson and Stuart Pocock. I have corrected a number of errors from the first edition, and I am grateful to colleagues who have pointed them out to me, in particular to Daniel Heitjan. I am very grateful to Douglas Altman and Janet Peacock, who have read the new draft. Their comments have made this a better book than it would otherwise have been; the faults which remain are mine. Special thanks are due to my head of department, Ross Anderson, for all his support, and to the staff of Oxford University Press. Most of all I thank Pauline Bland for her unfailing confidence and encouragement.

The book has been reset by me using LaTeX, so any errors which remain are definitely my own. All the graphs have been redrawn using Stata.

London, August 1994 M.B.

Contents

1 Introduction 1

1.1 Statistics and medicine 1
1.2 Statistics and mathematics 2
1.3 Statistics and computing 3
1.4 The scope of this book 3

2 The design of experiments 5

2.1 Comparing treatments 5
2.2 Random allocation 7
2.3 Methods of allocation without random numbers 10
2.4 Volunteer bias 13
2.5 Intention to treat 14
2.6 Cross-over designs 15
2.7 Selection of subjects for clinical trials 16
2.8 Response bias and placebos 18
2.9 Assessment bias and double blind studies 19
2.10 Laboratory experiments 20
2.11 Experimental units 22
2.12 Further points about trial design 23
2M Multiple choice questions 1 to 6 23
2E Exercise: The 'Know Your Midwife' trial 25

3 Sampling and observational studies 26

3.1 Observational studies 26
3.2 Censuses 26
3.3 Sampling 27
3.4 Random sampling 28
3.5 Sampling in clinical studies 32
3.6 Sampling in epidemiological studies 33
3.7 Cohort studies 35
3.8 Case-control studies 36
3.9 Questionnaire bias in observational studies 40
3M Multiple choice questions 7 to 13 42
3E Exercise: *Campylobacter jejuni* infection 43

4 Summarizing data 46
4.1 Types of data 46
4.2 Frequency distributions 47
4.3 Histograms and other frequency graphs 49
4.4 Shapes of frequency distribution 53
4.5 Medians and quantiles 54
4.6 The mean 56
4.7 Variance 58
4.8 Standard deviation 60
4A Appendix: The divisor for the variance 61
4B Appendix: Formulae for the sum of squares 63
4M Multiple choice questions 14 to 19 64
4E Exercise: Mean and standard deviation 65

5 Presenting data 67
5.1 Rates and proportions 67
5.2 Significant figures 68
5.3 Presenting tables 71
5.4 Pie charts 72
5.5 Bar charts 73
5.6 Scatter diagrams 75
5.7 Line graphs and time series 76
5.8 Misleading graphs 77
5.9 Logarithmic scales 80
5A Appendix: Logarithms 81
5M Multiple choice questions 20 to 24 83
5E Exercise: Creating graphs 85

6 Probability 86
6.1 Probability 86
6.2 Properties of probability 87
6.3 Probability distributions and random variables 87
6.4 The Binomial distribution 89
6.5 Mean and variance 91
6.6 Properties of means and variances 92
6.7 The Poisson distribution 94
6A Appendix: Permutations and combinations 96
6B Appendix: Expected value of a sum of squares 96
6M Multiple choice questions 25 to 31 98
6E Exercise: Probability and the life table 99

7 The Normal distribution 101
7.1 Probability for continuous variables 101

7.2 The Normal distribution 104
7.3 Properties of the Normal distribution 107
7.4 Variables which follow a Normal distribution 111
7.5 The Normal plot 112
7A Appendix: Chi-squared, t, and F 114
7M Multiple choice questions 32 to 37 117
7E Exercise: A Normal plot 118

8 Estimation 119
8.1 Sampling distributions 119
8.2 Standard error of a sample mean 120
8.3 Confidence intervals 123
8.4 Standard error of a proportion 125
8.5 The difference between two means 126
8.6 Comparison of two proportions 127
8.7 Standard error of a sample standard deviation 130
8M Multiple choice questions 38 to 43 130
8E Exercise: Means of large samples 131

9 Significance tests 133
9.1 Testing a hypothesis 133
9.2 An example: the sign test 134
9.3 Principles of significance tests 136
9.4 Significance levels and types of error 136
9.5 One and two sided tests of significance 137
9.6 Significant, real and important 138
9.7 Comparing the means of large samples 139
9.8 Comparison of two proportions 141
9.9 The power of a test 143
9.10 Multiple significance tests 144
9M Multiple choice questions 44 to 49 148
9E Exercise: Crohn's disease and cornflakes 149

10 Comparing the means of small samples 152
10.1 The t distribution 152
10.2 The one sample t method 155
10.3 The means of two independent samples 159
10.4 The use of transformations 161
10.5 Deviations from the assumptions of t methods 165
10.6 What is a large sample? 166
10.7 Serial data 166
10.8 Comparing two variances by the F test 169
10.9 Comparing several means 170

10.10 Assumptions of the analysis of variance 173
10.11 Comparison of means after analysis of variance 175
10A Appendix: The ratio mean/standard error 176
10M Multiple choice questions 50 to 56 177
10E Exercise: The paired t method 179

11 Regression and correlation 180
11.1 Scatter diagrams 180
11.2 Regression 181
11.3 The method of least squares 182
11.4 The regression of X on Y 185
11.5 The standard error of the regression coefficient 186
11.6 Using the regression line for prediction 188
11.7 Analysis of residuals 190
11.8 Deviations from assumptions in regression 191
11.9 Correlation 192
11.10 Significance test for the correlation coefficient 195
11.11 Uses of the correlation coefficient 197
11.12 Using repeated observations 197
11A Appendix: The least squares estimates 199
11B Appendix: Variance about the regression line 200
11C Appendix: The standard error of b 201
11M Multiple choice questions 57 to 61 202
11E Exercise: Comparing two regression lines 203

12 Methods based on rank order 205
12.1 Non-parametric methods 205
12.2 The Mann Whitney U Test 206
12.3 The Wilcoxon matched pairs test 212
12.4 Spearman's rank correlation coefficient, ρ 215
12.5 Kendall's rank correlation coefficient, τ 217
12.6 Continuity corrections 220
12.7 Parametric or non-parametric methods? 222
12M Multiple choice questions 62 to 66 223
12E Exercise: Application of rank methods 224

13 The analysis of cross-tabulations 225
13.1 The chi-squared test for association 225
13.2 Tests for 2 by 2 tables 227
13.3 The chi-squared test for small samples 229
13.4 Fisher's exact test 231
13.5 Yates' continuity correction for the 2 by 2 table 234
13.6 The validity of Fisher's and Yates' methods 234

13.7 The odds ratio 235
13.8 The chi-squared test for trend 238
13.9 McNemar's test for matched samples 241
13.10 The chi-squared goodness of fit test 244
13A Appendix: Why the chi-squared test works 245
13B Appendix: The formula for Fisher's exact test 247
13C Appendix: Standard error for the log odds ratio 248
13M Multiple choice questions 67 to 74 249
13E Exercise: Admissions to hospital in a heatwave 252

14 Choosing the statistical method 254
14.1 Method oriented and problem oriented teaching 254
14.2 Types of data 254
14.3 Comparing two groups 255
14.4 One sample and paired samples 257
14.5 Relationship between two variables 258
14M Multiple choice questions 75 to 80 260
14E Exercise: Choosing a statistical method 262

15 Clinical measurement 265
15.1 Making measurements 265
15.2 Repeatability and measurement error 266
15.3 Comparing two methods of measurement 269
15.4 Sensitivity and specificity 273
15.5 Reference interval or normal range 276
15.6 Survival data 279
15.7 Computer aided diagnosis 286
15.M Multiple choice questions 81 to 86 288
15.E Exercise: A reference interval 289

16 Mortality statistics and population structure 291
16.1 Mortality rates 291
16.2 Age standardization using the direct method 293
16.3 Age standardization by the indirect method 293
16.4 Demographic life tables 296
16.5 Vital statistics 300
16.6 The population pyramid 300
16M Multiple choice questions 87 to 92 303
16E Exercise: Deaths from volatile substance abuse 304

17 Multifactorial methods 306
17.1 Multiple regression 306
17.2 Significance tests in multiple regression 309

17.3 Interaction in multiple regression 311
17.4 Polynomial regression 313
17.5 Assumptions of multiple regression 314
17.6 Qualitative predictor variables 315
17.7 Multi-way analysis of variance 316
17.8 Logistic regression 317
17.9 Survival data using Cox regression 320
17.10 Stepwise regression 322
17.11 Meta-analysis: data from several studies 323
17.M Multiple choice questions 93 to 97 326
17.E Exercise: A multiple regression analysis 329

18 Determination of sample size 331
18.1 Estimation of a population mean 331
18.2 Estimation of a population proportion 332
18.3 Sample size for significance tests 332
18.4 Comparison of two means 334
18.5 Comparison of two proportions 337
18.6 Detecting a correlation 338
18.7 Accuracy of the estimated sample size 340
18.M Multiple choice questions 98 to 100 341
18.E Exercise: Estimation of sample sizes 341

19 Solutions to exercises 342

References 376

Index 384

1
Introduction

1.1 Statistics and medicine

The term 'statistics' means 'numerical data'. Statistics as an academic study is the science of assembling and interpreting numerical data. In clinical medicine, statistical methods are used to determine the accuracy of measurements, to compare measurement techniques, to assess diagnostic tests, to determine normal values, and to monitor patients. In the administration of medical services we are concerned with such things as bed use and perinatal mortality rates. It is in medical research, however, that statistics is most intimately involved, and it is with this area of application that this book is principally concerned. This is not to say that the book is addressed to the present or future researcher only. The medical profession is fond of research, but many doctors never try it. What nearly all doctors do is use the results of medical research, whether they are prescribing a new drug or advising a patient to give up smoking. In order to read the results of the enormous amount of research which pours into the medical journals, all doctors should have some understanding of ways in which studies are designed, and data are collected, analysed and interpreted. That is what this book is about.

In the past thirty years medical research has become deeply involved with the techniques of statistical inference. The work published in medical journals is full of statistical jargon and the results of statistical calculations. This acceptance of statistics, though gratifying to the medical statistician, may even have gone too far. More than once I have told a colleague that he did not need me to prove that his difference existed, as anyone could see it, only to be told in turn that without the magic of the P value he could not have his paper published.

Statistics has not always been so popular with the medical profession. Statistical methods were first used in medical research in the 19th century by workers such as Pierre-Charles-Alexandre Louis, William Farr and John Snow. Snow's studies of the modes of communication of cholera, for example, made use of epidemiological techniques upon which we have still made little improvement. Despite the work of these pioneers, however, statistical methods did not become widely used in clinical medicine until the middle of the twentieth century. It was then that the methods of randomized experimentation and statistical analysis based on sampling theory, which had been developed by Fisher and others, were introduced into medical research, notably by Bradford Hill. It rapidly became apparent that research in medicine raised many new problems in both design and analysis, and much work has been done since towards solving these by clinicians, statisticians and epidemiologists.

Although considerable progress has been made in such fields as the design of clinical trials, there remains much to be done in developing research methodology in medicine. It seems likely that this will always be so, for every research project is something new, something which has never been done before. Under these circumstances we make mistakes. No piece of research can be perfect and there will always be something which, with hindsight, we would have changed. Furthermore, it is often from the flaws in a study that we can learn most about research methods. For this reason, the work of several researchers is described in this book to illustrate the problems into which their designs or analyses led them. I do not wish to imply that these people were any more prone to error than the rest of the human race, or that their work was not a valuable and serious undertaking. Rather I want to learn from their experience of attempting something extremely difficult, trying to extend our knowledge, so that researchers and consumers of research may avoid these particular pitfalls in the future.

1.2 Statistics and mathematics

Many people are discouraged from the study of statistics by a fear of being overwhelmed by mathematics. It is true that many professional statisticians are also mathematicians, but not all are, and there are many very able appliers of statistics to their own fields. It is possible, though perhaps not very useful, to study statistics simply as a part of mathematics, with no concern for its application at all. Statistics may also be discussed without appearing to use any mathematics at all (e.g. Huff 1954).

The aspects of statistics described in this book can be understood and applied with the use of simple algebra. Only the algebra which is essential for explaining the most important concepts is given in the main text. This means that several of the theoretical results used are stated without a discussion of their mathematical basis. This is done when the derivation of the result would not aid much in understanding the application. For many readers the reasoning behind these results is not of great interest. For the reader who does not wish to take these results on trust, several chapters have appendices in which simple mathematical proofs are given. These appendices are designed to help increase the understanding of the more mathematically inclined reader and to be omitted by those who find that the mathematics serves only to confuse.

1.3 Statistics and computing

Practical statistics has always involved large amounts of calculation. When the methods of statistical inference were being developed in the first half of the twentieth century, calculations were done using pencil, paper, tables, slide rules and with luck a very expensive mechanical adding machine. Older books on statistics spend much time on the details of carrying out calculations and any reference to a 'computer' means a person who computes, not an electronic device. The development of the digital computer has brought changes to statistics as to many other fields. Calculations can be done quickly, easily and, we hope, accurately with a range of machines from pocket calculators with built-in statistical functions to powerful computers analysing data on many thousands of subjects. There is therefore no need to consider the problems of manual calculation in detail. The important thing is to know why particular calculations should be done and what the results of these calculations actually mean. Indeed, the danger in the computer age is not so much that people carry out complex calculations wrongly, but that they apply very complicated statistical methods without knowing why or what the computer output means. More than once I have been approached by a researcher bearing a two inch thick computer printout, and asking what it all means. Sadly, too often, the answer is that another tree has died in vain.

Computers are a great benefit to statistics in that calculations which would once have taken days can now be done in minutes, and statisticians use them a lot. Most of the calculations in this book were done using a computer and the graphs were produced with one. But the widespread availability of computers means that more calculations are being done, and being published, than ever before, and the chance of inappropriate statistical methods being applied may actually have increased. This misuse arises partly because people regard their data analysis problems as computing problems, not statistical ones, and seek advice from computer experts rather than statisticians. They often get good advice on how to do it, but rather poor advice about what to do, why to do it and how to interpret the results afterwards. It is therefore more important than ever that doctors, the consumers of research, understand something about the uses and limitations of statistical techniques.

1.4 The scope of this book

This book is intended as an introduction to some of the statistical ideas important to medicine. It does not tell you all you need to know to do medical research. Once you have understood the concepts discussed here, it is much easier to learn about the techniques of statistical analysis required to answer any particular question. There are several excellent standard works which describe the solutions to problems in the analysis of data (Armitage and Berry 1987, Snedecor and Cochran 1980, Altman 1991) and also more specialized books to which reference will be made where required.

What I hope the book will do is to give enough understanding of the statistical ideas commonly used in medicine to enable the doctor to read the medical literature competently and critically. It covers enough material for an undergraduate course in Statistics for medical students. At the time of writing, as far as can be established, it covers the material required to answer statistical questions set in the examinations of most of the Royal Colleges, except for the MRCPsych.

When working through a textbook, it is useful to be able to check your understanding of the material covered. Like most such books, this one has exercises at the end of each chapter, but to ease the tedium most of these are of the multiple choice type. There is also one long exercise, usually involving calculations, for each chapter. In keeping with the computer age, where laborious calculation would be necessary intermediate results are given to avoid this. Thus the exercises can be completed quite quickly and the reader is advised to try them. Solutions are given at the end of the book, in full for the long exercises and as brief notes with references to the relevant sections in the text for MCQs. Readers who would like more exercises are recommended to Osborn (1979).

Finally, a question many students of medicine ask as they struggle with statistics: is it worth it? As Altman (1982) has argued, bad statistics leads to bad research and bad research is unethical. Not only may it give misleading results, which can result in good therapies being abandoned and bad ones adopted, but it means that patients may have been exposed to potentially harmful new treatments for no good reason. Medicine is a rapidly changing field. In ten years' time, many of the therapies currently prescribed and many of our ideas about the causes and prevention of disease will be obsolete. They will be replaced by new therapies and new theories, supported by research studies and data of the kind described in this book, and probably presenting many of the same problems in interpretation. The doctor will be expected to decide for her- or himself what to prescribe or advise based on these studies. So a knowledge of medical statistics is one of the most useful things any doctor could acquire during her or his training.

2

The design of experiments

2.1 Comparing treatments

There are two broad types of study in medical research: observational and experimental. In observational studies, aspects of an existing situation are observed, as in a survey or a clinical case study. We then try to interpret our data to give an explanation of how the observed state of affairs has come about. In experimental studies, we do something, such as giving a drug, so that we can observe the result of our action. This chapter is concerned with the way statistical thinking is involved in the design of experiments. In particular, it deals with comparative experiments where we wish to study the difference between the effects of two or more treatments. These experiments may be carried out in the laboratory on animals or human volunteers, in the hospital or community on human patients, or, for trials of preventive interventions, on currently healthy people. We call the trials of treatments on human subjects **clinical trials**. The general principles of experimental design are the same, although there are special precautions which must be taken when experimenting with human subjects. The experiments whose results most concern clinicians are clinical trials, so the discussion will deal mainly with them. Suppose we want to know whether a new treatment is more effective than the present standard treatment. We could approach this in a number of ways:

(a) We could compare the results of the new treatment on new patients with records of previous results using the old treatment. This is seldom convincing, because there may be many differences between the patients who received the old treatment and the patients who will receive the new. As time passes, the general population from which patients come may become healthier, standards of ancillary treatment and nursing care may improve, or the social mix in the catchment area of the hospital may change. The nature of the disease itself may change. All these factors may produce changes in the patients' apparent response to treatment. For example, Christie (1979) showed this by studying the survival of stroke patients in 1978, after the introduction of a C-T head scanner, with that of patients treated in 1974, before the introduction of the scanner. He took

Table 2.1. Analysis of the difference in survival for matched pairs of stroke patients (Christie 1979)

	C-T scan in 1978		No C-T scan in 1978	
Pairs with 1978 better than 1974	9	(31%)	34	(38%)
Pairs with same outcome	18	(62%)	38	(43%)
Pairs with 1978 worse than 1974	2	(7%)	17	(19%)

the records of a group of patients treated in 1978, who received a C-T scan, and matched each of them with a patient treated in 1974 of the same age, diagnosis and level of consciousness on admission. As the first column of Table 2.1 shows, patients in 1978 clearly tended to have better survival than similar patients in 1974. The scanned 1978 patient did better than the unscanned 1974 patient in 31% of pairs, whereas the unscanned 1974 patient did better that the scanned 1978 patient in only 7% of pairs. However, he also compared the survival of patients in 1978 who did not receive a C-T scan with matched patients in 1974. These patients too showed a marked improvement in survival from 1974 to 1978 (Table 2.1). The 1978 patient did better in 38% of pairs and the 1974 patients in only 19% of pairs. There was a general improvement in outcome over a fairly short period of time. If we did not have the data on the unscanned patients from 1978 we might be tempted to interpret these data as evidence for the effectiveness of the C-T scanner. Historical controls like this are seldom very convincing, and usually favour the new treatment. We need to compare the old and new treatments concurrently.

(b) We could ask people to volunteer for the new treatment and give the standard treatment to those who do not volunteer. The difficulty here is that people who volunteer and people who do not volunteer are likely to be different in many ways apart from the treatments we give them. We will consider an example of the effects of volunteer bias in §2.4.

(c) We can allocate patients to the new treatment or the standard treatment and observe the outcome. The way in which patients are allocated to treatments can influence the results enormously, as the following example (Hill 1962) shows. Between 1927 and 1944 a series of trials of BCG vaccine were carried out in New York (Levine and Sackett 1946). Children from families where there was a case of tuberculosis were allocated to a vaccination group and given BCG vaccine, or to a control group who were not vaccinated. Between 1927 and 1932 physicians vaccinated half the children, the choice of which children to vaccinate being left to them. There was a clear advantage in survival for the BCG group (Table 2.2). However, there was also a clear tendency for the physician to vaccinate the children of more cooperative parents, and to leave those of less cooperative parents as controls. From 1933 allocation to treatment or control was done centrally,

Table 2.2. Results of studies of BCG vaccine in New York City (Hill 1962)

Period of trial	No. of children	No. of deaths from TB	Death rate	Average no. of visits to clinic during 1st year of follow-up	Proportion of parents giving good co-operation as judged by visiting nurses
1927–32 Selection made by physician:					
BCG group	445	3	0.67%	3.6	43%
Control group	545	18	3.30%	1.7	24%
1933–44 Alternate selection carried out centrally:					
BCG group	566	8	1.41%	2.8	40%
Control group	528	8	1.52%	2.4	34%

alternate children being assigned to control and vaccine. The difference in degree of cooperation between the parents of the two groups of children disappeared, and so did the difference in mortality. Note that these were a special group of children, from families where there was tuberculosis. In large trials using children drawn from the general population, BCG was shown to be effective in greatly reducing deaths from tuberculosis (Hart and Sutherland 1977)

Different methods of allocation to treatment can produce different results. This is because the method of allocation may not produce groups of subjects which are comparable, similar in every respect except the treatment. We need a method of allocation to treatments in which the characteristics of subjects will not affect their chance of being put into any particular group. This can be done using random allocation.

2.2 Random allocation

If we want to decide which of two people receive an advantage, in such a way that each has an equal chance of receiving it, we can use a simple, widely accepted method. We toss a coin. This is used to decide the way football matches begin, for example, and all appear to agree that it is fair. So if we want to decide which of two subjects should receive a vaccine, we can toss a coin. Heads and the first subject receives the vaccine, tails and the second receives it. If we do this for each pair of subjects we build up two groups which have been assembled without any characteristics of the subjects themselves being involved in the allocation. The only differences between the groups will be those due to chance. As we shall see later (Chapters 8 and 9), statistical methods enable us to measure the likely effects of chance. Any difference between the groups which is larger than this should be due to the treatment, since there will be no other differences between the groups. This method of dividing subjects into groups is called

Table 2.3. 1,000 random digits

	Column									
Row	1-4	5-8	9-12	13-16	17-20	21-24	25-28	29-32	33-36	37-40
1	36 45	88 31	28 73	59 43	46 32	00 32	67 15	32 49	54 55	75 17
2	90 51	40 66	18 46	95 54	65 89	16 80	95 33	15 88	18 60	56 46
3	98 41	90 22	48 37	80 31	91 39	33 80	40 82	38 26	20 39	71 82
4	55 25	71 27	14 68	64 04	99 24	82 30	73 43	92 68	18 99	47 54
5	02 99	10 75	77 21	88 55	79 97	70 32	59 87	75 35	18 34	62 53
6	79 85	55 66	63 84	08 63	04 00	18 34	53 94	58 01	55 05	90 99
7	33 53	95 28	06 81	34 95	13 93	37 16	95 06	15 91	89 99	37 16
8	74 75	13 13	22 16	37 76	15 57	42 38	96 23	90 24	58 26	71 46
9	06 66	30 43	00 66	32 60	36 60	46 05	17 31	66 80	91 01	62 35
10	92 83	31 60	87 30	76 83	17 85	31 48	13 23	17 32	68 14	84 96
11	61 21	31 49	98 29	77 70	72 11	35 23	69 47	14 27	14 74	52 35
12	27 82	01 01	74 41	38 77	53 68	53 26	55 16	35 66	31 87	82 09
13	61 05	50 10	94 85	86 32	10 72	95 67	88 21	72 09	48 73	03 97
14	11 57	85 67	94 91	49 48	35 49	39 41	80 17	54 45	23 66	82 60
15	15 16	08 90	92 86	13 32	26 01	20 02	72 45	94 74	97 19	99 46
16	22 09	29 66	15 44	76 74	94 92	48 13	75 85	81 28	95 41	36 30
17	69 13	53 55	35 87	43 23	83 32	79 40	92 20	83 76	82 61	24 20
18	08 29	79 37	00 33	35 34	86 55	10 91	18 86	43 50	67 79	33 58
19	37 29	99 85	55 63	32 66	71 98	85 20	31 93	63 91	77 21	99 62
20	65 11	14 04	88 86	28 92	04 03	42 99	87 08	20 55	30 53	82 24
21	66 22	81 58	30 80	21 10	15 53	26 90	33 77	51 19	17 49	27 14
22	37 21	77 13	69 31	20 22	67 13	46 29	75 32	69 79	39 23	32 43
23	51 43	09 72	68 38	05 77	14 62	89 07	37 89	25 30	92 09	06 92
24	31 59	37 83	92 55	15 31	21 24	03 93	35 97	84 61	96 85	45 51
25	79 05	43 69	52 93	00 77	44 82	91 65	11 71	25 37	89 13	63 87

random allocation or **randomization.**

Several methods of randomizing have been in use for centuries, including coins, dice, cards, lots, and spinning wheels. Some of the theory of probability which we shall use later to compare randomized groups was first developed as an aid to gambling. For large randomizations we use a different, non-physical randomizing method: random number tables. Table 2.3 provides an example, a table of 1 000 random digits. These are more properly called **pseudo-random numbers**, as they are generated by a mathematical process. They are available in tables (Kendall and Babington Smith 1971) or can be produced by computer and some calculators. We can use tables of random numbers in several ways to achieve random allocation. For example, let us randomly allocate 20 subjects to two groups, which I shall label A and B. We choose a random starting point in the table, using one of the physical methods described above. (I used decimal dice. These are 20-sided dice, numbered 0 to 9 twice, which fit our number system more conveniently than the traditional cube. Two such dice give a random number between 1 and 100, counting '0,0' as 100.) The random starting point was row 22, column 20, and the first twenty digits

Table 2.4. Allocation of 20 subjects to two groups

subject	digit	group	subject	digit	group
1	3	A	11	9	A
2	4	B	12	7	A
3	6	B	13	9	A
4	2	B	14	3	A
5	9	A	15	9	A
6	7	A	16	2	B
7	5	A	17	3	A
8	3	A	18	3	A
9	2	B	19	2	B
10	6	B	20	4	B

were 3, 4, 6, 2, 9, 7, 5, 3, 2, 6, 9, 7, 9, 3, 9, 2, 3, 3, 2 and 4. We now allocate subjects corresponding to odd digits to group A and those corresponding to even digits to B. The first digit, 3, is odd, so the first subject goes into group A. The second digit, 4, is even, so the second subject goes into group B, and so on. We get the allocation shown in Table 2.4. We could allocate into three groups by assigning to A if the digit is 1,2, or 3, to B if 4,5, or 6, and to C if 7,8, or 9, ignoring 0. There are many possibilities.

The system described above gave us unequal numbers in the two groups, 12 in A and 8 in B. We sometimes want the groups to be of equal size. One way to do this would be to proceed as above until either A or B has 10 subjects in it, all the remaining subjects going into the other groups. This is satisfactory in that each subject has an equal chance of being allocated to A or B, but it has a disadvantage. There is a tendency for the last few subjects all to have the same treatment. This characteristic sometimes worries researchers, who feel that the randomization is not quite right. In statistical terms the possible allocations are not equally likely. If we use this method for the random allocation described above, the 10th subject in group A would be reached at subject 15 and the last five subjects would all be in group B. We can ensure that all randomizations are equally likely by using the table of random numbers in a different way. For example, we can use the table to draw a random sample of 10 subjects from 20, as described in §3.4. These would form group A, and the remaining 10 group B. Another way is to put our subjects into small equal-sized groups, called **blocks**, and within each block to allocate equal numbers to A and B. This gives approximately equal numbers on the two treatments.

The use of random numbers and the generation of the random numbers themselves are simple mathematical operations well suited to the computers which are now readily available to researchers. It is very easy to program a computer to carry out random allocation, and once a program is available it can be used over and over again for further experiments.

The trial carried out by the Medical Research Council (MRC 1948) to

Table 2.5. Condition of patients on admission to trial of streptomycin (MRC 1948)

		Group	
		S	C
General condition	Good	8	8
	Fair	17	20
	Poor	30	24
Max. evening temperature in first week (°F)	98–98.9	4	4
	99–99.9	13	12
	100–100.9	15	17
	101+	24	19
Sedimentation rate	0–10	0	0
	11–20	3	2
	21–50	16	20
	51+	36	29

test the efficacy of streptomycin for the treatment of pulmonary tuberculosis is generally considered to have been the first randomized experiment in medicine. In this study the target population was patients with acute progressive bilateral pulmonary tuberculosis, aged 15–30 years. All cases were bacteriologically proved and were considered unsuitable for other treatments then available. The trial took place in three centres and allocation was by a series of random numbers, drawn up for each sex at each centre. The streptomycin group contained 55 patients and the control group 52 cases. The condition of the patients on admission is shown in Table 2.5. The frequency distributions of temperature and sedimentation rate were similar for the two groups; if anything the treated (S) group were slightly worse. However, this difference is no greater than could have arisen by chance, which, of course, is how it arose. The two groups are certain to be slightly different in some characteristics, especially with a fairly small sample, and we can take account of this in the analysis (Chapter 17).

After six months, 93% of the S group survived, compared to 73% of the control group. There was a clear advantage to the streptomycin group. The relationship of survival to initial condition is shown in Table 2.6. Survival was more likely for patients with lower temperatures, but the difference in survival between the S and C groups is clearly present within each temperature category where deaths occurred.

2.3 Methods of allocation without random numbers

In the second stage of the New York studies of BCG vaccine, the children were allocated to treatment or control alternately. Researchers often ask why this method cannot be used instead of randomization, arguing that the order in which patients arrive is random, so the groups thus formed

Table 2.6. Survival at six months in the MRC streptomycin trial, stratified by initial condition (MRC 1948)

Maximum evening temperature during first observation week	Outcome	Group	
		Streptomycin group	Control group
98–98.9°F	Alive	3	4
	Dead	0	0
99–99.9°F	Alive	13	11
	Dead	0	1
100–100.9°F	Alive	15	12
	Dead	0	5
101°F and above	Alive	20	11
	Dead	4	8

will be comparable. First, although the patients may appear to be in a random order, there is no guarantee that this is the case. We could never be sure that the groups are comparable. Second, this method is very susceptible to mistakes, or even to cheating in the patients' perceived interest. The experimenter knows what treatment the subject will receive before the subject is admitted to the trial. This knowledge may influence the decision to admit the subject, and so lead to biased groups. For example, an experimenter might be more prepared to admit a frail patient if the patient will be on the control treatment than if the patient would be exposed to the risk of the new treatment. This objection applies to using the last digit of the hospital number for allocation.

There are several examples reported in the literature of alterations to treatment allocations. Holten (1951) reported a trial of anticoagulant therapy for patients with coronary thrombosis. Patients who presented on even dates were to be treated and patients arriving on odd dates were to form the control group. The author reports that some of the clinicians involved found it 'difficult to remember' the criterion for allocation. Overall the treated patients did better than the controls (Table 2.7). Curiously, the controls on the even dates (wrongly allocated) did considerably better than control patients on the odd dates (correctly allocated) and even managed to do marginally better than those who received the treatment. The best outcome, treated or not, was for those who were incorrectly allocated. Allocation in this trial appears to have been rather selective.

Other methods of allocation set out to be random but can fall into this sort of difficulty. For example, we could use physical mixing to achieve randomization. This is quite difficult to do. As an experiment, take a deck of cards and order them in suits from ace of clubs to king of spades. Now shuffle them in the usual way and examine them. You will probably see many runs of several cards which remain together in order. Cards must be

Table 2.7. Outcome of a clinical trial using systematic allocation, with errors in allocation (Holten 1951)

	Even dates				Odd dates			
Outcome	Treated		Control		Treated		Control	
Survived	125		39		10		125	
Died	39	(25%)	11	(22%)	0	(0%)	81	(36%)
Total	164		50		10		206	

shuffled very thoroughly indeed before the ordering ceases to be apparent. The physical randomization method can be applied to an experiment by marking equal numbers on slips of paper with the names of the treatments, sealing them into envelopes and shuffling them. The treatment for a subject is decided by withdrawing an envelope. This method was used in another study of anticoagulant therapy by Carleton *et al.* (1960). These authors reported that in the latter stages of the trial some of the clinicians involved had attempted to read the contents of the envelopes by holding them up to the light, in order to allocate patients to their own preferred treatment.

Interfering with the randomization can actually be built into the allocation procedure, with equally disastrous results. In the Lanarkshire Milk Experiment, discussed by Student (1931), 10 000 school children received three quarters of a pint of milk per day and 10 000 children acted as controls. The children were weighed and measured at the beginning and end of the six month experiment. The object was to see whether the milk improved the growth of children. The allocation to the 'milk' or control group was done as follows:

> The teachers selected the two classes of pupils, those getting milk and those acting as controls, in two different ways. In certain cases they selected them by ballot and in others on an alphabetical system. In any particular school where there was any group to which these methods had given an undue proportion of well-fed or ill-nourished children, others were substituted to obtain a more level selection.

The result of this was that the control group had a markedly greater average height and weight than the milk group. Student interpreted this as follows:

> Presumably this discrimination in height and weight was not made deliberately, but it would seem probable that the teachers, swayed by the very human feeling that the poorer children needed the milk more than the comparatively well to do, must have unconsciously made too large a substitution for the ill-nourished among the (milk group) and too few among the controls and that this unconscious selection affected secondarily, both measurements.

Whether the bias was conscious or not, it spoiled the experiment, despite being from the best possible motives.

There is one non-random method which can be used successfully in clinical trials: minimization. In this method, new subjects are allocated to treatments so as to make the treatment groups as similar as possible in terms of the important prognostic factors. It is beyond the scope of this book, but see Pocock (1983) for a description.

2.4 Volunteer bias

One of the most interesting trials ever done was the field trial of Salk poliomyelitis vaccine carried out in 1954 in the USA (Meier 1977). This was carried out using two different designs simultaneously, due to a dispute about the correct method. In some districts, second grade school-children were invited to participate in the trial, and randomly allocated to receive vaccine or an inert saline injection. In other districts, all second grade children were offered vaccination and the first and third grade left unvaccinated as controls. The argument against this 'observed control' approach was that the groups may not be comparable, whereas the argument against the randomized control method was that the saline injection could provoke paralysis in infected children. The results are shown in Table 2.8. In the randomized control areas the vaccinated group clearly experienced far less polio than the control group. Since these were randomly allocated, the only difference between them should be the treatment, which is clearly preferable to saline. However, the control group also had more polio than those who had refused to participate in the trial. The difference between the control and not inoculated group is both in treatment (saline injection) and selection; they are self-selected as volunteers and refusers. The observed control areas enable us to distinguish between these two factors. The polio rates in the vaccinated children are very similar in both parts of the study, as are the rates in the not inoculated second grade children. It is the two control groups which differ. These were selected in different ways: in the randomized control areas they were volunteers, whereas in the observed controls areas they were everybody eligible, both potential volunteers and potential refusers. Now suppose that the vaccine were saline instead, and that the randomized vaccinated children had the same polio experience as those receiving saline. We would expect $200\,745 \times 57/100\,000 = 114$ cases. The total number of cases in the randomized areas would be $114 + 115 + 121 = 350$ and the rate per 100 000 would 47. This compares very closely with the rate of 46 in the observed control first and third grade group. Thus it seems that the principal difference between the saline control group of volunteers and the not inoculated group of refusers is selection, not treatment.

There is a simple explanation of this. Polio is a viral disease transmitted by the faecal-oral route. Before the development of vaccine almost everyone in the population was exposed to it at some time, usually in childhood. In the majority of cases, paralysis does not result and immunity is

Table 2.8. Result of the field trial of Salk poliomyelitis vaccine (Meier 1977)

Study group	Number in group	Paralytic Polio	
		Number of cases	Rate per 100 000
Randomized control:			
Vaccinated	200 745	33	16
Control	201 229	115	57
Not inoculated	338 778	121	36
Observed control:			
Vaccinated 2nd grade	221 998	38	17
Control 1st and 3rd grade	725 173	330	46
Unvaccinated 2nd grade	123 605	43	35

conferred without the child being aware of having been exposed to polio. In a small minority of cases, about one in 200, paralysis and occasionally death occurs and a diagnosis of polio is made. The older the exposed individual is, the greater the chance of paralysis developing. Hence, children who are protected from infection by high standards of hygiene are likely to be older when they are first exposed to polio than those children from homes with low standards of hygiene, and thus more likely to develop the clinical disease. There are many factors which may influence parents in their decision as to whether to volunteer or refuse their child for a vaccine trial. These may include education, personal experience, current illness, and others, but certainly include interest in health and hygiene. Thus in this trial the high risk children tended to be volunteered and the low risk children tended to be refused. The higher risk volunteer control children experienced 57 cases of polio per 100 000, compared to 36/100 000 among the lower risk refusers.

In most diseases, the effect of volunteer bias is opposite to this. Poor conditions are related both to refusal to participate and to high risk, whereas volunteers tend to be low risk. The effect of volunteer bias is then to produce an apparent difference in favour of the treatment. We can see that comparisons between volunteers and other groups can never be reliable indicators of treatment effects.

2.5 Intention to treat

In the observed control areas of the Salk trial (Table 2.8), quite apart from the non-random age difference, the vaccinated and control groups are not comparable. However, it is possible to make a reasonable comparison in this study by comparing all second grade children, both vaccinated and refused, to the control group. The rate in the second grade children is 23 per 100,000, which is less than the rate of 46 in the control group, demonstrating the effectiveness of the vaccine. The 'treatment' which we are evaluating is not vaccination itself, but a policy of offering vaccination and treating those who accept. A similar problem can arise in a randomized trial, for example in evaluating the effectiveness of health checkups (South-east London Screening Study Group 1977). Subjects were randomized to a screening group or to a control group. The screening group were invited to attend for an examination, some accepted and were screened and some refused. When comparing the results in terms of subsequent mortality, it was essential to compare the controls to the screening groups containing both screened and refusers. For example, the refusers may have included people who were already too ill to come for screening. The important point is that the random allocation procedure produces comparable groups and it is these we must compare, whatever selection may be made within them. We therefore analyse the data according to the way we intended to treat subjects, not the way in which they were actually treated. This analysis by **intention to treat**. The alternative, analysing by treatment actually received is called **on treatment** analysis.

2.6 Cross-over designs

Sometimes it is possible to use a subject as her or his own control. For example, when comparing analgesics in the treatment of arthritis, patients may receive in succession a new drug and a control treatment. The response to the two treatments can then be compared for each patient. These designs have the advantage of removing variability between subjects. We can carry out a trial with fewer subjects than would be needed for a two group trial.

Although all subjects receive all treatments, these trials must still be randomized. In the simplest case of treatment and control, patients may be given two different regimes: control followed by treatment or treatment followed by control. These may not give the same results, e.g. there may be a long term carry-over effect or time trend which makes treatment followed by control show less of a difference than control followed by treatment. Subjects are, therefore, assigned to a given order at random. It is possible in the analysis of cross-over studies to estimate the size of any carry-over effects which may be present.

As an example of the advantages of a cross-over trial, consider a trial of pronethalol in the treatment of angina pectoris (Pritchard *et al.* 1963). Angina pectoris is a chronic disease characterized by attacks of acute pain. Patients in this trial received either pronethalol or an inert control treatment (or placebo, see §2.8) in four periods of two weeks, two periods on the drug and two on the control treatment. These periods were in random order. The outcome measure was the number of attacks of angina experienced. These were recorded by the patient in a diary. Twelve patients took part in the trial. The results are shown in Table 2.9. The advantage

Table 2.9. Results of a trial of pronethalol for the treatment of angina pectoris (Pritchard *et al.* 1963)

Patient	Number of attacks while on		Difference
number	Placebo	Pronethalol	Placebo-Pronethalol
1	71	29	42
2	323	348	–25
3	8	1	7
4	14	7	7
5	23	16	7
6	34	25	9
7	79	65	14
8	60	41	19
9	2	0	2
10	3	0	3
11	17	15	2
12	7	2	5

in favour of pronethalol is shown by 11 of the 12 patients reporting fewer attacks of pain while on pronethalol than while on the control treatment. If we had obtained the same data from two separate groups of patients instead of the same group under two conditions, it would be far from clear that pronethalol is superior because of the huge variation between subjects. Using a two group design, we would need a much larger sample of patients to demonstrate the efficacy of the treatment.

Cross-over designs can be useful for laboratory experiments on animals or human volunteers. They should only be used in clinical trials where the treatment will not affect the course of the disease and where the patient's condition would not change appreciably over the course of the trial. A cross-over trial could be used to compare different treatments for the control of arthritis or asthma, for example, but not to compare different regimes for the management of myocardial infarction. However, a cross-over trial cannot be used to demonstrate the long term action of a treatment, as the nature of the design means that the treatment period must be limited. As most treatments of chronic disease must be used by the patient for a long time, a two sample trial of long duration is usually required to investigate fully the effectiveness of the treatment. Pronethalol, for example, was later found to have quite unacceptable side effects in long term use.

2.7 Selection of subjects for clinical trials

I have discussed the allocation of subjects to treatments at some length, but we have not considered where they come from. The way in which subjects are selected for an experiment may have an effect on its outcome. In practice, we are usually limited to subjects which are easily available to us. For example, in an animal experiment we must take the latest batch from the animal house. In a clinical trial of the treatment of myocardial infarction, we must be content with patients who are brought into the hospital. In experiments on human volunteers we sometimes have to use the researchers themselves.

As we shall see more fully in Chapter 3, this has important consequences for the interpretation of results. In trials of myocardial infarction, for example, we would not wish to conclude that, say, the survival rate with a new treatment in a trial in London would be the same as in a trial in Edinburgh. The patients may have a different history of diet, for example, and this may have a considerable effect on the state of their arteries and hence on their prognosis. Indeed, it would be very rash to suppose that we would get the same survival rate in a hospital a mile down the road. What we rely on is the comparison between randomized groups from the same population of subjects, and hope that if a treatment reduces mortality in London it will also do so in Edinburgh. This may be a reasonable supposition, and effects which appear in one population are likely to appear in another, but it cannot be proved on statistical grounds alone. Sometimes in extreme cases it turns out not to be true. BCG vaccine has been shown, by large, well conducted randomized trials, to be effective in reducing the incidence of tuberculosis in children in the UK. However, in India it appears to be far less effective (Lancet 1980). This may be because the amount of exposure to tuberculosis is so different in the two populations.

Given that we can use only the experimental subjects available to us, there are some principles which we use to guide our selection from them. As we shall see later, the lower the variability between the subjects in an experiment is, the better chance we have of detecting a treatment difference if it exists. This means that uniformity is desirable in our subjects. In an animal experiment this can be achieved by using animals of the same strain raised under controlled conditions. In a clinical trial we usually restrict our attention to patients of a defined age group and severity of disease. The Salk vaccine trial (§2.4) only used children in one school year. In the streptomycin trial (§2.2) the subjects were restricted to patients with acute bilateral pulmonary tuberculosis, bacteriologically proved, aged between 15 and 30 years, and unsuitable for other current therapy. Even with this narrow definition there was considerable variation among the patients, as Tables 2.5 and 2.6 show. Tuberculosis had to be bacteriologically proved because it is important to make sure that everyone has the disease we wish to treat. Patients with a different disease are not only potentially being wrongly treated themselves, but may make the results difficult to interpret. Restricting attention to a particular subset of patients, though useful, can lead to difficulties. For example, a treatment shown to be effective and safe in young people may not necessarily be so in the elderly. Trials have to be carried out on the sort of patients it is proposed to treat.

2.8 Response bias and placebos

The knowledge that she or he is being treated may alter a patient's response to treatment. This is called the **placebo effect**. A **placebo** is a pharmacologically inactive treatment given as if it were an active treatment. This effect may take many forms, from a desire to please the doctor to measurable biochemical changes in the brain. Mind and body are intimately connected, and unless the psychological effect is actually part of the treatment we usually try to eliminate such factors from treatment comparisons. This is particularly important when we are dealing with subjective assessments, such as of pain or well-being.

A fascinating example of the power of the placebo effect is given by Huskisson (1974). Three active analgesics, aspirin, Codis and Distalgesic, were compared with an inert placebo. Twenty two patients each received the four treatments in a cross-over design. The patients reported pain relief on a four point scale, from 0=no relief to 3=complete relief. All the treatments produced some pain relief, maximum relief being experienced after about two hours (Figure 2.1). The three active treatments were all superior to placebo, but not by very much. The four drug treatments were given in the form of tablets identical in shape and size, but each drug was given in four different colours. This was done so that patients could distinguish the drugs received, to say which they preferred. Each patient received four different colours, one for each drug, and the colour combinations were allocated randomly. Thus some patients received red placebos, some blue, and so on. As Figure 2.1 shows, red placebos were markedly more effective than other colours, and were just as effective as the active drugs! In this study not only is the effect of a pharmacologically inert placebo in producing reported pain relief demonstrated, but also the wide variability and unpredictability of this response. We must clearly take account of this in trial design. Incidentally, we should not conclude that red placebos always work best. There is, for example, some evidence that patients being treated for anxiety prefer tablets to be in a soothing green, and depressive symptoms respond best to a lively yellow (Schapira *et al.* 1970).

In any trial involving human subjects it is desirable that the subjects should not be able to tell which treatment is which. In a study to compare two or more treatments this should be done by making the treatments as similar as possible. Where subjects are to receive no treatment an inactive placebo should be used if possible. Sometimes when two very different active treatments are compared a double placebo can be used. For example, when comparing a drug given a single dose with a drug taken daily for seven days, subjects on the single dose drug may receive a daily placebo and those on the daily dose a single placebo at the start. Placebos are not always

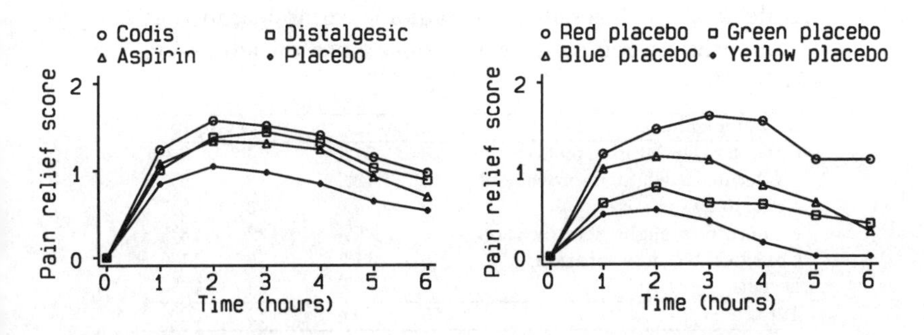

Fig. 2.1. Pain relief in relation to drug and to colour of placebo (after Huskisson 1974)

possible or ethical. In the MRC trial of streptomycin, where the treatment involved several injections a day for several months, it was not regarded as ethical to do the same with an inert saline solution and no placebo was given. In the Salk vaccine trial, the inert saline injections were placebos. It could be argued that paralytic polio is not likely to respond to psychological influences, but how could we be really sure of this? The certain knowledge that a child had been vaccinated may have altered the risk of exposure to infection as parents allowed the child to go swimming, for example. Finally, the use of a placebo may also reduce the risk of assessment bias as we shall see in §2.9.

2.9 Assessment bias and double blind studies

The response of subjects is not the only thing affected by knowledge of the treatment. The assessment by the researcher of the response to treatment may also be influenced by the knowledge of the treatment.

Some outcome measures do not allow for much bias on the part of the assessor. For example, if the outcome is survival or death, there is little possibility that unconscious bias may affect the observation. However, if we are interested in an overall clinical impression of the patient's progress, or in changes in an X-ray picture, the measurement may be influenced by our desire (or otherwise) that the treatment should succeed. It is not enough to be aware of this danger and allow for it, as we may have the similar problem of 'bending over backwards to be fair'. Even such an apparently objective measure as blood pressure can be influenced by the expectations of the experimenter, and special measuring equipment has been devised to avoid this (Rose *et al.* 1964).

We can avoid the possibility of such bias by using **blind assessment**, that is, the assessor does not know which treatment the subject is receiving. If a clinical trial cannot be conducted in such a way that the clinician in

Table 2.10. Assessment of radiological appearance at six months as compared with appearance on admission (MRC 1948)

Radiological assessment	S Group		C Group	
Considerable improvement	28	51%	4	8%
Moderate or slight improvement	10	18%	13	25%
No material change	2	4%	3	6%
Moderate or slight deterioration	5	9%	12	23%
Considerable deterioration	6	11%	6	11%
Deaths	4	7%	14	27%
Total	55	100%	52	100%

charge does not know the treatment, blind assessment can still be carried out by an external assessor. When the subject does not know the treatment and blind assessment is used, the trial is said to be **double blind**.

Placebos may be just as useful in avoiding assessment bias as for response bias. The subject is unable to tip the assessor off as to treatment, and there is likely to be less material evidence to indicate to an assessor what it is. In the anticoagulant study by Carleton *et al.* (1960) described above, the treatment was supplied through an intravenous drip. Control patients had a dummy drip set up, with a tube taped to the arm but no needle inserted, primarily to avoid assessment bias. In the Salk trial, the injections were coded and the code for a case was only broken after the decision had been made as to whether the child had polio and if so of what severity.

In the streptomycin trial, one of the outcome measures was radiological change. X-ray plates were numbered and then assessed by two radiologists and a clinician, none of whom knew to which patient and treatment the plate belonged. The assessment was done independently, and they only discussed a plate if they had not all come to the same conclusion. Only when a final decision had been arrived at was the link between plate and patient made. The results are shown in Table 2.10. The clear advantage of streptomycin is shown in the considerable improvement of over half the S group, compared to only 8% of the controls.

2.10 Laboratory experiments

So far we have looked at clinical trials, but exactly the same principles apply to laboratory research on animals. It may well be that in this area the principles of randomization are not so well understood and even more critical attention is needed from the reader of research reports. One reason for this may be that great effort has been put into producing genetically similar animals, raised in conditions as close to uniform as is practicable. The researcher using such animals as subjects may feel that the resulting animals show so little biological variability that any natural differences between them will be dwarfed by the treatment effects. This is not necessarily so, as the following examples illustrate.

A colleague was looking at the effect of tumour growth on macrophage counts in rats. The only significant difference was between the initial values in tumour induced and non-induced rats, that is, before the tumour-inducing treatment was given. There was a simple explanation for this surprising result. The original design had been to give the tumour-inducing treatment to each of a group of rats. Some would develop tumours and others would not, and then the macrophage counts would be compared between the two groups thus defined. In the event, all the rats developed tumours. In an attempt to salvage the experiment my colleague obtained a second batch of animals, which he did not treat, to act as controls. The difference between the treated and untreated animals was thus due to differences in parentage or environment, not to treatment.

That problem arose by changing the design during the course of the experiment. Problems can arise from ignoring randomization in the design of a comparative experiment. Another colleague wanted to know whether a treatment would affect weight gain in mice. Mice were taken from a cage one by one and the treatment given, until half the animals had been treated. The treated animals were put into smaller cages, 5 to a cage, which were placed together in a constant environment chamber. The control mice were in cages also placed together in the constant environment chamber. When the data were analysed, it was discovered that the mean initial weights was greater in the treated animals than in the control group. In a weight gain experiment this could be quite important! It may have been that when picking up the animals to apply the treatment, the larger animals were easier to pick up. What that experimenter should have done was to place the mice in the boxes, give each box a place in the constant environment chamber, then allocate the boxes to treatment or control at random. We would then have two groups which were comparable, both in initial values and in any environmental differences which may exist in the constant environment chamber.

These examples are given to show that even when the experimental material is as uniform as laboratory animals biological variability is still present. Randomization was devised to cope with this and it is the only effective method we have.

2.11 Experimental units

In the weight gain experiment described above, each box of mice contained five animals. These animals were not independent of one another, but interacted. In a box the other four animals formed part of the environment of the fifth, and so might influence its growth. The box of five mice is called an **experimental unit**. An experimental unit is the smallest group of subjects in an experiment whose response cannot be affected by other subjects. We need to know the amount of natural variation which exists between experimental units before we can decide whether the treatment effect is distinguishable from this natural variation. In the weight gain experiment, the mean weight gain in each box should be calculated, and the mean difference estimated using the two sample t method (§10.3).

The question of the experimental unit arises when the treatment is applied to the provider of care rather than to the patient directly. For example, White *et al.* (1989) compared three randomly allocated groups of GPs, the first given an intensive programme of small group education to improve their treatment of asthma, the second a lesser intervention, and the third no intervention at all. For each GP, a sample of her or his asthmatic patients was selected. These patients received questionnaires about their symptoms, the research hypothesis being that the intensive programme would result in fewer symptoms among their patients. The experimental unit was the GP, not the patient. The asthma patients treated by an individual GP were use to monitor the effect of the intervention on that GP. The proportion of patients who reported symptoms was used as a measure of the GP's effectiveness, and the mean of these proportions was compared between the groups using one-way analysis of variance (§10.9). Another example would be a trial of population screening for a disease (§15.3), where screening centres were set up in some health districts and not in others. We should find the mortality rate for each district separately and then compare the mean rate in the group of screening districts with that in the group of control districts.

The most extreme case arises when there is only one experimental unit per treatment. For example, consider a health education experiment involving two schools. In one school a special health education programme was mounted, aimed to discourage children from smoking. Both before and afterwards, the children in each school completed questionnaires about cigarette smoking. In this example the school is the experimental unit. There is no reason to suppose that two schools should have the same proportion of smokers among their pupils, or that two schools which do have equal proportions of smokers will remain so. The experiment would be much more convincing if we had several schools and randomly allocated them to receive the health education programme or to be controls. We would then look for a consistent difference between the treated and control schools, using the proportion of smokers in the school as the variable.

2.12 Further points about trial design

There are many aspects of experimental design which we have not yet discussed. These include experiments to compare several factors at once. For example, we might wish to study the effect of a drug at different doses in the presence or absence of a second drug, with the subject standing or supine. This is usually designed as a factorial experiment, where every possible combination of treatments is used. These designs are unusual in clinical research but are sometimes used in laboratory work. They are described in more advanced texts (Armitage and Berry 1987; Snedecor and Cochran 1980).

The trials described above all had a fixed sample size, decided at the start of the experiment. Because in medicine it is desirable to expose as few patients as possible to potentially hazardous treatments, sequential designs have been developed in which the data are analysed as they are collected. As soon as the difference between treatments is large enough to be convincing, the trial is stopped (Armitage 1975).

Finally, I have yet to mention the ethics of clinical trials. The objection to randomized experimentation may be made that we are withholding a potentially beneficial treatment from patients. However, any biologically active treatment is potentially harmful, and we are surely not justified in giving potentially harmful treatments to patients before the benefits have been demonstrated conclusively. Without properly conducted controlled clinical trials to support it, each administration of a treatment to a patient becomes an uncontrolled experiment, whose outcome, good or bad, cannot be predicted.

For the choice of sample size see Chapter 18. For accounts of the theory and practice of clinical trials, see Pocock (1983) and Johnson and Johnson (1977).

2M Multiple choice questions 1 to 6

(Each branch is either true or false)

1. When testing a new medical treatment, suitable control groups include patients who:

(a) are treated by a different doctor at the same time;

(b) are treated in a different hospital;

(c) are not willing to receive the new treatment;

(d) were treated by the same doctor in the past;

(e) are not suitable for the new treatment.

2. In an experiment to compare two treatments, subjects are allocated using random numbers so that:

(a) the sample may be referred to a known population;

(b) when deciding to admit a subject to the trial, we do not know which treatment that subject would receive;

(c) the subjects will get the treatment best suited to them;

(d) the two groups will be similar, apart from treatment;

(e) treatments may be assigned according to the characteristics of the subject.

3. In a double blind clinical trial:

(a) the patients do not know which treatment they receive;

(b) each patient receives a placebo;

(c) the patients do not know that they are in a trial;

(d) each patient receives both treatments;

(e) the clinician making assessment does not know which treatment the patient receives.

4. In a trial of a new vaccine, children were assigned at random to a 'vaccine' and a 'control' group. The 'vaccine' group were offered vaccination, which two thirds accepted:

(a) the group which should be compared to the controls is all children who accepted vaccination;

(b) those refusing vaccination should be included in the control group;

(c) the trial is double blind;

(d) those refusing vaccination should be excluded;

(e) the trial is useless because not all the treated group were vaccinated.

5. Cross-over designs for clinical trials:

(a) may be used to compare several treatments;

(b) involve no randomization;

(c) require fewer patients than do designs comparing independent groups;

(d) are useful for comparing treatments intended to alleviate chronic symptoms;

(e) use the patient as his own control.

6. Placebos are useful in clinical trials:

(a) when two apparently similar active treatments are to be compared;

(b) to guarantee comparability in non-randomized trials;

(c) because the fact of being treated may itself produce a response;

(d) because they may help to conceal the subject's treatment from assessors;

(e) when an active treatment is to be compared to no treatment.

Table 2.11. Method of delivery in the KYM study

Method of delivery	Accepted KYM		Refused KYM		Control women	
	%	n	%	n	%	n
Normal	80.7	352	69.8	30	74.8	354
Instrumental	12.4	54	14.0	6	17.8	84
Caesarian	6.9	30	16.3	7	7.4	35

2E Exercise: The 'Know Your Midwife' trial

The Know Your Midwife (KYM) scheme was a method of delivering maternity care for low-risk women. A team of midwives ran a clinic, and the same midwife would give all antenatal care for a mother, deliver the baby, and give postnatal care. The KYM scheme was compared to standard antenatal care in a randomized trial (Flint and Poulengeris 1986). It was thought that the scheme would be very attractive to women and that if they knew it was available they might be reluctant to be randomized to standard care. Eligible women were randomized without their knowledge to KYM or to the control group, who received the standard antenatal care provided by St. George's Hospital. Women randomized to KYM were sent a letter explaining the KYM scheme and inviting them to attend. Some women declined and attended the standard clinic instead. The mode of delivery for the women is shown in Table 2.11. Normal obstetric data were recorded on all women, and the women were asked to complete questionnaires (which they could refuse) as part of a study of antenatal care, though they were not told about the trial.

1. The women knew what type of care they were receiving. What effect might this have on the outcome?
2. What comparison should be made to test whether KYM has any effect on method of delivery?
3. Do you think it was ethical to randomize women without their knowledge?

3

Sampling and observational studies

3.1 Observational studies

In this chapter we shall be concerned with observational studies. Instead of changing something and observing the result, as in an experiment or clinical trial, we observe the existing situation and try to understand what is happening. Studying people in the wild, as it were, can be extremely difficult and it is often impossible to draw unequivocal conclusions. We shall start by considering how to get descriptive information about populations in which we are interested. We shall go on to the problem of using such information to study disease processes and the possible causes of disease.

3.2 Censuses

One simple question we can ask about any group of interest is how many members it has. For example, we need to know how many people live in a country and how many of them are in various age and sex categories, in order to monitor the changing pattern of disease and to plan medical services. We can obtain it by a **census**. In a census, the whole of a defined population is counted. In the United Kingdom, as in many developed countries, a population census is held every ten years. This is done by dividing the entire country into small areas called enumeration districts, usually containing between 100 and 200 households. It is the responsibility of an enumerator to identify every household in the district and ensure that a census form is completed, listing all members of the household and providing a few simple pieces of information. Even though completion of the census form is compelled by law, and enormous effort goes into ensuring that every household is included, there are undoubtedly some who are missed. The final data, though extremely useful, are not totally reliable.

The medical profession takes part in a massive, continuing census of deaths, by registering for each death which occurs not only the name of the deceased and cause of death, but also details of age, sex, place of residence and occupation. Census methods are not restricted to national populations. They can be used for more specific administrative purposes too. For example, we might want to know how many patients are in a particular hospital at a particular time, how many of them are in different diagnostic groups, in different age/sex groups, and so on. We can then use this information together with estimates of the death and discharge rates to estimate how many beds these patients will occupy at various times in the future (Bewley *et al.* 1975; 1981).

3.3 Sampling

A census of a single hospital can only give us reliable information about that hospital. We cannot easily generalize our results to hospitals in general. If we want to obtain information about the hospitals of the United Kingdom, two courses are open to us: we can study every hospital, or we can take a representative sample of hospitals and use that to draw conclusions about hospitals as a whole.

Most statistical work is concerned with using samples to draw conclusions about some larger population. In the clinical trials described in Chapter 2, the patients act as a sample from a larger population consisting of all similar patients and we do the trial to find out what would happen to this larger group were we to give them a new treatment.

The word 'population' is used in common speech to mean 'all the people living in an area', frequently of a country. In statistics, we define the term more widely. A **population** is any collection of individuals in which we may be interested, where these individuals may be anything, and the number of individuals may be finite or infinite. Thus, if we are interested in some characteristics of the British people, the population is 'all people in Britain'. If we are interested in the treatment of diabetes the population is 'all diabetics'. If we are interested in the blood pressure of a particular patient, the population is 'all possible measurements of blood pressure in that patient'. If we are interested in the toss of two coins, the population is 'all possible tosses of two coins'. The first two examples are finite populations and could in theory if not practice be completely examined; the second two are infinite populations and could not. We could only ever look at a **sample**, which we will define as being a group of individuals taken from a larger population and used to find out something about that population.

How should we choose a sample from a population? The problem of getting a representative sample is similar to that of getting comparable groups of patients discussed in §2.1-3. We want our sample to be representative, in some sense, of the population. We want it to have all the characteristics in terms of the proportions of individuals with particular qualities as has the whole population. In a sample from a human population, for example, we want the sample to have about the same proportion of men and women as in the population, the same proportions in different age groups, in occupational groups, with different diseases, and so on. In addition, if we use a sample to estimate the proportion of people with a disease, we want to know how reliable this estimate is, how far from the proportion in the whole population the estimate is likely to be.

It is not sufficient to choose the most convenient group. For example, if we wished to predict the results of an election, we would not take as our sample people waiting in bus queues. These may be easy to interview, at least until the bus comes, but the sample would be heavily biased towards those who cannot afford cars and thus towards lower income groups. In the same way, if we wanted a sample of medical students we would not take the front two rows of the lecture theatre. They may be unrepresentative in having an unusually high thirst for knowledge, or poor eyesight.

How can we choose a sample which does not have a built-in bias? We might divide our population into groups, depending on how we think various characteristics will affect the result. To ask about an election, for example, we might group the population according to age, sex and social class. We then choose a number of people in each group by knocking on doors until the quota is made up, and interview them. Then, knowing the distributions of these categories in the population (from census data, etc.) we can get a far better picture of the views of the population. This is called **quota sampling**. In the same way we could try to choose a sample of rats by choosing given numbers of each weight, age, sex, etc. There are difficulties with this approach. First, it is rarely possible to think of all the relevant classifications. Second, it is still difficult to avoid bias within the classifications, by picking interviewees who look friendly, or rats which are easy to catch. Third, we can only get an idea of the reliability of findings by repeatedly doing the same type of survey, and of the representativeness of the sample by knowing the true population values (which we can actually do in the case of elections), or by comparing the results with a sample which does not have these drawbacks. This method can be quite effective when similar surveys are made repeatedly as in opinion polls or market research. It is less useful for medical problems, where we are continually asking new questions. We need a method where bias is avoided and where we can estimate the reliability of the sample from the sample itself. As in §2.2, we use a random method: random sampling.

3.4 Random sampling

The problem of obtaining a sample which is representative of a larger population is very similar to that of allocating patients into two comparable groups. We want a way of choosing members of the sample which does not depend on their own characteristics. The only way to be sure of this is to select them at random, so that whether or not each member of the population is chosen for the sample is purely a matter of chance.

For example, to take a random sample of five students from a class of eighty, we could write all the names on pieces of paper, mix them thoroughly in a hat or other suitable container, and draw out five. All students have the same probability, 5/80, of being chosen, and so we have a random sample. All samples of 5 students are equally likely, too, because each student is chosen quite independently of the others. This method is called **simple random sampling**.

As we have seen in §2.2, physical methods of randomizing are often not very suitable for statistical work. We usually use tables of random digits, such as Table 2.3, or random numbers generated by a computer program. We could use Table 2.3 to draw our sample of 5 from 80 students in several ways. For example, we could list the students, numbered from 1 to 80. This list from which the sample is to be drawn is called the **sampling frame**. We choose a starting point in the random number table (Table 2.3), say row 20, column 5. This gives us the following pairs of digits:

14 04 88 86 28 92 04 03 42 99 87 08

We could use these pairs of digits directly as subject numbers. We choose subjects numbered 14 and 4. There is no subject 88 or 86, so the next chosen is number 28. There is no 92, so the next is 4. We already have this subject in the sample, so we carry on to the next pair of digits, 03. The final member of the sample is number 42. Our sample of 5 students is thus numbers 3, 4, 14, 28 and 42.

There appears to be some pattern in this sample. Two numbers are adjacent (3 and 4) and three are divisible by 14 (14, 28 and 42). Random numbers often appear to us to have pattern, perhaps because the human mind is always looking for it. On the other hand, if we try to make the sample 'more random' by replacing either 3 or 4 by a subject near the end of the list, we are imposing a pattern of uniformity on the sample and destroying its randomness. All groups of five are equally likely and may happen, even 1, 2, 3, 4, 5.

This method of using the table is fine for drawing a small sample, but it can be tedious for drawing large samples, because of the need to check for duplicates. There are many other ways of doing it. For example, we can drop the requirement for a sample of fixed size, and only require that each member of the population will have a fixed probability of being in the sample. We could draw a $5/80 = 1/16$ sample of our class by using the digits in groups to give a decimal number, say,

.1404 .8886 .2892 .0403 .4299 .8708

We then choose the first member of the population if 0.1404 is less than 1/16. It is not, so we do not include this member, nor the second, corresponding to 0.8886, nor the third, corresponding to 0.2892. The fourth corresponds to 0.0403, which is less than 1/16 (0.0625) and so the fourth member is chosen as a member of the sample, and so on. This method is only suitable for fairly large samples, as the size of the sample obtained can be very variable in small sampling problems. In the example there is a better than 1 in 10 chance of finishing with a sample of 2 or fewer.

Random sampling ensures that the only ways in which the sample differs from the population will be those due to chance. It has a further advantage, because the sample is random, we can apply the methods of probability theory to the data obtained. As we shall see in Chapter 8, this enables us to estimate how far from the population value the sample value is likely to be.

The problem with random sampling is that we must have a list of the population from which the sample is to be drawn. Lists of populations may be hard to find, or they may be very cumbersome. For example, to sample the adult population in the UK, we could use the electoral roll. But a list of some 40 000 000 names would be difficult to handle, and in practice we would first take a random sample of electoral wards, and then a random sample of electors within these wards. This is, for obvious reasons, a **multi-stage random sample**. This approach contains the element of randomness, and so samples will be representative of the populations from which they are drawn. However, not all samples have an equal chance of being chosen, so it is not the same as simple random sampling.

We can also carry out sampling without a list of the population itself, provided we have a list of some larger units which contain all the members of the population. For example, we can obtain a random sample of school children in an area by starting with a list of schools, which is quite easy to come by. We then draw a simple random sample of schools and all the children within our chosen schools form the sample of children. This is called a **cluster sample**, because we take a sample of clusters of individuals.

Sometimes it is desirable to divide the population into different strata, for example into age and sex groups, and take random samples within these. This is rather like quota sampling, except that within the strata we choose at random. If the different strata have different values of the quantity we are measuring, this **stratified random sampling** can increase our precision considerably. There are many complicated sampling schemes for use in different situations.

In §2.3 I looked at the difficulties which can arise using methods of allocation which appear random but do not use random numbers. In sampling, two such methods are often suggested by researchers. One is to take every tenth subject from the list, or whatever fraction is required. The other is to use the last digit of some reference number, such as the hospital number, and take as the sample subjects where this is, say, 3 or 4. These sampling methods are **systematic** or **quasi-random**. It is not usually obvious why they should not give 'random' samples, and it may be that in many cases they would be just as good as random sampling. They are certainly easier. To use them, we must be very sure that there is no pattern to the list which could produce an unrepresentative group. If it is possible, random sampling seems safer.

Volunteer bias can be as serious a problem in sampling studies as it is in trials (§2.4). Having drawn the sample, if we can only obtain data from a subset of them this subset will not be a random sample of the population. Its members will be self selected. It is often very difficult to get data from every member of a sample. The proportion for whom data is obtained is called the **response rate** and in a sample survey of the general population is likely to be between 70% and 80%. The possibility that those lost from the sample are different in some way must be considered. For example, they may tend to be ill, which can be a serious problem in disease prevalence studies. In a study of cigarette smoking and respiratory disease in Derbyshire schoolchildren, we drew a random sample of schools, and our sample of children was all children in the first secondary school year (Banks *et al.* 1978). We thus had a random cluster sample. The response rate to our survey was 80%, most of those lost being absent from school on the day. Now, some of these absentees were ill and some were truants. Our sample may thus lead us to underestimate the prevalence of respiratory symptoms, by omitting sufferers with current acute disease, and the prevalence of cigarette smoking by omitting those who have gone for a quick smoke behind the bike sheds.

One of the most famous sampling disasters, the *Literary Digest* poll of 1936, illustrates these dangers (Bryson 1976). This was a poll of voting intentions in the 1936 U.S. presidential election, fought by Roosevelt and Landon. The sample was a complex one. In some cities every registered voter was included, in others one in two, and for the whole of Chicago one in three. Ten million sample ballots were mailed to prospective voters, but only 2.3 million, less than a quarter, were returned. Still, two million is a lot of Americans, and these predicted a 60% vote to Landon. In fact, Roosevelt won with 62% of the vote. The response was so poor that the sample was most unlikely to be representative of the population, no matter how carefully the original sample was drawn. Two million Americans can be wrong! It is not the mere size of the sample, but its representativeness which is important. Provided the sample is truly representative, 2 000 voters is all you need to estimate voting intentions to within two per cent, which is enough for election prediction if they tell the truth and don't change their minds (see §18E).

3.5 Sampling in clinical studies

Having extolled the virtues of random sampling and cast doubt on all other sampling methods, I must admit that most medical data are not obtained in this way. This is partly because the practical difficulties are immense. To obtain a reasonable sample of the population of the UK, anyone can get a list of electoral wards, take a random sample of them, buy copies of the electoral rolls for the chosen wards and then take a random sample of names from it. But suppose you want to obtain a sample of patients with carcinoma of the bronchus, to see how many smoke cigarettes. You could get a list of hospitals easily enough and get a random sample of them, but then things would become difficult. The names of patients will only be released by the consultant in charge should he so wish, and you will need his permission before approaching them. Any study of human patients requires ethical approval, and you will need this from the ethics committee of each of your chosen hospitals. Getting the cooperation of so many people is a task to daunt the hardiest, and obtaining ethical approval alone can take more than a year.

The result of this is that clinical studies are done on the patients to hand. I have touched on this problem in the context of clinical trials (§2.7) and the same applies to other types of clinical study. In a clinical trial we are concerned with the comparison of two treatments and we hope that the superior treatment in Stockport will also be the superior treatment in Southampton. If we are studying clinical measurement, we can hope that a measurement method which is repeatable in Middlesbrough will be repeatable in Maidenhead, and that two different methods giving similar results in one place will give similar results in another. Studies which are not comparative give more cause for concern. The natural history of a disease described in one place may differ in unpredictable ways from that in another, due to differences in the environment and the genetic make up of the local population. Reference ranges for quantities of clinical interest, the limits within which values from most healthy people will lie, may well differ from place to place, yet they are often determined on groups of subjects which are quite unrepresentative even of the local population.

Studies based on local groups of patients are not without value. This is particularly so when we are concerned with comparisons between groups, as in a clinical trial, or relationships between different variables. However, we must always bear the limitations of the sampling method in mind when interpreting the results of such studies.

In general, most medical research has to be carried out using samples drawn from populations which are much more restricted than those about which we wish to draw conclusions. We may have to use patients in one hospital instead of all patients, or the population of a small area rather than that of the whole country or planet. We may have to rely on volunteers for studies of normal subjects, given most people's dislike of having needles pushed into them and disinclination to spend hours hooked up to batteries of instruments. Groups of normal subjects contain medical students, nurses and laboratory technicians far more often than would be expected by chance. In animal research the problem is even worse, for not only does one batch of one strain of mice have to represent the whole species, it often has to represent members of a different order, namely humans.

Findings from such studies can only apply to the population from which the sample was drawn. Any conclusion which we come to about wider populations, such as all patients with the disease in question, depends on evidence which is not statistical and often unspecified, namely our general experience of natural variability and experience of similar studies. This may let us down, and results established in one population may not apply to another. We have seen this in the use of BCG vaccine in India (§2.7). It is very important wherever possible that studies should be repeated by other workers on other populations, so that we can sample the larger population at least to some extent.

3.6 Sampling in epidemiological studies

One of the most important and difficult tasks in medicine is to determine the causes of disease, so that we may devise methods of prevention. We are working in an area where experiments are often neither possible nor ethical. For example, to determine that cigarette smoking caused cancer, we could imagine a study in which children were randomly allocated to a 'twenty cigarettes a day for fifty years' group and a 'never smoke in your life' group. All we would have to do then would be to wait for the death certificates. However, we could not persuade our subjects to stick to the treatment and deliberately setting out to cause cancer is not ethical. We must therefore observe the disease process as best we can, by watching people in the wild state rather than under laboratory conditions.

When we do this we must face the fact that the disease effect and putative cause do not exist in isolation but in a complex interplay of many intervening factors. We must do our best to assure ourselves that the relationship we observe is not the result of some other factor acting on both 'cause' and 'effect'. For example, it was once thought that the African fever tree, the yellow-barked acacia, caused malaria, because those unwise enough to camp under them were likely to develop the disease. This tree grows by water where mosquitos breed, and provides an ideal day-time resting place for these insects, whose bite transmits the plasmodium parasite which produces the disease. It was the water and the mosquitos which were the important factors, not the tree. Indeed, the name 'malaria' comes from a similar incomplete observation. It means 'bad air' and comes from the belief that the disease was caused by the air in marshy places, where the mosquitos bred. Epidemiological study designs must try to deal with the complex interrelationships between different factors in order to deduce the true mechanism of disease causation. We also use a number of different approaches to the study of these problems, to see whether all produce the same answer.

One method is to use differences in mortality rates between countries or changes over time. Here the data are whole population census data, so there is no sampling problem. The problem is rather to do with variations in diagnostic fashion and with the intervention of other variables. For example, it has been observed that countries with a high consumption of animal fat tend to have high mortality from coronary artery disease. However, such countries tend to have low consumption of dietary fibre also, so we must try to disentangle the effects of one from those of the other, which may not be possible.

Another approach is the **cross-sectional study**. We take some sample or whole population and observe whether or not they have either disease or possible cause. For example, Banks *et al.* (1978) wanted to know whether smoking causes respiratory symptoms in school children. We gave questionnaires to all first year secondary school boys in a random sample of schools in Derbyshire. Among boys who had never smoked, 3% reported a cough first thing in the morning, compared to 19% of boys who said that they smoked one or more cigarettes per week. We have a problem. The sample was representative of boys of this age in Derbyshire who answer questionnaires, but we want our conclusions to apply at least to the United Kingdom, if not the developed world or the whole planet. We argue that although the prevalence of symptoms and the strength of the relationship may vary between populations, the existence of the relationship is unlikely only to occur in the population studied. We also have the problem that smoking and respiratory symptoms may not be directly related, but may both be related to some other factor. For example, children whose parents smoke may be more likely to develop respiratory symptoms, because of passive inhalation of their parent's smoke, and also be more influenced to try smoking themselves. We can test this by looking separately at the relationship between the child's smoking and symptoms for those whose parents are not smokers, and for those whose parents are smokers. As Figure 3.1 shows, this relationship in fact persisted (§17.8) and there was no reason to suppose that a third causal factor was at work. A third problem is that the respondents may not be telling the truth, which we shall tackle in §3.9.

Most diseases are not suited to this simple cross-sectional approach, because they are rare events. For example, lung cancer accounts for 9%

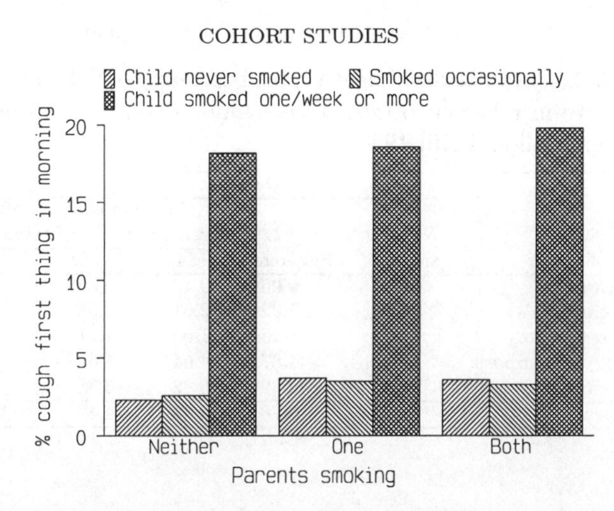

Fig. 3.1. Prevalence of self-reported morning cough in Derbyshire schoolboys, by their own and their parents' cigarette smoking (Bland *et al.* 1978)

of male deaths in the UK (OPCS, DH2 No.7), and so is a very important disease. However the proportion of people who are known to have the disease at any given time, the **prevalence**, is quite low. Most deaths from lung cancer take place after the age of 45, so we will consider a sample of men aged 45 and over. The average remaining life span of these men, in which they could contract lung cancer, will be about 30 years. The average time from diagnosis to death is about half a year so of those who will contract lung cancer only $1/30 \times 1/2$ will have been diagnosed when the sample is drawn. Only 9% of the sample will develop lung cancer anyway, so the proportion with the disease at any time is $1/30 \times 1/2 \times 9\% = 0.2\%$ or 2 per thousand. We would need a very large sample indeed to get a worthwhile number of lung cancer cases.

3.7 Cohort studies

One way of getting round the problem of the small proportion of people with the disease of interest is the **cohort study**. This is a **prospective** design, as we start with the possible cause and see whether this leads to the disease in the future. We take a group of people, the **cohort**, and observe whether they have the suspected causal factor. We then follow them over time and observe whether they develop the disease. A cohort study usually takes a long time, as we must wait for the future event to occur. It involves keeping track of large numbers of people, sometimes for many years, and often very large numbers must be included in the sample to ensure sufficient numbers will develop the disease between those with and without the factor to enable comparisons to be made.

Table 3.1. Standardized death rates per year per 1 000 men aged 35 or more in relation to most recent amount smoked, 53 months follow-up (Doll and Hill 1956)

	Death rate among				
			Men smoking a daily average weight of tobacco of		
Cause of death	Non-smokers	Smokers	1–14g	15–24g	25g+
Lung cancer	0.07	0.90	0.47	0.86	1.66
Other cancer	2.04	2.02	2.01	1.56	2.63
Other respiratory	0.81	1.13	1.00	1.11	1.41
Coronary thrombosis	4.22	4.87	4.64	4.60	5.99
Other causes	6.11	6.89	6.82	6.38	7.19
All causes	13.25	15.78	14.92	14.49	18.84

A noted cohort study of mortality in relation to cigarette smoking was carried out by Doll and Hill (1956). They sent a questionnaire to all members of the medical profession in the UK, who were asked to give their name, address, age and details of current and past smoking habits. The deaths among this group were recorded. Only 60% of doctors co-operated, so in fact the cohort does not represent all doctors. The results for the first 53 months are shown in Table 3.1.

The cohort represents doctors willing to return questionnaires, not people as a whole. We cannot use the death rates as estimates for the whole population, or even for all doctors. What we can say is that, in this group, smokers were far more likely than non-smokers to die from lung cancer. It would be surprising if this relationship were only true for doctors, but we cannot definitely say that this would be the case for the whole population, because of the way the sample has been chosen.

We also have the problem of other intervening variables. Doctors were not allocated to be smokers or non-smokers as in a clinical trial; they chose for themselves. The decision to begin smoking may be related to many things (social factors, personality factors, genetic factors) which may also be related to lung cancer. We must consider these alternative explanations very carefully before coming to any conclusion about the causes of cancer. In this study there were no data to test such hypotheses, a common problem in cohort studies. Because the sample was so large, only a little information was collected on each member of it.

3.8 Case-control studies

Another solution to the problem of the small number of people with the disease of interest is the **case-control study**. In this we take a group of people with the disease, the cases, and a second group without the disease, the controls. We then find the exposure of each subject to the possible causative factor and see whether this differs between the two groups. Before

Table 3.2. Numbers of smokers and non-smokers among lung cancer patients and age and sex matched controls with diseases other than cancer (Doll and Hill 1950)

	Non-smokers		Smokers		Total
Males:					
Lung cancer patients	2	(0.3%)	647	(99.7%)	649
Control patients	27	(4.2%)	622	(95.8%)	649
Females:					
Lung cancer patients	19	(31.7%)	41	(68.3%)	60
Control patients	32	(53.3%)	28	(46.7%)	60

their cohort study, Doll and Hill (1950) carried out a case-control study into the aetiology of lung cancer. Twenty London hospitals notified all patients admitted with carcinoma of the lung, the cases. An interviewer visited the hospital to interview the case, and, at the same time, selected a patient with diagnosis other than cancer, of the same sex and within the same 5 year age group as the case, in the same hospital at the same time, as a control. When more than one suitable patient was available, the patient chosen was the first in the ward list considered by the ward sister to be fit for interview. Table 3.2 shows the relationship between smoking and lung cancer for these patients. A smoker was anyone who had smoked as much as one cigarette a day for as much as one year. It appears that cases were more likely than controls to smoke cigarettes. Doll and Hill concluded that smoking is an important factor in the production of carcinoma of the lung.

The case-control study is an attractive method of investigation, because of its relative speed and cheapness compared to other approaches. However, there are difficulties in the selection of cases, the selection of controls, and obtaining the data. Because of these, case-control studies sometimes produce contradictory and conflicting results.

The first problem is the selection of cases. This usually receives little consideration beyond a definition of the type of disease and a statement about the confirmation of the diagnosis. This is understandable, as there is usually little else that the investigators can do about it. They start with the available set of patients. However, these patients do not exist in isolation. They are the result of some process which has led to them being diagnosed as having the disease and thus being available for study. For example, suppose we suspect that oral contraceptives might cause cancer of the breast. We have a group of patients diagnosed as having cancer of the breast. We must ask ourselves whether any of these were detected at a medical examination which took place because the woman was seeing a doctor to receive a prescription. If this were so, the pill would be associated with the detection of the disease rather than its cause.

Far more difficulty is caused by the selection of controls. We want a group of people who do not have the disease in question, but who are otherwise comparable to our cases. We must first decide the population from which they are to be drawn. There are two sources of controls: the general population and patients with other diseases. The latter is usually preferred because of its accessibility. Now these two populations are clearly not the same. For example, Doll and Hill gave the current smoking habits of 1014 men and women with diseases other than cancer, 14% of whom were currently non-smokers. They commented that there was no difference between smoking in the disease groups respiratory disease, cardiovascular disease, gastro-intestinal disease and others. However, in the general population the percentage of current non-smokers was 18% for men and 59% for women (Todd 1972). The smoking rate in the patient group as a whole was high. Since their report, of course, smoking has been associated with diseases in each group. Smokers get more disease and are more likely to be in hospital than non-smokers.

Intuitively, the comparison we want to make is between people with the disease and healthy people, not people with a lot of other diseases. We want to find out how to prevent disease, not how to choose one disease or another! However, it is much easier to use hospital patients as controls. There may then be a bias because the factor of interest may be associated with other diseases. Suppose we want to investigate the relationship between a disease and cigarette smoking using hospital controls. Should we exclude patients with lung cancer from the control group? If we include them, our controls may have more smokers than the general population, but if we exclude them we may have fewer. This problem is usually resolved by choosing specific patient groups, such as fracture cases, whose illness is thought to be unrelated to the factor being investigated. In case-control studies using cancer registries, controls are sometimes people with other forms of cancer. Sometimes more than one control group is used.

Having defined the population we must choose the sample. There are many factors which affect exposure to risk factors, such as age and sex. The most straightforward way is to take a large random sample of the control population, ascertain all the relevant characteristics, and then adjust for differences during the analysis, using methods described in Chapter 17. The alternative is to try to match a control to each case, so that for each case there is a control of the same age, sex, etc. Having done this, then we can compare our cases and controls knowing that the effects of these intervening variables are automatically adjusted for. If we wish to exclude a case we must exclude its control, too, or the groups will no longer be comparable. We can have more than one control per case, but the analysis becomes complicated.

Matching on some variables does not ensure comparability on all. Indeed, if it did there would be no study. Doll and Hill matched on age, sex and hospital. They recorded area of residence and found that 25% of their cases were from outside London, compared to 14% of controls. If we want to see whether this influences the smoking and lung cancer relationship we must make a statistical adjustment anyway. (Doll's and Hill's solution was to restrict attention to 98 pairs from district hospitals in London.) What should we match for? The more we match for, the fewer intervening variables there are to worry about. On the other hand, it becomes more and more difficult to find matches. Even matching on age and sex, Doll and Hill could not always find a control in the same hospital, and had to look elsewhere. Matching for more than age and sex can be very difficult.

Having decided on the matching variables we then find in the control population all the possible matches. If there are more matches than we need, we should choose the number required at random. Other methods, such as that used by Doll and Hill who allowed the ward sister to choose, have obvious problems of potential bias. If no suitable control can be found, we can do two things. We can widen the matching criteria, say age to within ten years rather than five, or we can exclude the case.

There are difficulties in interpreting the results of case-control studies. One is that the case-control design is usually **retrospective**, that is, we are starting with the present disease state, e.g. lung cancer, and relating it to the past, e.g. history of smoking. We may have to rely on the unreliable memories of our subjects. There is a problem of assessment bias in such studies, just as in clinical trials (§2.9). Interviewers will very often know whether the interviewee is a case or control and this may well affect the way questions are asked. The same problem arises in the recall of past events by the case. For example, the mother of a handicapped child may be more likely than the mother of a normal child to remember events in pregnancy which may have caused damage. These and other considerations make case-control studies extremely difficult to interpret. The evidence from such studies can be useful, but data from other types of investigation must be considered, too, before any firm conclusions are drawn.

There are many problems in using these observational designs, and the medical consumer of such research must be aware of them. We have no better way to tackle these questions and so we must make the best of them and look for consistent relationships which stand up to the most severe examination. We can also look for confirmation of our findings indirectly, from animal models and from dose-response relationships in the human population. However, we must accept that perfect proof is impossible in these issues and it is unreasonable to demand it. Sometimes, as with smoking and health, we must act on the balance of the evidence.

Table 3.3. Replies to two similar questions about ill health, by age (Hedges 1978)

	Age (years)			
	16–34	35–54	55+	Total
Can do something (a)	75%	64%	56%	65%
Can do something (b)	45%	49%	50%	49%

3.9 Questionnaire bias in observational studies

We have already looked at response bias in clinical trials (§2.8) and the same problems arise in observational studies. This is often further complicated because so much data have to be supplied by the subjects themselves. The way in which a question is asked may influence the reply. Sometimes the bias in a question is obvious. Compare these:

(a) Do you think people should be free to provide the best medical care possible for themselves and their families, free of interference from a State bureaucracy?
(b) Should the wealthy be able to buy a place at the head of the queue for medical care, pushing aside those with greater need, or should medical care be shared solely on the basis of need for it?

Version (a) expects the answer yes, version (b) expects the answer no. We would hope not to be misled by such blatant manipulation, but the effects of question wording can be much more subtle than this. Hedges (1978) reports several examples of the effects of varying the wording of questions. He asked two groups of about 800 subjects one of the following:

(a) Do you feel you take enough care of your health, or not?
(b) Do you feel you take enough care of your health, or do you think you could take more care of your health?

In reply to question (a), 82% said that they took enough care, whereas only 68% said this in reply to question (b). Even more dramatic was the difference between this pair:

(a) Do you think a person of your age can do anything to prevent ill-health in the future or not?
(b) Do you think a person of your age can do anything to prevent ill-health in the future, or is it largely a matter of chance?

Not only was there a difference in the percentage who replied that they could do something, but as Table 3.3 shows this answer was related to age for version (a) but not for version (b). Here version (b) is ambiguous, as it is quite possible to think that health is largely a matter of chance but that there is still something one can do about it. Only if it is totally a matter of chance is there nothing one can do.

Sometimes the respondents may interpret the question in a different way from the questioner. For example, when asked whether they usually coughed first thing in the morning, 3.7% of the Derbyshire schoolchildren replied that they did. When their parents were asked about the child's symptoms 2.4% replied positively, not a dramatic difference. Yet when asked about cough at other times in the day or at night 24.8% of children said yes, compared to only 4.5% of their parents (Bland *et al.* 1979). These symptoms all showed relationships to the child's smoking and other potentially causal variables, and also to one another. We are forced to admit that we are measuring something, but that we are not sure what!

Another possibility is that respondents may not understand the question at all, especially when it includes medical terms. In an earlier study of cigarette smoking by children, we found that 85% of a sample agreed that smoking caused cancer, but that 41% agreed that smoking was not harmful (Bewley *et al.* 1974). There are at least two possible explanations for this: being asked to agree with the negative statement 'smoking is not harmful' may have confused the children, or they may not see cancer as harmful. We have evidence for both of these possibilities. In a repeat study in Kent we asked a further sample of children whether they agreed that smoking caused cancer and that 'smoking is bad for your health' (Bewley and Bland 1976). In this study 90% agreed that smoking causes cancer and 91% agreed that smoking is bad for your health. In another study (Bland *et al.* 1975), we asked children what was meant by the term 'lung cancer'. Only 13% seemed to us to understand and 32% clearly did not, often saying 'I don't know'. They nearly all knew that lung cancer was caused by smoking, however.

The setting in which a question is asked may also influence replies. Opinion pollsters International Communications and Market Research conducted a poll in which half the subjects were questioned by interviewers about their voting preference and half were given a secret ballot (McKie 1992). By each method 33% chose 'Labour', but 28% chose 'Conservative' at interview and 7% would not say, whereas 35% chose 'Conservative' by secret ballot and only 1% would not say. Hence the secret method produced a Conservative majority, as at the then recent general election, and the open interview a Labour majority. For another example, Sibbald *et al.* (1994) compared two random samples of GPs. One sample were approached by post and then by telephone if they did not reply after two reminders, and the other were contacted directly by telephone. Of the predominatly postal sample, 19% reported that they provided counselling themselves, compared to 36% of the telephone sample, and 14% reported that their health visitor provided counselling compared to 30% of the telephone group. Thus the method of asking the question influenced the answer. One must be very cautious when interpreting questionnaire replies.

Often the easiest and best method, if not the only method, of obtaining data about people is to ask them. When we do it, we must be very careful to ensure that questions are straightforward, unambiguous and in language the respondents will understand. If we do not do this then disaster is likely to follow.

3M Multiple choice questions 7 to 13

(Each branch is either true or false)

7. In statistical terms, a population:

(a) consists only of people;

(b) may be finite;

(c) may be infinite;

(d) can be any set of things in which we are interested;

(e) may consist of things which do not actually exist.

8. A one day census of in-patients in a psychiatric hospital could:

(a) give good information about the patients in that hospital at that time;

(b) give reliable estimates of seasonal factors in admissions;

(c) enable us to draw conclusions about the psychiatric hospitals of Britain;

(d) enable us to estimate the distribution of different diagnoses in mental illness in the local area;

(e) tell us how many patients there were in the hospital.

9. In simple random sampling:

(a) each member of the population has an equal chance of being chosen;

(b) adjacent members of the population must not be chosen;

(c) likely errors cannot be estimated;

(d) each possible sample of the given size has an equal chance of being chosen;

(e) the decision to include a subject in the sample depends only on the subject's own characteristics.

10. Advantages of random sampling include:

(a) it can be applied to any population;

(b) likely errors can be estimated;

(c) it is not biassed;

(d) it is easy to do;

(e) the sample can be referred to a known population.

11. In a study of hospital patients, 20 hospitals were chosen at random from a list of all hospitals. Within each hospital, 10% of patients were chosen at random:

(a) the sample of patients is a random sample;
(b) all hospitals had an equal chance of being chosen;
(c) all patients had an equal chance of being chosen;
(d) the sample could be used to make inferences about all hospital patients at that time;
(e) all possible samples of patients had an equal chance of being chosen.

12. To examine the relationship between alcohol consumption and cancer of the oesophagus, feasible studies include:

(a) questionnaire survey of a random sample from the electoral role;
(b) comparison of history of alcohol consumption between a group of oesophageal cancer patients and a group of healthy controls matched for age and sex;
(c) comparison of current oesophageal cancer rates in a group of alcoholics and a group of teetotallers;
(d) comparison by questionnaire of history of alcohol consumption between a group of oesophageal cancer patients and a random sample from the electoral role in the surrounding district;
(e) comparison of death rates due to cancer of the oesophagus in a large sample of subjects whose alcohol consumption has been determined in the past.

13. In a case-control study to investigate whether eczema in children is related to cigarette smoking by their parents:

(a) parents would be asked about their smoking habits at the child's birth and the child observed for subsequent development of eczema;
(b) children of a group of parents who smoke would be compared to children of a group of parents who are non-smokers;
(c) parents would be asked stop to smoking to see whether their children's eczema was reduced;
(d) The smoking habits of the parents of a group of children with eczema would be compared to the smoking habits of the parents of a group of children without eczema;
(e) parents would be randomly allocated to smoking or non-smoking groups.

3E Exercise: *Campylobacter jejuni* infection

Campylobacter jejuni is a bacterium causing gastro-intestinal illness, spread by the faecal-oral route. It infects many species, and human infection has been recorded from handling pet dogs and cats, handling and eating chicken and other meats, and via milk and water supplies. Treatment is by antibiotics.

In May, 1990, there was a fourfold rise in the isolation rate of *C. je-*

Table 3.4. Doorstep delivery of milk bottles and exposure to bird attack

	No. (%) exposed			
	Cases		Controls	
Doorstep milk delivery	29	(91%)	47	(73%)
Previous milk bottle attack by birds	26	(81%)	25	(39%)
Milk bottle attack in week before illness	26	(81%)	5	(8%)
Protective measures taken	6	(19%)	14	(22%)
Handling attacked milk bottle in week before illness	17	(53%)	5	(8%)
Drinking milk from attacked bottle in week before illness	25	(80%)	5	(8%)

Table 3.5. Frequency of bird attacks on milk bottles

Number of days of week when attacks took place	Cases	Controls
0	3	42
1–3	11	3
4–5	5	1
6–7	10	1

juni in the Ogwr District, Mid-Glamorgan. The mother of a young boy admitted to hospital with febrile convulsions resulting from *C. jejuni* infection reported that her milk bottles had been attacked by birds during the week before her son's illness, a phenomenon which had been associated with campylobacter infection in another area. This observation, with the rise in *C. jejuni*, prompted a case-control study (Southern *et al.* 1990).

A 'case' was defined as a person with laboratory confirmed *C. jejuni* infection with onset between May 1 and June 1 1990, resident in an area with Bridgend at its centre. Cases were excluded if they had spent one or more nights away from this area in the week before onset, if they could have acquired the infection elsewhere, or were members of a household in which there had been a case of diarrhea in the preceding four weeks.

The controls were selected from the register of the general practice of the case, or in a few instances from practices serving the same area. Two controls were selected for each case, matched for sex, age (within 5 years), and area of residence.

Cases and controls were interviewed by means of a standard questionnaire at home or by telephone. Cases were asked about their exposure to various factors in the week before the onset of illness. Controls were asked the same questions about the corresponding week for their matched cases. Before we approached a control for an interview, we wrote explaining the purpose of the investigation. If a control or member of his or her family had had diarrhea lasting more than 3 days in the week before or during the illness of the respective case, or had spent any nights during that week away from home, another control was found. Evidence of bird attack included the pecking or tearing off of milk bottle tops. A history of bird attack was defined as a previous attack at that house.

55 people with campylobacter infection resident in the area were reported during the study period. Of these, 19 were excluded and 4 could not be interviewed, leaving 32 cases and 64 matched controls. There was no difference in milk consumption between cases and controls, but more cases than controls reported doorstep delivery of bottled milk, previous milk bottle attack by birds, milk bottle attack by birds in the index week, and handling or drinking milk from an attacked bottle (Table 3.4). Cases reported bird attacks more frequently than controls (Table 3.5). Controls were more likely to have protected their milk bottles from attack or to have discarded milk from attacked bottles. Almost all subjects whose milk bottles had been attacked mentioned that magpies and jackdaws were common in their area, though only 3 had actually witnessed attacks and none reported bird droppings near bottles.

None of the other factors investigated (handling raw chicken; eating chicken bought raw; eating chicken, beef or ham bought cooked; eating out; attending barbecue; cat or dog in the house; contact with other cats or dogs; and contact with farm animals) were significantly more common in controls than cases. Bottle attacks seemed to have ceased when the study was carried out, and no milk could be obtained for analysis.

1. What problems were there in selecting cases?
2. What problems were there in the selection of controls?
3. Are there any problems about data collection?
4. From the above, do you think there is convincing evidence that bird attacks on milk bottles cause campylobacter infection?
5. What further studies might be carried out?

4
Summarizing data

4.1 Types of data

In Chapters 2 and 3 we looked at ways in which data are collected. In this chapter we shall see how data can be summarized to help to reveal information they contain. We do this by calculating numbers from the data which extract the important material. These numbers are called **statistics**. A statistic is anything calculated from the data alone.

It is often useful to distinguish between three types of data: qualitative, discrete quantitative and continuous quantitative. **Qualitative** data arise when individuals may fall into separate classes. These classes may have no numerical relationship with one another at all, e.g. sex: male, female; types of dwelling: house, maisonette, flat, lodgings; eye colour: brown, grey, blue, green, etc. **Quantitative** data are numerical, arising from counts or measurements. If the values of the measurements are integers (whole numbers), like the number of people in a household, or number of teeth which have been filled, those data are said to be **discrete**. If the values of the measurements can take any number in a range, such as height or weight, the data are said to be **continuous**. In practice there is overlap between these categories. Most continuous data are limited by the accuracy with which measurements can be made. Human height, for example, is difficult to measure more accurately than to the nearest millimetre and is more usually measured to the nearest centimetre. So only a finite set of possible measurements is actually available, although the quantity 'height' can take an infinite number of possible values, and the measured height is really discrete. However, the methods described below for continuous data will be seen to be those appropriate for its analysis.

We shall refer to qualities or quantities such as sex, height, age, etc. as **variables**, because they vary from one member of a sample to another. A qualitative variable is also termed a **categorical variable** or an **attribute**. We shall use these terms interchangeably.

Table 4.1. Principle diagnosis of patients in Tooting Bec Hospital

Diagnosis	Number of patients
Schizophrenia	474
Affective disorders	277
Organic brain syndrome	405
Subnormality	58
Alcoholism	57
Other and not known	196
Total	1 467

4.2 Frequency distributions

When data are purely qualitative, the simplest way to deal with them is to count the number of cases in each category. For example, in the analysis of the census of a psychiatric hospital population (§3.2), one of the variables of interest was the patient's principal diagnosis (Bewley *et al.* 1975). To summarize these data, we count the number of patients having each diagnosis. The results are shown in Table 4.1. The count of individuals having a particular quality is called the **frequency** of that quality. For example, the frequency of schizophrenia is 474. The proportion of individuals having the quality is called the **relative frequency** or **proportional frequency**. The relative frequency of schizophrenia is $474/1467 = 0.32$ or 32%. The set of frequencies of all the possible categories is called the **frequency distribution** of the variable.

In this census we assessed whether patients were 'likely to be discharged', 'possibly to be discharged' or 'unlikely to be discharged'. The frequencies of these categories are shown in Table 4.2. Likelihood of discharge is a qualitative variable, like diagnosis, but the categories are ordered. This enables us to use another set of summary statistics, the cumulative frequencies. The **cumulative frequency** for a value of a variable is the number of individuals with values less than or equal to that value. Thus, if we order likelihood of discharge from 'unlikely', through 'possibly' to 'likely' the cumulative frequencies are 871, 1 210 (= 871 + 339) and 1 467. The **relative cumulative frequency** for a value is the proportion of individuals in the sample with values less than or equal to that value. For the example they are 0.59 (= 871/1 467), 0.82 and 1.00. Thus we can see that the proportion of patients for whom discharge was not thought likely was 0.82 or 82%.

As we have noted, likelihood of discharge is a qualitative variable, with ordered categories. Sometimes this ordering is taken into account in analysis, sometimes not. Although the categories are ordered these are not quantitative data. There is no sense in which the difference between 'likely' and 'possibly' is the same as the difference between 'possibly' and 'unlikely'.

Table 4.2. Likelihood of discharge of patients in Tooting Bec Hospital

Discharge:	Frequency	Relative frequency	Cumulative frequency	Relative cumulative frequency
unlikely	871	0.59	871	0.59
possible	339	0.23	1210	0.82
likely	257	0.18	1467	1.00
Total	1467	1.00	1467	1.00

Table 4.3. Parity of 125 women attending antenatal clinics at St. George's Hospital

Parity	Frequency	Relative frequency (per cent)	Cumulative frequency	Relative cumulative frequency (per cent)
0	59	47.2	59	47.2
1	44	35.2	103	82.4
2	14	11.2	117	93.6
3	3	2.4	120	96.0
4	4	3.2	124	99.2
5	1	0.8	125	100.0
Total	125	100.0	125	100.0

Table 4.3 shows the frequency distribution of a quantitative variable, parity. This shows the number of previous pregnancies for a sample of women booking for delivery at St. George's Hospital. Only certain values are possible, as the number of pregnancies must be an integer, so this variable is discrete. The frequency of each separate value is given.

Table 4.4 shows a continuous variable, forced expiratory volume in one second (FEV1) in a sample of male medical students. As most of the values occur only once, to get a useful frequency distribution we need to divide the FEV1 scale into class intervals, e.g. from 3.0 to 3.5, from 3.5 to 4.0, and so on, and count the number of individuals with FEV1s in each class interval. The class intervals should not overlap, so we must decide which interval contains the boundary point to avoid it being counted twice. It is usual to put the lower boundary of an interval into that interval and the higher boundary into the next interval. Thus the interval starting at 3.0 and ending at 3.5 contains 3.0 but not 3.5. We can write this as '3.0 – ' or '3.0 – 3.5^- or '3.0 – 3.499'

If we take a starting point of 2.5 and an interval of 0.5 we get the frequency distribution shown in Table 4.5. Note that this is not unique. If we take a starting point of 2.4 and an interval of 0.2 we get a different set of frequencies.

The frequency distribution can be calculated easily and accurately using a computer. Manual calculation is not so easy but must be done carefully

Table 4.4. FEV1 (litres) of 57 male medical students

2.85	3.19	3.50	3.69	3.90	4.14	4.32	4.50	4.80	5.20
2.85	3.20	3.54	3.70	3.96	4.16	4.44	4.56	4.80	5.30
2.98	3.30	3.54	3.70	4.05	4.20	4.47	4.68	4.90	5.43
3.04	3.39	3.57	3.75	4.08	4.20	4.47	4.70	5.00	
3.10	3.42	3.60	3.78	4.10	4.30	4.47	4.71	5.10	
3.10	3.48	3.60	3.83	4.14	4.30	4.50	4.78	5.10	

Table 4.5. Frequency distribution of FEV1 in 57 male medical students

FEV1	Frequency	Relative frequency (per cent)
2.0	0	0.0
2.5	3	5.3
3.0	9	15.8
3.5	14	24.6
4.0	15	26.3
4.5	10	17.5
5.0	6	10.5
5.5	0	0.0
Total	57	100.0

Table 4.6. Tally system for finding the frequency distribution of FEV1

FEV1		Frequency
2.0		0
2.5	///	3
3.0	///// ////	9
3.5	///// ///// ////	14
4.0	///// ///// /////	15
4.5	///// /////	10
5.0	///// /	6
5.5		0
Total		57

and systematically. One way recommended by many texts (e.g. Hill 1977) is to set up a tally system, as in Table 4.6. We go through the data and for each individual make a tally mark by the appropriate interval. We then count up the number in each interval. In practice this is very difficult to do accurately, and it needs to be checked and double-checked. Hill (1977) recommends writing each number on a card and dealing the cards into piles corresponding to the intervals. It is then easy to check that each pile contains only those cases in that interval and count them. This is undoubtedly superior to the tally system. My own preferred method is to order the observations from lowest to highest before marking the interval boundaries and counting, or to use the stem and leaf plot described below.

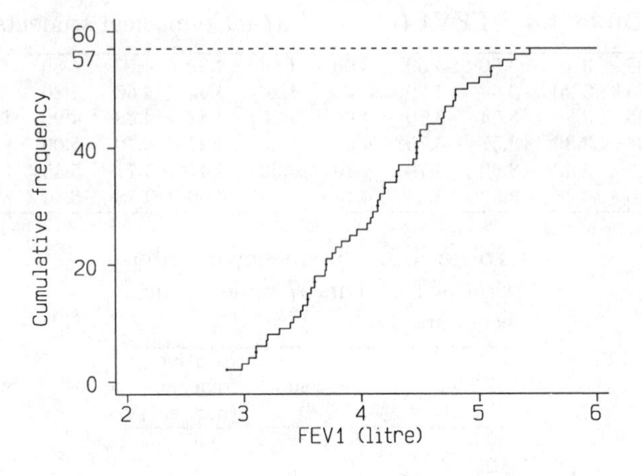

Fig. 4.1. Cumulative frequency distribution of FEV1 in a sample of male medical students

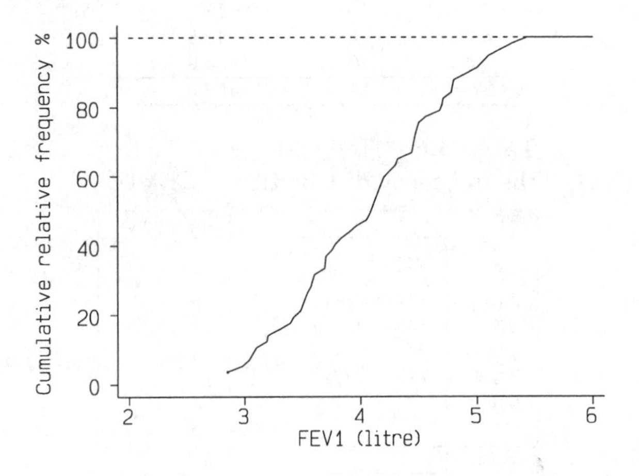

Fig. 4.2. Cumulative frequency polygon of FEV1

4.3 Histograms and other frequency graphs

Graphical methods are very useful for examining frequency distributions. Figure 4.1 shows a graph of the cumulative frequency distribution for the FEV1 data. This is called a step function. We can smooth this by joining successive points where the cumulative frequency changes by straight lines, to give a **cumulative frequency polygon**. Figure 4.2 shows this for the cumulative relative frequency distribution of FEV1. This plot is very useful for calculating some of the summary statistics referred to in §4.5.

The most common way of depicting a frequency distribution is by a

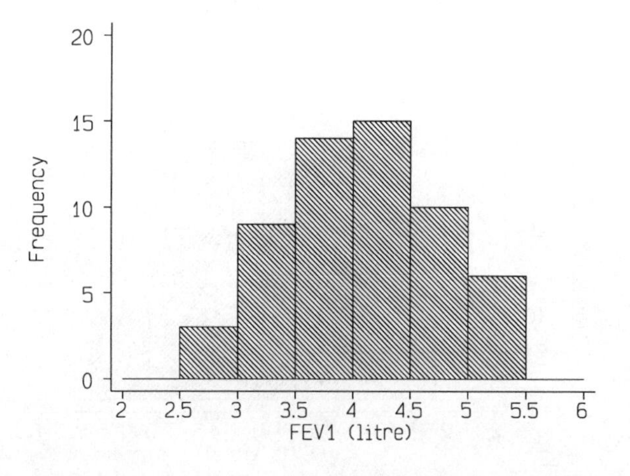

Fig. 4.3. Histogram of FEV1: frequency scale

histogram. This is a diagram where the class intervals are on an axis and rectangles with heights or areas proportional to the frequencies erected on them. Figure 4.3 shows the histogram for the FEV1 distribution in Table 4.5. The vertical scale shows frequency, the number of observations in each interval. Figure 4.4 shows a histogram for the same distribution, with frequency per unit FEV1 (or frequency density) shown on the vertical axis. The distributions appear identical and we may well wonder whether it matters which method we choose. We see that it does matter when we consider a frequency distribution with unequal intervals, as in Table 4.7. If we plot the histogram using the heights of the rectangles to represent relative frequency in the interval we get Figure 4.5, whereas if we use the relative frequency per year we get Figure 4.6. These histograms tell different stories. Figure 4.5 suggests that the most common age for accident victims is between 15 and 44 years, whereas Figure 4.6 suggests it is between 0 and 4. Figure 4.6 is correct, Figure 4.5 being distorted by the unequal class intervals. It is therefore preferable in general to use the frequency per unit rather than per class interval when plotting a histogram. The frequency for a particular interval is then represented by the area of the rectangle on that interval. Only when the class intervals are all equal can the frequency for the class interval be represented by the height of the rectangle.

A different version of the histogram has been developed by Tukey (1977), the **stem and leaf plot** (Figure 4.7). The rectangles are replaced by the numbers themselves. The 'stem' is the first digit or digits of the number and the 'leaf' the trailing digit. The first row of Figure 4.7 represents the numbers 2.8, 2.8, and 2.9, which in the data are 2.85, 2.85, and 2.98. The plot provides a good summary of data structure while at the

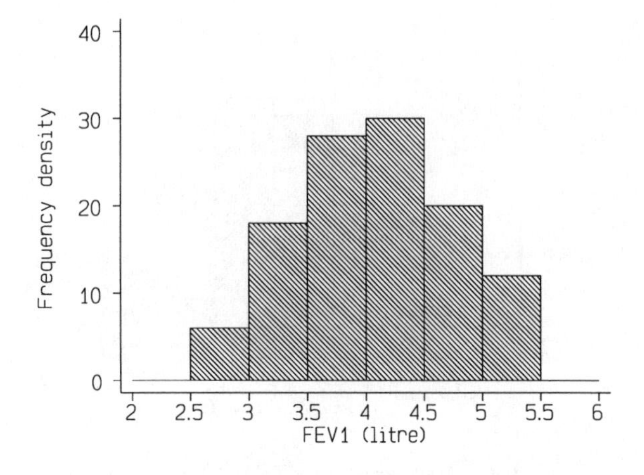

Fig. 4.4. Histogram of FEV1: frequency per unit FEV1 or frequency density scale

Table 4.7. Distribution of age in people suffering accidents in the home

Age group	Relative frequency (per cent)	Relative frequency per year (per cent)
0–4	25.3	5.06
5–14	18.9	1.89
15–44	30.3	1.01
45–64	13.6	0.68
65+	11.7	0.33

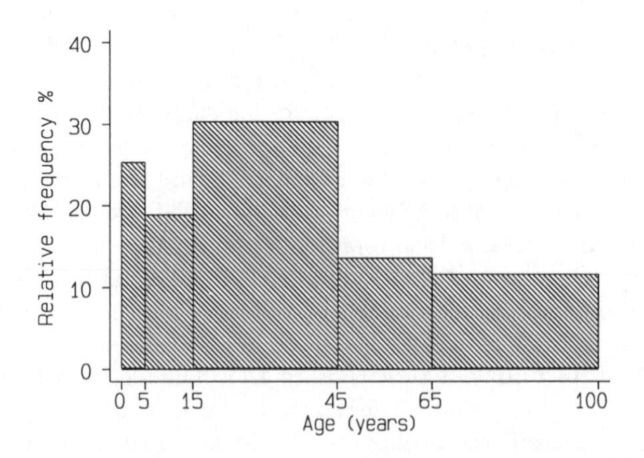

Fig. 4.5. Age distribution of home accident victims: relative frequency scale

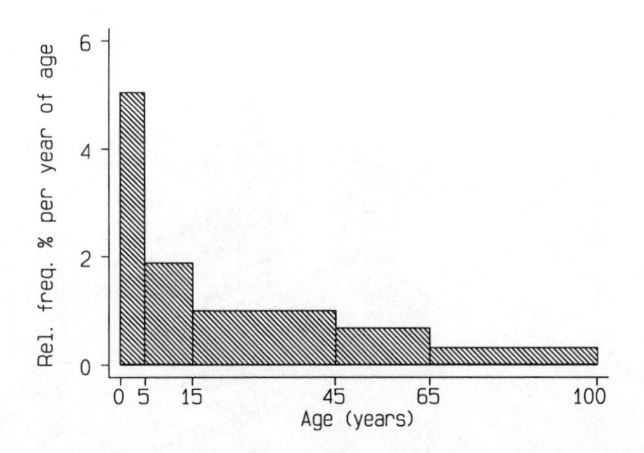

Fig. 4.6. Age distribution of home accident victims: relative frequency density scale

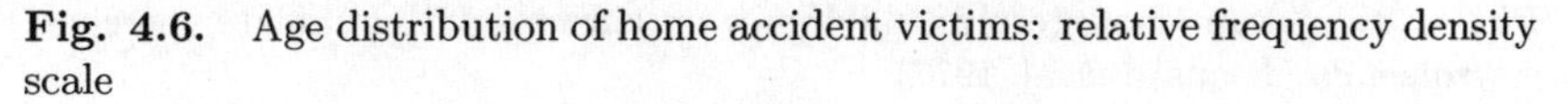

Fig. 4.7. Stem and leaf plot for the FEV1 data, rounded down to one decimal place

same time we can see other characteristics such as a tendency to prefer some trailing digits to others, called digit preference (§15.2). It is also easy to construct and much less prone to error than the tally method of finding a frequency distribution.

4.4 Shapes of frequency distribution

Figure 4.3 shows a frequency distribution of a shape often seen in medical data. The distribution is roughly symmetrical about its central value and has frequency concentrated about one central point. The most common value is called the **mode** of the distribution and Figure 4.3 has one such point. It is **unimodal**. Figure 4.8 shows a very different shape. Here there are two distinct modes, one near 5 and the other near 8.5. This distribution is **bimodal**. We must be careful to distinguish between the unevenness in the histogram which results from using a small sample to represent a large population and those which result from genuine bimodality in the data. The trough between 6 and 7 in Figure 4.8 is very marked and might represent

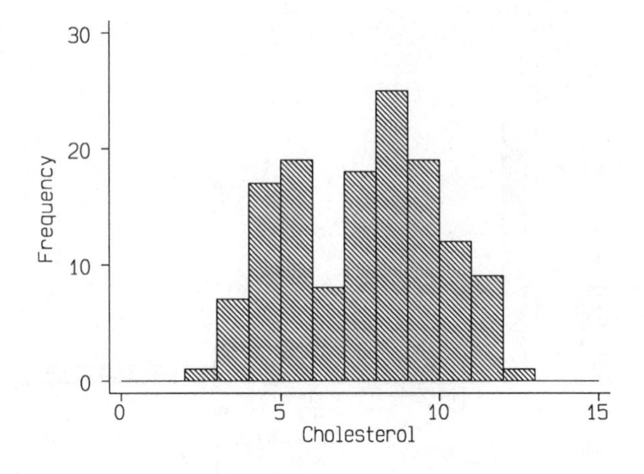

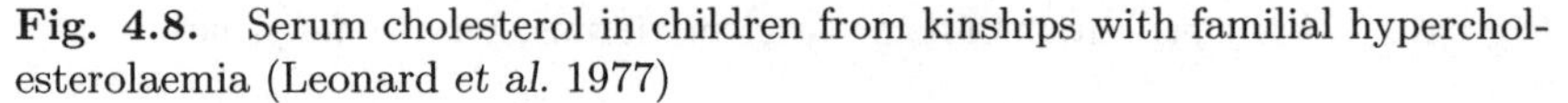

Fig. 4.8. Serum cholesterol in children from kinships with familial hypercholesterolaemia (Leonard *et al.* 1977)

a genuine bimodality. In this case we have children, some of whom have a condition which raises the cholesterol level and some of whom do not. We actually have two separate populations represented with some overlap between them. However, almost all distributions encountered in medical statistics are unimodal.

Figure 4.9 differs from Figure 4.3 in a different way. We have already noted that the distribution of FEV1 is symmetrical. The distribution of serum triglyceride is **skew**, that is, the distance from the central value to the extreme is much greater on one side than it is on the other. The parts of the histogram near the extremes are called the **tails** of the distribution. If the tail on the right is longer than the tail on the left as in Figure 4.9, the distribution is **skew to the right** or **positively skew**. If the tail on the left is longer, the distribution is **skew to the left** or **negatively skew**. If the tails are equal the distribution is **symmetrical**. Most distributions encountered in medical work are symmetrical or skew to the right, for reasons we shall discuss later (§7.4).

4.5 Medians and quantiles

We often want to summarize a frequency distribution in a few numbers, for ease of reporting or comparison. The most direct method is to use quantiles. The **quantiles** are values which divide the distribution such that there is a given proportion of observations below the quantile. For example, the median is a quantile. The **median** is the central value of the distribution, such that half the points are less than or equal to it and half are greater than or equal to it. We can estimate any quantiles easily

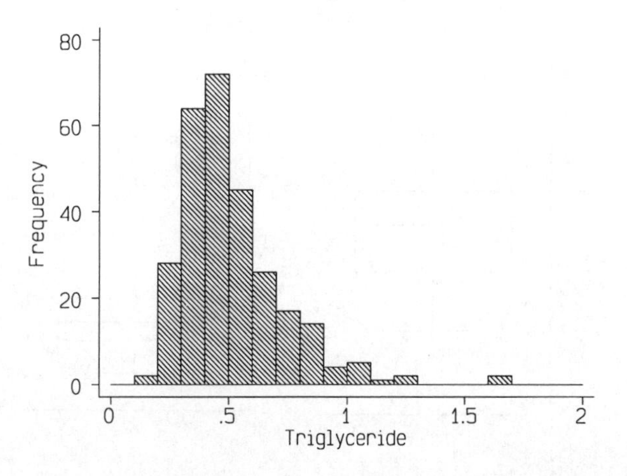

Fig. 4.9. Serum triglyceride in cord blood from 282 babies

from the cumulative frequency distribution or a stem and leaf plot. For the FEV1 data the median is 4.1, the 29th value in Table 4.4. If we have an even number of points, we choose a value midway between the two central values.

In general, we estimate the q quantile, the value such that a proportion q will be below it, as follows. We have n ordered observations which divide the scale into $n+1$ parts: below the lowest observation, above the highest and between each adjacent pair. The proportion of the distribution which lies below the ith observation is estimated by $i/(n+1)$. We set this equal to q and get $i = q(n+1)$. If i is an integer, the ith observation is the required quantile estimate. If not, let j be the integer part of i, the part before the decimal point. The quantile will lie between the jth and $j+1$th observations. We estimate it by

$$x_j + (x_{j+1} - x_j) \times (i - j)$$

For the median, for example, the 0.5 quantile, $i = q(n+1) = 0.5 \times (57+1) = 29$, the 29th observation as before.

Other quantiles which are particularly useful are the **quartiles** of the distribution. The quartiles divide the distribution into four equal parts, called **fourths**. The second quartile is the median. For the FEV1 data the first and third quartiles are 3.54 and 4.53. For the first quartile, $i = 0.25 \times 58 = 14.5$. The quartile is between the 14th and 15th observations, which are both 3.54. For the third quartile, $i = 0.75 \times 58 = 43.5$, so the quartile lies between the 42nd and 43rd observations, which are 4.50 and 4.56. The quantile is given by $4.50 + (4.56 - 4.50) \times (43.5 - 43) = 4.53$. We often divide the distribution into 100 **centiles** or **percentiles**. The median

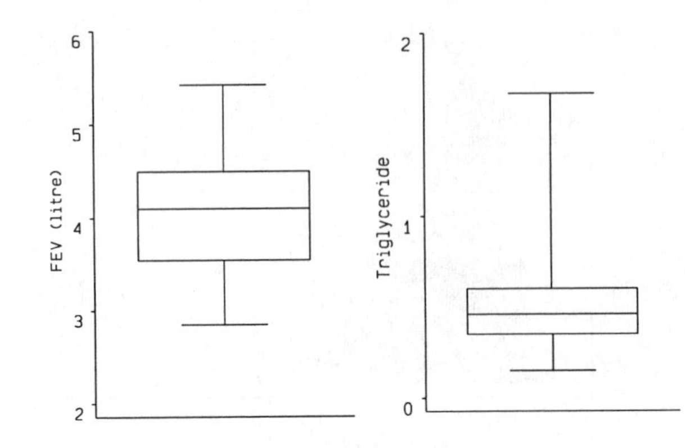

Fig. 4.10. Box and whisker plots for FEV1 and for serum triglyceride

is thus the 50th centile. For the 20th centile of FEV1, $i = 0.2 \times 58 = 11.6$, so the quantile is between the 11th and 12th observation, 3.42 and 3.48, and can be estimated by $3.42 + (3.48 - 3.42) \times (11.6 - 11) = 3.46$. We can estimate these easily from Figure 4.2 by finding the position of the quantile on the vertical axis, e.g. 0.2 for the 20th centile or 0.5 for the median, drawing a horizontal line to intersect the cumulative frequency polygon, and reading the quantile off the horizontal axis.

Tukey (1977) used the median, quartiles, maximum and minimum as a convenient five figure summary of a distribution. He also suggested a neat graph, the **box and whisker plot**, which represents this (Figure 4.10). The box shows the distance between the quartiles, with the median marked as a line, and the 'whiskers' show the extremes. The different shapes of the FEV1 and serum triglyceride distributions is clear from the graph. An observation a long way from the rest of the data may be shown separately. The plot is useful for showing the comparison of several groups (Figure 4.11).

4.6 The mean

The median is not the only measure of central value for a distribution. Another is the **arithmetic mean** or **average**, usually referred to simply as the **mean**. This is found by taking the sum of the observations and dividing by their number. For example, consider the following hypothetical data:

2 3 9 5 4 0 6 3 4

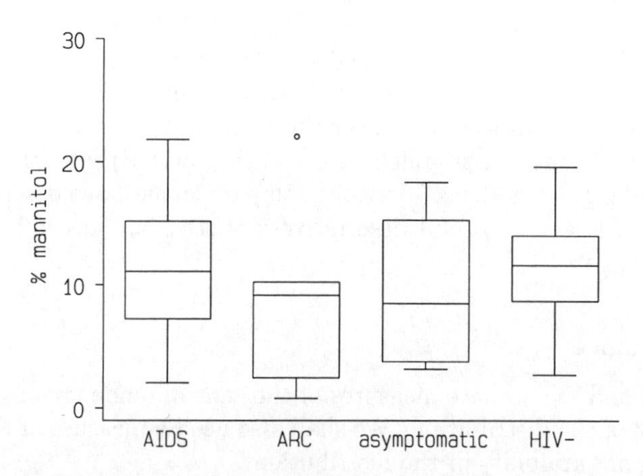

Fig. 4.11. Box plots showing a roughly symmetrical variable in four groups, with an extreme point (data in Table 10.8)

The sum is 36 and there are 9 observation, so the mean is $36/9 = 4.0$. At this point we will need to introduce some algebraic notation, widely used in statistics. We denote the observations by

$$x_1, x_2, ..., x_i, ..., x_n$$

There are n observations and the ith of these is x_i. For the example, $x_4 = 5$ and $n = 9$. The sum of all the x_i is

$$\sum_{i=1}^{n} x_i$$

The summation sign is an upper case Greek letter, sigma, the Greek S. When it is obvious that we are adding the values of x_i for all values of i, which runs from 1 to n, we abbreviate this to $\sum x_i$ or simply to $\sum x$. The mean of the x_i is denoted by $\bar{x}$ ('x bar'), and

$$\bar{x} = \frac{1}{n}\sum x_i = \frac{\sum x_i}{n}$$

The sum of the 57 FEV1s is 231.51 and hence the mean is $231.51/57 = 4.06$. This is very close to the median, 4.1, so the median is within 1% of the mean. This is not so for the triglyceride data. The median triglyceride is 0.46 but the mean is 0.51, which is higher. The median is 10% away from the mean. If the distribution is symmetrical the sample mean and median will be about the same, but in a skew distribution they will not.

If the distribution is skew to the right, as for serum triglyceride, the mean will be greater, if it is skew to the left the median will be greater. This is because the values in the tails affect the mean but not the median.

The sample mean has much nicer mathematical properties than the median and is thus more useful for the comparison methods described later. The median is a very useful descriptive statistic, but not much used for other purposes.

4.7 Variance

The mean and median are measures of the central tendency or position of the middle of the distribution. We shall also need a measure of the spread, dispersion or variability of the distribution.

One obvious measure is the **range**, the difference between the highest and lowest values. This is a useful descriptive measure, but has two disadvantages. Firstly, it depends only on the extreme values and so can vary a lot from sample to sample. Secondly, it depends on the sample size. The larger the sample is, the further apart the extremes are likely to be. We can see this if we consider a sample of size 2. If we add a third member to the sample the range will only remain the same if the new observation falls between the other two, otherwise the range will increase. We can get round the second of these problems by using the **interquartile range**, the differences between the first and third quartiles. However, the interquartile range is quite variable from sample to sample and is also mathematically intractable. Although a useful descriptive measure, it is not the one preferred for purposes of comparison.

The most commonly used measures of dispersion are the variance and standard deviation. We start by calculating the difference between each observation and the sample mean, called the **deviations from the mean**, Table 4.8. If the data are widely scattered, many of the observations x_i will be far from the mean $\bar{x}$ and so many deviations $x_i - \bar{x}$ will be large. If the data are narrowly scattered, very few observations will be far from the mean and so few deviations $x_i - \bar{x}$ will be large. We need some kind of average deviation to measure the scatter. If we add all the deviations together, we get zero, because $\sum(x_i - \bar{x}) = \sum x_i - \sum \bar{x} = \sum x_i - n\bar{x}$ and $n\bar{x} = \sum x_i$. Instead we square the deviations and then add them, as shown in Table 4.8. This removes the effect of sign; we are only measuring the size of the deviation not the direction. This gives us $\sum(x_i - \bar{x})^2$, in the example equal to 52, called the **sum of squares about the mean**, usually abbreviated to **sum of squares**.

Clearly, the sum of squares will depend on the number of observations as well as the scatter. We want to find some kind of average squared deviation. This leads to a difficulty. Although we want an average squared

Table 4.8. Deviations from the mean of 9 observations

Observations x_i	Deviations from the mean $x_i - \bar{x}$	Squared deviations $(x_i - \bar{x})^2$
2	−2	4
3	−1	1
9	5	25
5	1	1
4	0	0
0	−4	16
6	2	4
3	−1	1
4	0	0
36	0	52

deviation, we divide the sum of squares by $n-1$, not n. This is not the obvious thing to do and puzzles many students of statistical methods. The reason is that we are interested in estimating the scatter of the population, rather than the sample, and the sum of squares about the sample mean is proportional to $n-1$ (§4A, §6B). Dividing by n would lead to small samples producing lower estimates of variability than large samples. The minimum number of observations from which the variability can be estimated is 2, a single observation cannot tell us how variable the data are. If we used n as our divisor, for $n = 1$ the sum of squares would be zero, giving a variance of zero. With the correct divisor of $n-1$, $n = 1$ gives the meaningless ratio 0/0, reflecting the impossibility of estimating variability from a single observation. The estimate of variability is called the **variance**, defined by

$$\text{variance} = \frac{1}{n-1}\sum(x_i - \bar{x})^2$$

We have already said that $\sum(x_i - \bar{x})^2$ is called the sum of squares. The quantity $n-1$ is called the **degrees of freedom** of the variance estimate (§7A). We have:

$$\text{variance} = \frac{\text{sum of squares}}{\text{degrees of freedom}}$$

We shall usually denote the variance by s^2. In the example, the sum of squares is 52 and there are 9 observations, giving 8 degrees of freedom. Hence $s^2 = 52/8 = 6.5$.

The formula $\sum(x_i - \bar{x})^2$ gives us a rather tedious calculation. There is another formulae for the sum of squares, which make the calculation easier to carry out. This is simply an algebraic manipulation of the first form and give exactly the same answers. We thus have two formulae for variance:

$$s^2 = \frac{1}{n-1}\sum(x_i - \bar{x}^2)$$

$$s^2 = \frac{1}{n-1}\left(\sum x_i^2 - \frac{\left(\sum x_i\right)^2}{n}\right)$$

These are simply algebraic manipulations of the first form and give exactly the same answers. The algebra is quite simple and is given in §4B. For example, using the second formula for the nine observations, we have:

$$\begin{aligned}
\sum x_i^2 &= 2^2+3^2+9^2+5^2+4^2+0^2+6^2+3^2+4^2 \\
&= 4+9+81+25+16+0+36+9+16 \\
&= 196 \\
\sum x_i &= 36 \\
s^2 &= \frac{1}{n-1}\left(\sum x_i^2 - \frac{(\sum x_i)^2}{n}\right) \\
&= \frac{1}{9-1}\left(196 - \frac{36^2}{9}\right) \\
&= \frac{1}{8}(196-144) \\
&= 52/8 \\
&= 6.5
\end{aligned}$$

as before. On a calculator this is a much easier formula than the first, as the numbers need only be put in once. It can be inaccurate, because we subtract one large number from another to get a small one. For this reason the first formula would be used in a computer program.

4.8 Standard deviation

The variance is calculated from the squares of the observations. This means that it is not in the same units as the observations, which limits its use as a descriptive statistic. The obvious answer to this is to take the square root, which will then have the same units as the observations and the mean. The square root of the variance is called the **standard deviation**, usually denoted by s. Thus,

$$\begin{aligned}
s &= \sqrt{\frac{1}{n-1}\sum (x_i-\bar{x})^2} \\
&= \sqrt{\frac{1}{n-1}\left(\sum x_i^2 - \frac{\left(\sum x_i\right)^2}{n}\right)}
\end{aligned}$$

Returning to the FEV data, we calculate the variance and standard deviation as follows. We have $n = 57$, $\sum x_i = 231.51$, $\sum x_i^2 = 965.45$.

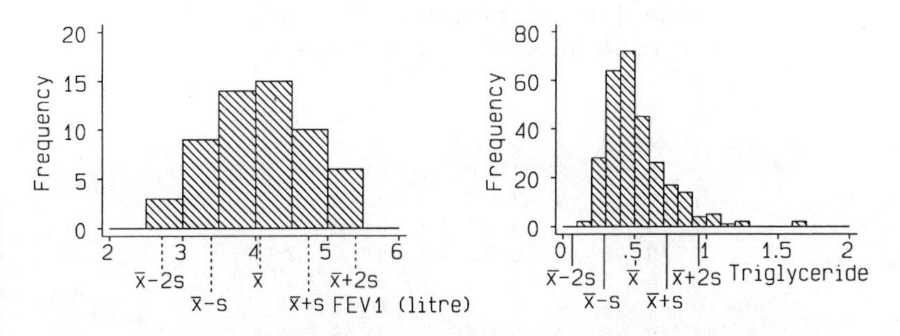

Fig. 4.12. Histograms of FEV1 and triglyceride with mean and standard deviation

$$\begin{aligned}\text{Sum of squares} &= \sum x_i^2 - \frac{(\sum x_i)^2}{n} \\ &= 965.45 - \frac{231.51^2}{57} \\ &= 965.45 - 940.296 \\ &= 25.154\end{aligned}$$

$$\begin{aligned}s^2 &= \frac{\text{sum of squares}}{n-1} \\ &= \frac{25.154}{57-1} \\ &= 0.449\end{aligned}$$

The standard deviation is $s = \sqrt{s^2} = \sqrt{0.449} = 0.67$ litres.

Figure 4.12 shows the relationship between mean, standard deviation and frequency distribution. For FEV1, we see that the majority of observations are within one standard deviation of the mean, and nearly all within two standard deviations of the mean. There is a small part of the histogram outside the $\bar{x} - 2s$ to $\bar{x} + 2s$ interval, on either side of this symmetrical histogram. As Figure 4.12 also shows, this is true for the highly skew triglyceride data, too. In this case, the outlying observations are all in one tail of the distribution, however. In general, we expect roughly 2/3 of observations to lie within one standard deviation of the mean and 95% to lie within two standard deviations of the mean.

4A Appendix: The divisor for the variance

The variance is found by dividing the sum of squares about the sample mean by $n-1$, not by n. This is because we want the scatter about the population mean, and the scatter about the sample mean is always less.

Table 4.9. Population of 100 random digits for a sampling experiment

9	1	0	7	5	6	9	5	8	8
1	8	8	8	5	2	4	8	3	1
2	8	1	8	5	8	4	0	1	9
1	9	7	9	7	2	7	7	0	8
7	0	2	8	8	7	2	5	4	1
1	0	5	7	6	5	0	2	2	2
6	5	5	7	4	1	7	3	3	3
2	1	6	9	4	4	7	6	1	7
1	6	3	8	0	5	7	4	8	6
8	6	8	3	5	8	2	7	2	4

The sample mean is 'closer' to the data points than is the population mean. We shall try a little sampling experiment to show this. Table 4.9 shows a set of 100 random digits which we shall take as the population to be sampled. They have mean 4.74 and the sum of squares about the mean is 811.24. Hence the average squared difference from the mean is 8.1124. We can take samples of size two at random from this population using a pair of decimal dice, which will enable us to choose any digit numbered from 00 to 99. The first pair chosen was 5 and 6 which has mean 5.5. The sum of squares about the population mean 4.74 is $(5-4.74)^2+(6-4.74)^2 = 1.655$. The sum of squares about the sample mean is $(5-5.5)^2+(6-5.5)^2 = 0.5$.

The sum of squares about the population mean is greater than the sum of squares about the sample mean, and this will always be so. Table 4.10 shows this for 20 such samples of size two. The average sum of squares about the population mean is 13.6, and about the sample mean it is 5.7. Hence dividing by the sample size ($n = 2$) we have mean square differences of 6.8 about the population mean and 2.9 about the sample mean. Compare this to 8.1 for the population as a whole. We see that the sum of squares about the population mean is quite close to 8.1, while the sum of squares about the sample mean is much less. However, if we divide the sum of squares about the sample mean by $n-1$, i.e. 1, instead of n we have 5.7, which is not much different to the 6.8 from the sum of squares about the population mean.

Table 4.11 shows the results of a similar experiment with more samples being taken. The Table shows the two average variance estimates using n and $n-1$ as the divisor of the sum of squares, for sample sizes 2, 3, 4, 5 and 10. We see that the sum of squares about the sample mean divided by n increases steadily with sample size, but if we divide it by $n-1$ instead of n the estimate does not change as the sample size increases. The sum of squares about the sample mean is proportional to $n-1$.

Table 4.10. Sampling pairs from Table 4.9

Sample		$\sum(x_i - \mu)^2$	$\sum(x_i - \bar{x})^2$
5	6	1.655	0.5
8	8	21.255	0.0
6	1	15.575	12.5
9	3	21.175	18.0
5	5	0.135	0.0
7	7	10.215	0.0
1	7	19.095	18.0
9	8	28.775	0.5
3	3	6.055	0.0
5	1	14.055	8.0
8	3	13.655	12.5
5	7	5.175	2.0
5	2	5.575	4.5
5	7	5.175	2.0
8	8	21.255	0.0
3	2	10.535	0.5
0	4	23.015	8.0
9	3	21.175	18.0
5	2	7.575	4.5
6	9	19.735	4.5
Mean		13.6432	5.7

Table 4.11. Mean sums of squares about the sample mean for sets of 100 random samples from Table 4.10

Number in	Mean variance estimates	
sample, n	$\frac{1}{n}\sum(x_i - \bar{x})^2$	$\frac{1}{n-1}\sum(x_i - \bar{x})^2$
2	4.5	9.1
3	5.4	8.1
4	5.9	7.9
5	6.2	7.7
10	7.2	8.0

4B Appendix: Formulae for the sum of squares

The different formulae for sums of squares are derived as follows:

$$\begin{aligned}
\text{sum of squares} &= \sum(x_i - \bar{x})^2 \\
&= \sum(x_i^2 - 2x_i\bar{x} + \bar{x}^2) \\
&= \sum x_i^2 - \sum 2x_i\bar{x} + \sum \bar{x}^2 \\
&= \sum x_i^2 - 2\bar{x}\sum x_i + n\bar{x}^2
\end{aligned}$$

because $\bar{x}$ has the same value for each of the n observations. Now, $\sum x_i = n\bar{x}$, so

$$\begin{aligned}\text{sum of squares} &= \sum x_i^2 - 2\bar{x}n\bar{x} + n\bar{x}^2 \\ &= \sum x_i^2 - 2n\bar{x}^2 + n\bar{x}^2 \\ &= \sum x_i^2 - n\bar{x}^2\end{aligned}$$

and putting $\bar{x} = \frac{1}{n}\sum x_i$

$$\begin{aligned}\text{sum of squares} &= \sum x_i^2 - n\left(\frac{1}{n}\sum x_i\right)^2 \\ &= \sum x_i^2 - \frac{\left(\sum x_i\right)^2}{n}\end{aligned}$$

We thus have three formulae for variance:

$$\begin{aligned}s^2 &= \frac{1}{n-1}\sum(x_i - x)^2 \\ &= \frac{1}{n-1}\left(\sum x_i^2 - n\bar{x}^2\right) \\ &= \frac{1}{n-1}\left(\sum x_i^2 - \frac{(\sum x_i)^2}{n}\right)\end{aligned}$$

4M Multiple choice questions 14 to 19

(Each branch is either true or false)

14. Which of the following are qualitative variables:

(a) sex;

(b) parity;

(c) diastolic blood pressure;

(d) diagnosis;

(e) height.

15. Which of the following are continuous variables:

(a) blood glucose;

(b) peak expiratory flow rate;

(c) age last birthday;

(d) exact age;

(e) family size.

16. When a distribution is skew to the right:

(a) the median is greater than the mean;

(b) the distribution is unimodal;

(c) the tail on the left is shorter than the tail on the right;

(d) the standard deviation is less than the variance;

(e) the majority of observations are less than the mean.

17. The shape of a frequency distribution can be described using:

(a) a box and whisker plot;

(b) a histogram;

(c) a stem and leaf plot;

(d) mean and variance;

(e) a table of frequencies.

18. For the sample 3, 1, 7, 2, 2:

(a) the mean is 3;

(b) the median is 7;

(c) the mode is 2;

(d) the range is 1;

(e) the variance is 5.5.

19. Diastolic blood pressure has a distribution which is slightly skew to the right. If the mean and standard deviation were calculated for the diastolic pressures of a random sample of men:

(a) there would be fewer observations below the mean than above it;

(b) the standard deviation would be approximately equal to the mean;

(c) the majority of observations would be more than one standard deviation from the mean;

(d) the standard deviation would estimate the accuracy of blood pressure measurement;

(e) About 95% of observations would be expected to be within two standard deviations of the mean.

4E Exercise: Mean and standard deviation

This exercise gives some practice in one of the most fundamental calculations in statistics, that of the sum of squares and standard deviation. It also shows the relationship of the standard deviation to the frequency distribution. Table 4.12 shows blood glucose levels obtained from a group of medical students.

1. Make a stem and leaf plot for these data.
2. Find the minimum, maximum and quartiles and sketch a box and

Table 4.12. Random blood glucose levels from a group of first year medical students (mmol/litre)

4.7	3.6	3.8	2.2	4.7	4.1	3.6	4.0	4.4	5.1
4.2	4.1	4.4	5.0	3.7	3.6	2.9	3.7	4.7	3.4
3.9	4.8	3.3	3.3	3.6	4.6	3.4	4.5	3.3	4.0
3.4	4.0	3.8	4.1	3.8	4.4	4.9	4.9	4.3	6.0

whisker plot.
3. Find the frequency distribution, using a class interval of 0.5.
4. Sketch the histogram of this frequency distribution. What term best describes the shape: symmetrical, skew to the right or skew to the left?
5. For the first column only, i.e. for 4.7, 4.2, 3.9, and 3.4, calculate the standard deviation using the deviations from the mean formula

$$s = \sqrt{\frac{1}{n-1}\sum(x_i - \bar{x})^2}$$

First calculate the mean. Calculate the deviations from the mean, i.e. the differences between the observations and the mean. Now square these and add to give the sum of squares about the mean. How many degrees of freedom are there for these four observations? Hence calculate the variance and the standard deviation.
6. For the same four numbers, calculate the standard deviation using the formula

$$s = \sqrt{\frac{1}{n-1}\left(\sum x_i^2 - \frac{(\sum x_i)^2}{n}\right)}$$

First calculate the sum of the observations and the sum of the observations squared. Hence calculate the sum of squares about the mean. Is this the same as that found in 4 above? Hence calculate the variance and the standard deviation.
7. Use the following summations for the whole sample: $\sum x_i = 162.2$, $\sum x_i^2 = 676.74$. Calculate the mean of the sample, the sum of squares about the mean, the degrees of freedom for this sum of squares, and hence estimate the variance and standard deviation.
8. Calculate the mean $\pm$ one standard deviation and mean $\pm$ two standard deviations. Indicate these points and the mean on the histogram. What do you observe about their relationship to the frequency distribution?

5
Presenting data

5.1 Rates and proportions

Having collected our data as described in Chapters 2 and 3 and extracted information from it using the methods of Chapter 4, we must find a way to convey this information to others. In this chapter we shall look at some of the methods of doing that. We begin with rates and proportions.

When we have data in the form of frequencies, we often need to compare the frequency with certain conditions in groups containing different totals. In Table 2.1, for example, two groups of patient pairs were compared, 29 where the later patient had a C-T scan and 89 where neither had a C-T scan. The later patient did better in 9 of the first group and 34 of the second group. To compare these frequencies we compare the proportions 9/29 and 34/89. These are 0.31 and 0.38, and so we can conclude that there is little difference. In Table 2.1, these were given as percentages, that is, the proportion out of 100 rather than out of 1, to avoid the decimal point. In Table 2.8, the Salk vaccine trial, the proportions contracting polio were presented as the number per 100 000 for the same reason.

A **rate** expresses the frequency of the characteristic of interest per 1000 (or per 100 000, etc.) of the population. For example, in Table 3.1, the results of the study of smoking by doctors, the data were presented as the number of deaths per 1 000 doctors per year. This is not a proportion, as a further adjustment has been made to allow for the time period observed. Furthermore, the rate has been adjusted to take account of any differences in the age distributions of smokers and non-smokers (§16.2). Sometimes the actual denominator for a rate may be continually changing. The number of deaths from lung cancer among men in England and Wales for 1983 was 26 502. The denominator for the death rate, the number of males in England and Wales, changed throughout 1983, as some died, some were born, some left the country and some entered it. The death rate is calculated by using a representative number, the estimated population at the end of June 1983, the middle of the year. This was 24 175 900, giving a death rate of 26 502/24 175 900, which equals 0.001 096, or 109.6 deaths per 100 000 at risk per year. A number of the rates used in medical statistics are described in §16.5.

The use of rates and proportions enables us to compare frequencies obtained from unequal sized groups, base populations or time periods, but we must beware of their use when their bases or denominators are not given. Victora (1982) reported a drug advertisement sent to doctors which described the antibiotic phosphomycin as being '100% effective in chronic urinary infections'. This is very impressive. How could we fail to prescribe a drug which is 100% effective? The study on which this was based used 8 patients, after excluding 'those whose urine contained phosphomycin-resistant bacteria'. If the advertisement has said the drug was effective in 100% of 8 cases, we would have been less impressed. Had we known that it worked in 100% of 8 cases selected because it might work in them, we would have been still less impressed. The same paper quotes an advertisement for a cold remedy, where 100% of patients showed improvement. This was out of 5 patients! As Victora remarked, such small samples are understandable in the study of very rare diseases, but not for the common cold.

Sometimes we can fool ourselves as well as others by omitting denominators. I once carried out a study of the distribution of the soft tissue tumour Kaposi's sarcoma in Tanzania (Bland *et al.* 1977), and while writing it up I came across a paper setting out to do the same thing (Schmid 1973). One of the factors studied was tribal group, of which there are over 100 in Tanzania. This paper reported 'the tribal incidence in the Wabende, Wambwe and Washirazi is remarkable ... These small tribes, each with fewer than 90 000 people, constitute the group in which a tribal factor can be suspected'. This is based on the following rates of tumours per 10 000 population: national, 0.1; Wabende, 1.3; Wambwe, 0.7; Washirazi, 1.3. These are very big rates compared to the national, but the populations on which they are based are small, 8 000, 14 000 and 15 000 respectively (Egero and Henin 1973). To get a rate of 1.3/10 000 out of 8 000 Wabende people we must have $8\,000 \times 1.3/10\,000 = 1$ case! Similarly we have 1 case among the 14 000 Wambwe and 2 among the 15 000 Washirazi. We can see that there is not enough data to draw the conclusions which the author has done. Rates and proportions are powerful tools and we must beware of them becoming detatched from the original data.

5.2 Significant figures

When we calculated the death rate due to lung cancer among men in 1983 we quoted the answer as 0.001 096 or 109.6 per 100 000 per year. This is an approximation. The rate to the greatest number of figures my calculator will give is 0.001 096 215 653 and this number would probably go on indefinitely, turning into a recurring series of digits. The decimal system of representing numbers cannot in general represent fractions exactly. We know that $1/2 = 0.5$, but $1/3 = 0.333\,333\,33\ldots$, recurring infinitely. This does not usually worry us, because for most applications the difference between 0.333 and 1/3 is too small to matter. Only the first few non-zero digits of the number are important and we call these the **significant digits** or **significant figures**. There is usually little point in quoting statistical data to more than three significant figures. After all, it hardly matters whether the lung cancer mortality rate is 0.001 096 or 0.001 097. The value 0.001 096 is given to 4 significant figures. The leading zeros are not significant, the first significant digit in this number being '1'. To three significant figures we get 0.001 10, because the last digit is 6 and so the 9 which preceeds it is rounded up to 10. Note that significant figures are not the same as decimal places. The number 0.00110 is given to 5 decimal places, the number of digits after the decimal point. When rounding to the nearest digit, we leave the last significant digit, 9 in this case, if what follows it is less than 5, and increase by one if what follows is greater than 5. When we have exactly 5, I would always round up, i.e. 1.5 goes to 2. This means that 0, 1, 2, 3, 4 go down and 5, 6, 7, 8, 9 go up, which seems unbiased. Some writers take the view that 5 should go up half the time and down half the time, since it is exactly midway between the preceeding digit and that digit plus one. Various methods are suggested for doing this but I do not recommend them myself. In any case, it is usually a mistake to round to so few significant figures that this matters.

How many significant figures we need depends on the use to which the number is to be put and on how accurate it is anyway. For example, if we have a sample of 10 sublingual temperatures measured to the nearest half degree, there is little point in quoting the mean to more than 3 significant figures. What we should *not* do is to round numbers to a few significant figures before we have completed our calculations. In the lung cancer mortality rate example, suppose we round the numerator and denominator to two significant figures. We have $27\,000/24\,000\,000 = 0.001\,125$ and the answer is only correct to two figures. This can spread through calculations causing errors to build up. We always try to retain several more significant figures than we required for the final answer.

Consider Table 5.1. This shows mortality data in terms of the exact numbers of deaths in one year. The table is taken from a much larger table (OPCS 1991) which shows the numbers dying from every cause of death in the International Classification of Diseases (ICD), which gives numerical codes to many hundreds of causes of death. The full table, which also gives deaths by age group, covers 70 A4 pages. Table 5.1 shows deaths for broad groups of diseases called ICD chapters. This table is not a good way to present these data if we want to get an understanding of the frequency distribution of cause of death, and the differences between

Table 5.1. Deaths by sex and cause, England and Wales, 1989 (OPCS 1991, DH2 No. 10)

I.C.D. chapter and type of disease		Number of deaths	
		Males	Females
I	Infectious and parasitic	1 246	1 297
II	Neoplasms (cancers)	75 172	69 948
III	Endocrine, nutritional and metabolic diseases and immunity disorders	4 395	5 758
IV	Blood and blood forming organs	1 002	1 422
V	Mental disorders	4 493	9 225
VI	Nervous system and sense organs	5 466	5 990
VII	Circulatory system	127 435	137 165
VIII	Respiratory system	33 489	33 223
IX	Digestive system	7 900	10 779
X	Genitourinary system	3 616	4 156
XI	Complications of pregnancy, childbirth and the puerperium	0	56
XII	Skin and subcutaneous tissues	250	573
XIII	Musculo-skeletal system and connective tissues	1 235	4 139
XIV	Congenital anomalies	897	869
XV	Certain conditions originating in the perinatal period	122	118
XVI	Signs, symptoms and ill-defined conditions	1 582	3 082
XVII	Injury and poisoning	11 073	6 427
Total		279 373	294 227

causes in men and women. This is even more true of the 70 page original. This is not the purpose of the table, of course. It is a source of data, a reference document from which users extract information for their own purposes. Let us see how Table 5.1 can be simplified. First, we can reduce the number of significant figures. Let us be extreme and reduce the data to one significant figure (Table 5.2). This makes comparisons rather easier, but it is still not obvious which are the most important causes of death. We can improve this by re-ordering the table to put the most frequent cause, diseases of the circulatory system, first (Table 5.3). We can also combine a lot of the smaller categories into an 'others' group. I did this arbitrarily, by combining all those accounting for less than 2% of the total. Now it is clear at a glance that the most important causes of death in England and Wales are diseases of the circulatory system, neoplasms and diseases of the respiratory system, and that these dwarf all the others. Of course, mortality is not the only indicator of the importance of a disease. ICD chapter XIII, diseases of the musculo-skeletal system and connective tissues, are easily seen from Table 5.2 to be only minor causes of death, but this group includes arthritis and rheumatism, the most important illness in its effects on daily activity.

Table 5.2. Deaths by sex and cause, England and Wales, 1989, rounded to one significant figure

I.C.D. chapter and type of disease		Number of deaths	
		Males	Females
I	Infectious and parasitic	1 000	1 000
II	Neoplasms (cancers)	80 000	70 000
III	Endocrine, nutritional and metabolic diseases and immunity disorders	4 000	6 000
IV	Blood and blood forming organs	1 000	1 000
V	Mental disorders	4 000	9 000
VI	Nervous system and sense organs	5 000	6 000
VII	Circulatory system	100 000	100 000
VIII	Respiratory system	30 000	30 000
IX	Digestive system	8 000	10 000
X	Genitourinary system	4 000	4 000
XI	Complications of pregnancy, childbirth and the puerperium	0	60
XII	Skin and subcutaneous tissues	300	600
XIII	Musculo-skeletal system and connective tissues	1000	4 000
XIV	Congenital anomalies	900	900
XV	Certain conditions originating in the perinatal period	100	100
XVI	Signs, symptoms and ill-defined conditions	2 000	3 000
XVII	Injury and poisoning	10 000	6 000
Total		300 000	300 000

Table 5.3. Deaths by sex, England and Wales, 1989, for major causes

I.C.D. chapter and type of disease	Number of deaths	
	Males	Females
Circulatory system (VII)	100 000	100 000
Neoplasms (cancers) (II)	80 000	70 000
Respiratory system (VIII)	30 000	30 000
Injury and poisoning (XVII)	10 000	6 000
Digestive system (IX)	8 000	10 000
Others	20 000	20 000
Total	300 000	300 000

5.3 Presenting tables

Tables 5.1 to 5.3 illustrate a number of useful points about the presentation of tables. Like all the tables in this book, they are designed to stand alone from the text. There is no need to refer to material buried in some paragraph to interpret the table. A table is intended to communicate information, so it should be easy to read and understand. A table should have a clear title, stating clearly and unambiguously what the table represents. The rows and columns must also be labelled clearly.

Table 5.4. Calculations for a pie chart of the distribution of cause of death

Cause of death	Frequency	Relative frequency	Angle (degrees)
Circulatory system	137 165	0.466 19	168
Neoplasms (cancers)	69 948	0.237 73	86
Respiratory system	33 223	0.112 92	41
Injury and poisoning	6 427	0.021 84	8
Digestive system	10 779	0.036 63	13
Nervous system	5 990	0.020 36	7
Others	30 695	0.104 32	38
Total	294 227	1.000 00	361

When proportions, rates or percentages are used in a table together with frequencies, they must be easy to distinguish from one another. This can be done, as in Table 2.9, by adding a '%' symbol, or by including a place of decimals. The addition in Table 2.9 of the 'total' row and the '100%' makes it clear that the percentages are calculated from the number in the treatment group, rather than the number with that particular outcome or the total number of patients.

5.4 Pie charts

It is often convenient to present data pictorially. Information can be conveyed much more quickly by a diagram than by a table of numbers. This is particularly useful when data are being presented to an audience, as here the information has to be got across in a limited time. It can also help a reader get the salient points of a table of numbers. Unfortunately, unless great care is taken, diagrams can also be very misleading and should be treated only as an addition to numbers, not a replacement.

We have already discussed methods of illustrating the frequency distribution of a qualitative variable. We will now look at the equivalent of the histogram for qualitative data, the **pie chart** or **pie diagram**. This shows the relative frequency for each category by dividing a circle into sectors, the angles of which are proportional to the relative frequency. We thus multiply each relative frequency by 360, to give the corresponding angle in degrees.

Table 5.4 shows the calculation for drawing a pie chart to represent the distribution of cause of death for females, using the data of Tables 5.1 and 5.3. (The total degrees are 361 rather than 360 because of rounding errors in the calculations.) The resulting pie chart is shown in Figure 5.1. This diagram is said to resemble a pie cut into pieces for serving, hence the name.

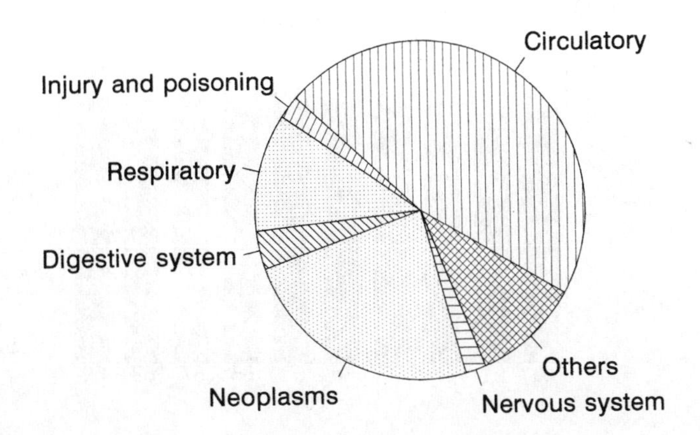

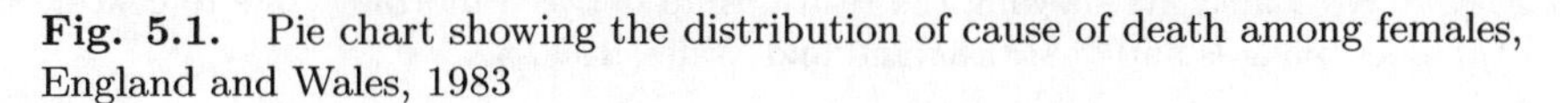

Fig. 5.1. Pie chart showing the distribution of cause of death among females, England and Wales, 1983

Table 5.5. Cancer of the oesophagus: standardised mortality rate per 100 000 per year, England and Wales, 1960–1969

Year	Mortality rate	Year	Mortality rate
60	5.1	65	5.4
61	5.0	66	5.4
62	5.2	67	5.6
63	5.2	68	5.8
64	5.2	69	6.0

5.5 Bar charts

Histograms and pie charts depict the distribution of a single variable. A **bar chart** or **bar diagram** shows the relationship between two variables, one being quantitative and the other either qualitative or a quantitative variable which is grouped, such as time in years. The values of the first variable are shown by the heights of bars, one bar for each category of the second variable. Table 5.5 shows the mortality due to cancer of the oesophagus in England and Wales over a 10 year period. There appears from the table to be an increase in mortality over this period. Figure 5.2 shows this relationship, the heights of the bars being proportional to the mortality.

Bar charts can be used to represent relationships between more than two variables. Figure 5.3 shows the relationship between children's reports of breathlessness and cigarette smoking by themselves and their parents. We can see quickly that the prevalence of the symptom increases both with the child's smoking and with that of their parents.

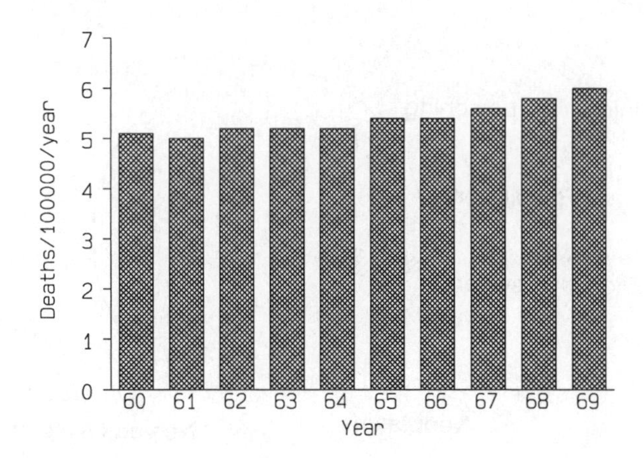

Fig. 5.2. Bar chart showing the relationship between mortality due to cancer of the oesophagus and year, England and Wales, 1960-69

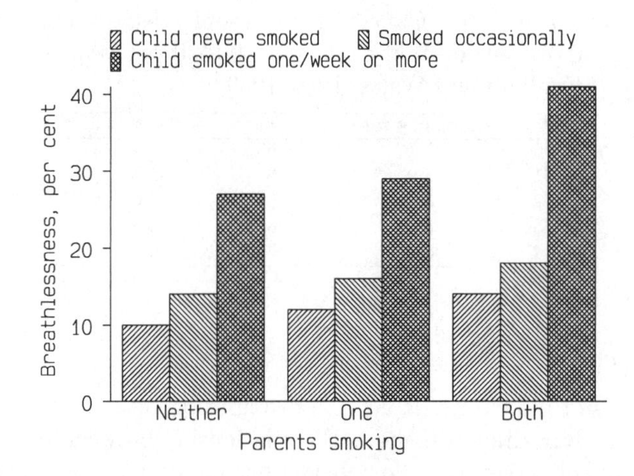

Fig. 5.3. Bar chart showing the relationship between the prevalence of self-reported breathlessness among school children and two possible causative factors

In the published paper reporting these respiratory symptom data (Bland *et al.* 1978) the bar chart was not used; the data were given in the form of tables. It was thus available for other researchers to compare to their own or to carry out calculations upon. The bar chart was used to present the results during a conference, where the most important thing was to convey an outline of the analysis quickly.

Table 5.6. Vital capacity (VC) and height for 44 female medical students

Height (cm)	VC (litres)	Height (cm)	VC (litres)	Height (cm)	VC (litres)
155.0	2.20	163.0	2.72	168.0	2.78
155.0	2.65	163.0	2.82	168.0	3.63
155.4	3.06	163.0	3.40	169.4	2.80
158.0	2.40	164.0	2.90	170.0	3.88
160.0	2.30	165.0	3.07	171.0	3.38
160.2	2.63	166.0	3.03	171.0	3.75
161.0	2.56	166.0	3.50	171.5	2.99
161.0	2.60	166.0	3.66	172.0	2.83
161.0	2.80	166.0	3.69	172.0	4.47
161.0	2.90	166.6	3.06	174.0	4.02
161.0	3.40	167.0	3.48	174.2	4.27
161.2	3.39	167.0	3.72	176.0	3.77
162.0	2.88	167.0	3.80	177.0	3.81
162.0	2.96	167.6	3.06	180.6	4.74
162.0	3.12	167.8	3.70		

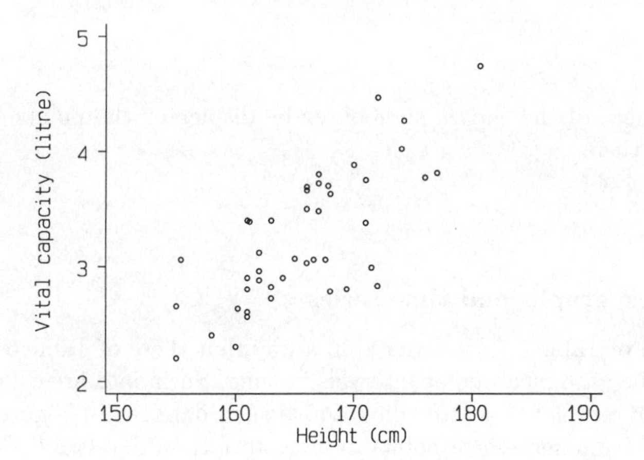

Fig. 5.4. Scatter diagram showing the relationship between vital capacity and height for a group of female medical students

5.6 Scatter diagrams

The bar chart would be a rather clumsy method for showing the relationship between two continuous variables, such as vital capacity and height (Table 5.6). For this we use a **scatter diagram** or **scattergram** (Figure 5.4). This is made by marking the scales of the two variables along horizontal and vertical axes. Each pair of measurements is plotted with a cross, circle, or some other suitable symbol at the point indicated by using the measurements as coordinates. If there is more than one observation at the some coordinate we can indicate this by using the number of observa-

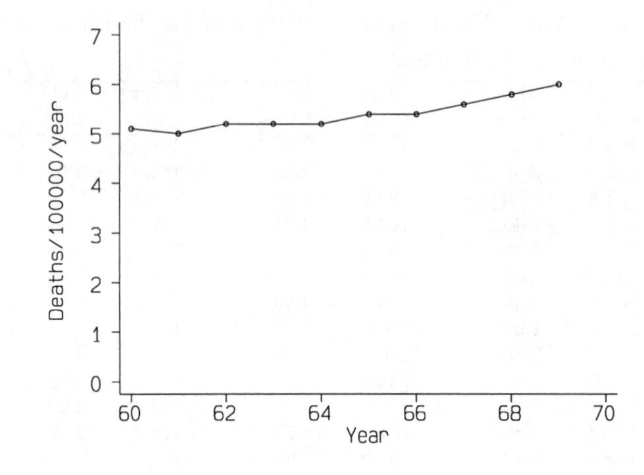

Fig. 5.5. Line graph showing changes in cancer of the oesophagus mortality over time

tion in place of the chosen symbol, or by displacing the points slightly to separate them.

5.7 Line graphs and time series

The data of Table 5.5 are ordered in a way that those of Table 5.6 are not, in that they are recorded at intervals in time. Such data are called a **time series**. If we plot a scatter diagram of such data, as in Figure 5.5, it is natural to join successive points by lines to form a line graph. We do not even need to mark the points at all; all we need is the line. This would not be sensible in Figure 5.4, as the observations are independent of one another and quite unrelated, whereas in Figure 5.5 there is likely to be a relationship between adjacent points. Here the mortality rate recorded for cancer of the oesophagus will depend on a number of things which vary over time including possibly causal factors, such as tobacco and alcohol consumption, and clinical factors, such as better diagnostic techniques and methods of treatment.

Line graphs are particularly useful when we want to show the change of more than one quantity over time. Figure 5.6 shows levels of zidovudine (AZT) in the blood of AIDS patients at several times after administration of the drug, for patients with normal fat absorption and with fat malabsorption (§10.8). The difference in response to the two treatments is very clear.

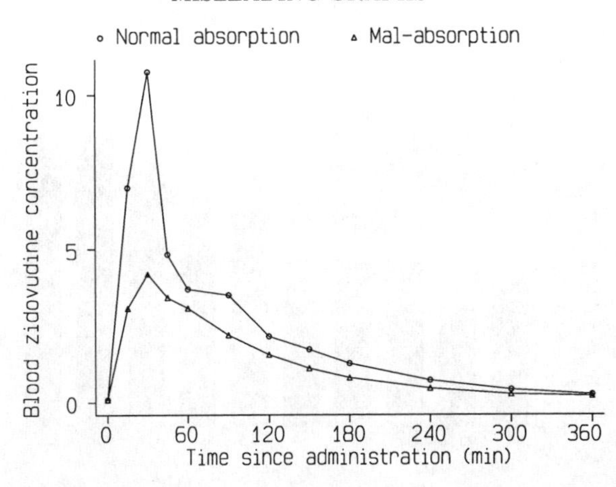

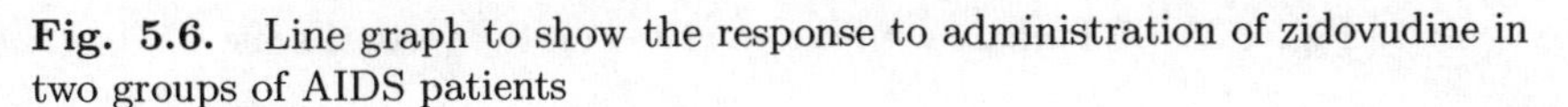

Fig. 5.6. Line graph to show the response to administration of zidovudine in two groups of AIDS patients

5.8 Misleading graphs

Figure 5.2 is clearly titled and labelled and can be read independently of the surrounding text. The principles of clarity outlined for tables apply equally here. After all, a diagram is a method of conveying information quickly and this object is defeated if the reader or audience has to spend time trying to sort out exactly what a diagram really means. Because the visual impact of diagrams can be so great, further problems arise in their use.

The first of these is the missing zero. Figure 5.7 shows a second bar chart representing the data of Table 5.5. This chart appears to show a very rapid increase in mortality, compared to the gradual increase shown in Figure 5.2. Yet both show the same data. Figure 5.7 omits most of the vertical scale, and instead stretches that small part of the scale where the change takes place. Even when we are aware of this, it is difficult to look at this graph and not think that it shows a large increase in mortality. It helps if we visualize the baseline as being somewhere near the bottom of the page.

There is no zero on the horizontal axis in Figures 5.2 and 5.7, either. There are two reasons for this. There is no practical 'zero time' on the calendar; we use an arbitrary zero. Also, there is an unstated assumption that mortality rates vary with time and not the other way round.

The zero is omitted in Figure 5.4. This is almost always done in scatter diagrams, yet if we are to gauge the importance of the relationship between vital capacity and height by the relative change in vital capacity over the height range we need the zero on the vital capacity scale. The origin is

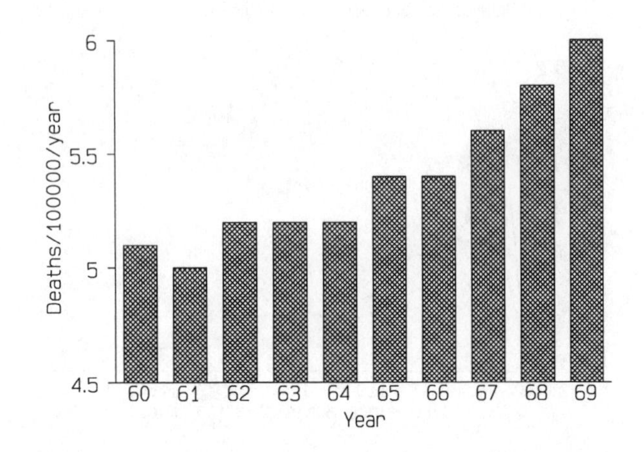

Fig. 5.7. Bar chart with zero omitted on the vertical scale

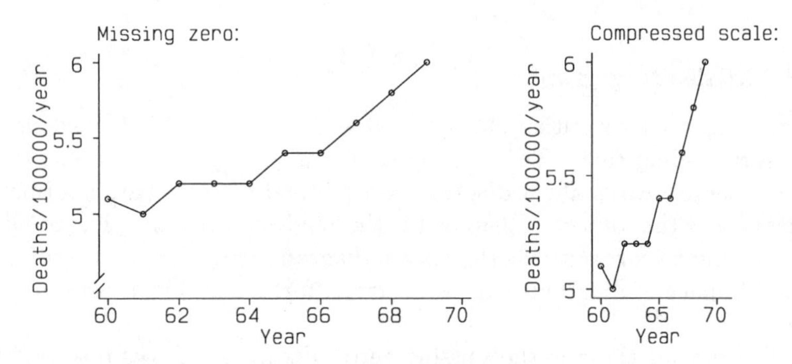

Fig. 5.8. Line graphs with a missing zero and with a stretched vertical and compressed horizontal scale, a 'gee whiz' graph

often omitted on scatter diagrams because we are usually concerned with the existence of a relationship and the distributions followed by the observations, rather than its magnitude. We estimate the latter in a different way, described in Chapter 11.

Line graphs are particularly at risk of undergoing the sort of distortion of missing zero described in §5.8. Many computer programs resist drawing barcharts like Figure 5.7, but will produce a line graph with a truncated scale as the default. Figure 5.8 shows a line graph with a truncated scale, corresponding to Figure 5.7. Just as there, the message of the graph is a dramatic increase in mortality, which the data themselves do not really support. We can make this even more dramatic by stretching the vertical scale and compressing the horizontal scale. The effect is now really impres-

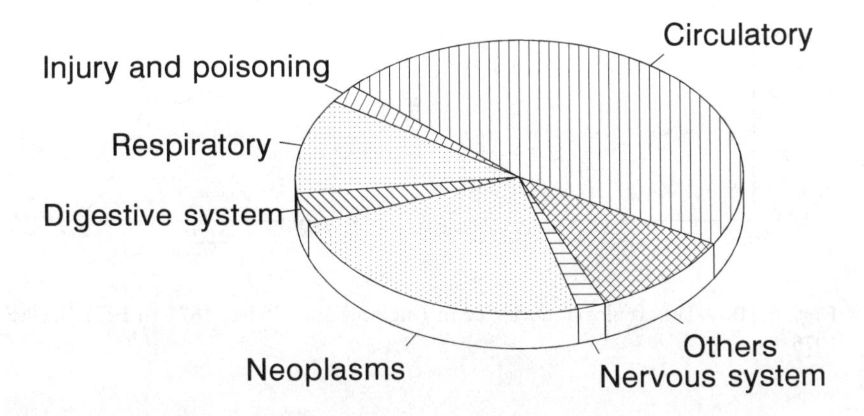

Fig. 5.9. Figure 5.1 with three-dimensional effects

sive and looks much more likely than Figure 5.5 to attract research funds, Nobel prizes and interviews on television. Huff (1954) aptly names such horrors 'gee whiz graphs'. They are even more dramatic if we omit the scales altogether and show only the soaring line.

This is not to say that authors who show only part of the scale are deliberately trying to mislead. There are often good arguments against graphs with vast areas of boring blank paper. In Figure 5.4, we are not interested in vital capacities near zero and can feel quite justified in excluding them. In Figure 5.8 we certainly are interested in zero mortality; it is surely what we are aiming for. The point is that graphs can so easily mislead the unwary reader, so let the reader beware.

The advent of powerful personal computers led to an increase in the ability to produce complicated graphics. Simple charts, such as Figure 5.1, are informative but not visually exciting. One way of decorating such graphs is make them appear three-dimensional. Figure 5.9 shows the effect. The angles are no longer proportional to the numbers which they represent. The areas are, but because they are different shapes it is difficult to compare them. This defeats the primary object of conveying information quickly and accurately. Another approach to decorating diagrams is to turn them into pictures. In a **pictogram** the bars of the bar chart are replaced by pictures. Pictograms can be highly misleading, as the height of a picture, drawn with three-dimensional effect, is proportional to the number represented, but what we see is the volume. Such decorated graphs are like the illuminated capitals of medieval manuscripts: nice to look at but hard to read. I think they should be avoided.

Huff (1954) recounts that the president of a chapter of the American

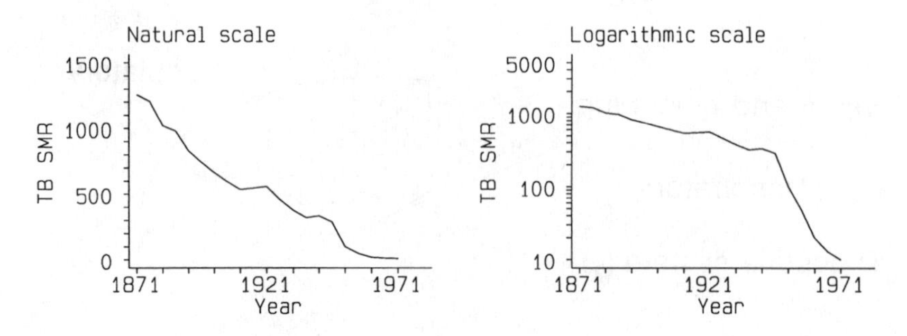

Fig. 5.10. Tuberculosis mortality in England and Wales, 1871 to 1971 (DHSS 1976)

Statistical Association criticized him for accusing presenters of data of trying to deceive. The statistician argued that incompetence was the problem. Huff's reply was that diagrams frequently sensationalize by exageration and rarely minimize anything, that presenters of data rarely distort those data to make their case appear weaker than it is. The errors are too one-sided for us to ignore the possibility that we are being fooled. When presenting data, especially graphically, be very careful that the data are shown fairly. When on the receiving end, beware!

5.9 Logarithmic scales

Figure 5.10 shows a line graph representing the fall in tuberculosis mortality in England and Wales over 100 years (DHSS 1976). We can see a rather unsteady curve, showing the continuing decline in the disease. Figure 5.10 also shows the mortality plotted on a logarithmic (or log) scale. A **logarithmic scale** is one where two pairs of points will be the same distance apart if their ratios are equal, rather than their differences. Thus the distance between 1 and 10 is equal to that between 10 and 100, not to that between 10 and 19. (See §5A if you do not understand this.) The logarithmic line shows a clear kink in the curve about 1950, the time when a number of effective anti-TB measures, chemotherapy with streptomycin, BCG vaccine and mass screening with X rays were introduced. If we consider the properties of logarithms (§5A), we can see how the log scale for the tuberculosis mortality data produced such sharp changes in the curve. If the relationship is such that the mortality is falling with a constant proportion, such as 10% per year, the absolute fall each year depends on the absolute level in the preceding year:

$$\text{mortality in 1960} = \text{constant} \times \text{mortality in 1959}$$

So if we plot mortality on a log scale we get:

$$\log(\text{mortality in 1960}) = \log(\text{constant}) + \log(\text{mortality in 1959})$$

For mortality in 1961, we have

$$\begin{aligned}\log(\text{mortality in 1961}) &= \log(\text{constant}) + \log(\text{mortality in 1960}) \\ &= \log(\text{constant}) + \log(\text{constant}) \\ &\quad + \log(\text{mortality in 1959}) \\ &= 2 \times \log(\text{constant}) + \log(\text{mortality in 1959})\end{aligned}$$

Hence we get a straight line relationship between log mortality and time t:

$$\log(\text{mortality after } t \text{ years}) = t \times \log(\text{constant}) + \log(\text{mortality as start})$$

When the constant proportion changes, the slope of the straight line formed by plotting log(mortality) against time changes and there is a very obvious kink in the line.

Log scales are very useful analytic tools. However, a graph on a log scale can be very misleading if the reader does not allow for the nature of the scale. The log scale in Figure 5.10 shows the increased rate of reduction in mortality associated with the anti-TB measures quite plainly, but it gives the impression that these measures were important in the decline of TB. This is not so. If we look at the corresponding point on the natural scale, we can see that all these measures did was to accelerate a decline which had been going on for a long time (see Radical Statistics Health Group 1976).

5A Appendix: Logarithms

Logarithms are not simply a method of calculation dating from before the computer age, but a set of fundamental mathematical functions. Because of their special properties they are much used in statistics. We shall start with logarithms (or logs for short) to base 10, the common logarithms used in calculations. The log to base 10 of a number x is y where

$$x = 10^y$$

We write $y = \log_{10}(x)$. Thus for example $\log_{10}(10) = 1$, $\log_{10}(100) = 2$, $\log_{10}(1\,000) = 3$, $\log_{10}(10\,000) = 4$, and so on. If we multiply two numbers, the log of the product is the sum of their logs:

$$\log(xy) = \log(x) + \log(y)$$

For example,

$$100 \times 1\,000 = 10^2 \times 10^3 = 10^{2+3} = 10^5 = 100\,000$$

Or in log terms:

$$\log_{10}(100 \times 1\,000) = \log_{10}(100) + \log_{10}(1\,000) = 2 + 3 = 5$$

Hence, $100 \times 1\,000 = 10^5 = 100\,000$. This means that any multiplicative relationship of the form

$$y = a \times b \times c \times d$$

can be made additive by a log transformation:

$$\log(y) = \log(a) + \log(b) + \log(c) + \log(d)$$

This is the process underlying the fit to the Lognormal Distribution described in §7.4.

There is no need to use 10 as the base for logarithms. We can use any number. The log of a number x to base b can be found from the log to base a by a simple calculation:

$$\log_b(x) = \frac{\log_a(x)}{\log_a(b)}$$

Ten is convenient for arithmetic using log tables, but for other purposes it is less so. For example, the gradient, slope or differential of the curve $y = \log_{10}(x)$ is $\log_{10}(e)/x$, where $e = 2.718\,281\ldots$ is a constant which does not depend on the base of the logarithm. This leads to awkward constants spreading through formulae. To keep this to a minimum we use logs to the base e, called natural or Napierian logarithms after the mathematician John Napier. This is the logarithm usually produced by LOG(X) functions in computer languages.

Figure 5.11 shows the log curve for three different bases, 2, e and 10. The curves all go through the point (1,0), i.e. $\log(1) = 0$. As x approaches 0, $\log(x)$ becomes a larger and larger negative number, tending towards minus infinity as x tends to zero. There are no logs of negative numbers. As x increases from 1, the curve becomes flatter and flatter. Though $\log(x)$ continues to increase, it does so more and more slowly. The curves all go through $(base, 1)$ i.e. $\log(base) = 1$. The curve for log to the base 2 goes through (2,1), (4,2), (8,3) because $2^1 = 2$, $2^2 = 4$, $2^3 = 8$. We can see that the effect of replacing data by their logs will be to stretch out the scale at the lower end and contract it at the upper.

We often work with logarithms of data rather than the data themselves. This may have several advantages. Multiplicative relationships may become additive, curves may become straight lines and skew distributions may become symmetrical.

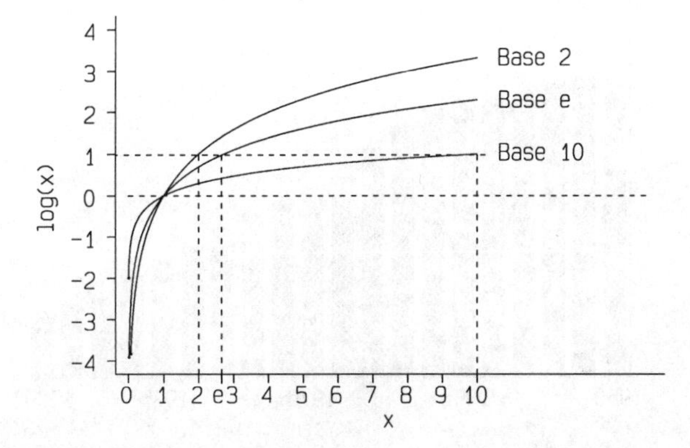

Fig. 5.11. Logarithmic curves to three different bases

We transform back to the natural scale using the **antilogarithm** or **antilog**. If $y = \log_{10}(x)$, $x = 10^y$ is the antilog of y. If $z = \log_e(x)$, $x = e^z$ or $x = \exp(z)$ is the antilog of z. If your computer program doesn't transform back, most calculators have e^x and 10^x functions for this purpose.

5M Multiple choice questions 20 to 24

(Each branch is either true or false)

20. 'After treatment with Wondermycin, 66.67% of patients made a complete recovery'

(a) Wondermycin is wonderful;

(b) This statement may be misleading because the denominator is not given;

(c) The number of significant figures used suggest a degree of precision which may not be present;

(d) Some control information is required before we can draw any conclusions about Wondermycin;

(e) There might be only a very small number of patients.

21. The number 1 729.543 71:

(a) to two significant figures is 1 700;

(b) to three significant figures is 1 720;

(c) to six decimal places is 1 729.54;

(d) to three decimal places is 1 729.544;

(e) to five significant figures is 1 729.5.

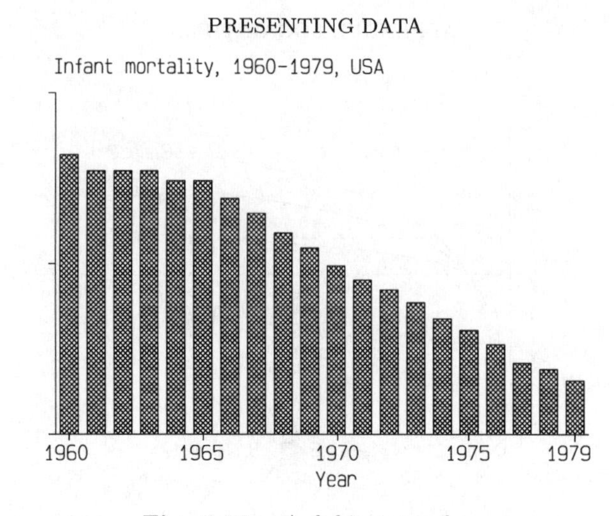

Fig. 5.12. A dubious graph

22. Figure 5.12:

(a) shows a histogram;

(b) should have the vertical axis labelled;

(c) should show the zero on the vertical axis;

(d) should show the zero on the horizontal axis;

(e) should show the units for the vertical axis.

23. Logarithmic scales used in graphs showing time trends:

(a) show changes in the trend clearly;

(b) often produce straight lines;

(c) give a clear idea of the magnitude of changes;

(d) should show the zero point from the original scale;

(e) compress intervals between large numbers compared to those between small numbers.

24. The following methods can be used to show the relationship between two variables:

(a) histogram;

(b) pie chart;

(c) scatter diagram;

(d) bar chart;

(e) line graph.

Table 5.7. Weekly geriatric admissions in Wandsworth Health District from May to September, 1982 and 1983 (Fish *et al.* 1985)

Week	1982	1983	Week	1982	1983
1	24	20	12	11	25
2	22	17	13	6	22
3	21	21	14	10	26
4	22	17	15	13	12
5	24	22	16	19	33
6	15	23	17	13	19
7	23	20	18	17	21
8	21	16	19	10	28
9	18	24	20	16	19
10	21	21	21	24	13
11	17	20	22	15	29

5E Exercise: Creating graphs

In this exercise we shall display graphically some of the data we have studied so far.

1. Table 4.1 shows diagnoses of patients in a hospital census. Display these data as a graph.
2. Table 2.7 shows the paralytic polio rates for several groups of children. Construct a bar chart for the results from the randomized control areas.
3. Table 3.1 shows some results from the study of mortality in British doctors. Show these graphically.
4. Table 4.3 shows the parity of a group of women. Show these graphically.
5. Table 5.7 shows the numbers of geriatric admissions in Wandsworth Health District for each week from May to September in 1982 and 1983. Show these data graphically. Why do you think the two years were different?

6

Probability

6.1 Probability

We use data from a sample to draw conclusions about the population from which it is drawn. For example, in a clinical trial we might observe that a sample of patients given a new treatment respond better than patients given an old treatment. We want to know whether the improvement would be seen in the whole population of patients, or if it could be due to chance. The theory of probability enables us to link samples and populations, and to draw conclusions about populations from samples. We shall start the discussion of probability with some simple randomizing devices, such as coins and dice, but the relevance to medical problems should soon became apparent.

We first ask what exactly is meant by 'probability'. There are several different approaches to this in statistics. We shall take the frequency definition. The **probability** that an event will happen under given circumstances may be defined as the proportion of repetitions of those circumstances in which the event would occur in the long run. For example, if we toss a coin it comes down either heads or tails. Before we toss it, we have no way of knowing which will happen, but we do know that it will either be heads or tails. After we have tossed it, of course, we know exactly what the outcome is. If we carry on tossing our coin, we should get several heads and several tails. If we go on doing this for long enough, then we would expect to get as many heads as we do tails. So the probability of a head being thrown is half, because in the long run a head should occur on half of the throws. The number of heads which might arise in several tosses of the coin is called a **random variable**, that is, a variable which can take more than one value with given probabilities. In the same way, a thrown die can show six faces, numbered one to six, with equal probability. We can investigate random variables such as the number of sixes in a given number of throws, the number of throws before the first six, and so on.

The same definition of probability applies to continuous measurement, such as human height. For example, suppose the median height in a population of women is 168 cm. Then half the women are above 168 cm in height. If we choose women at random (i.e. without the characteristics of the woman influencing the choice) then in the long run half the women chosen will have heights above 168 cm. The probability of a woman having height above 168 cm is one half. Similarly, if 1/10 of the women have height greater than 180 cm, a woman chosen at random will have height greater than 180 cm with probability 1/10. In the same way we can find the probability of height being between any given values. When we measure a continuous quantity we are always limited by the method of measurement, and so when we say a woman's height is 170 cm we mean that it is between, say, 169.5 cm and 170.5 cm, depending in the accuracy with which we measure. So what we are interested in is the probability of the random variable taking values between certain limits rather than particular values.

6.2 Properties of probability

The following simple properties follow from the definition of probability.

1. A probability lies between 0.0 and 1.0. When the event never happens the probability is 0.0, when it always happens the probability is 1.0.

2. **Addition rule**. Suppose two events are **mutually exclusive**, i.e. when one happens the other cannot happen. Then the probability that one or the other happens is the sum of their probabilities. For example, a thrown die may show a one or a two, but not both. The probability that it shows a one or a two $= 1/6 + 1/6 = 2/6$.

3. **Multiplication rule**. Suppose two events are **independent**, i.e. knowing one has happened tells us nothing about whether the other happens. Then the probability that both happen is the product of their probabilities. For example, suppose we toss a coin twice. Then the second toss is independent of the first toss, and the probability of two heads occurring is $1/2 \times 1/2 = 1/4$. Consider two independent events A and B. The proportion of times A happens in the long run is the probability of A. Since A and B are independent, of those times when A happens, a proportion, equal to probability of B, will have B happen also. Hence the proportion of times that A and B happen together is the probability of A multiplied by the probability of B.

6.3 Probability distributions and random variables

Suppose we have a set of events which are mutually exclusive and which includes all the events which can possibly happen. The sum of their probabilities is 1.0. The set of these probabilities make up a **probability distribution**. For example, if we toss a coin the two possibilities, head

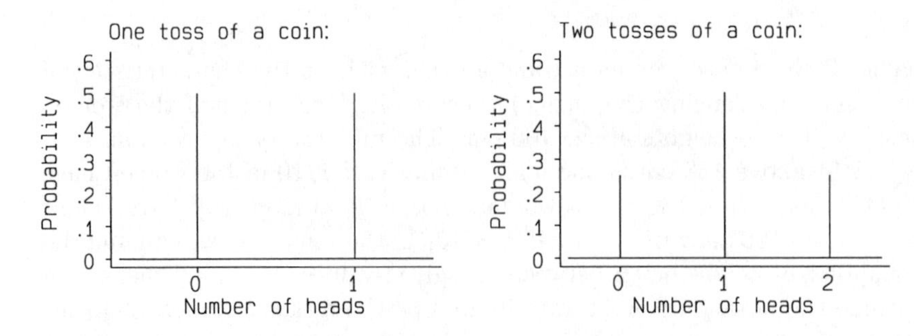

Fig. 6.1. Probability distribution for the number of heads shown in one toss and in two tosses of a coin

or tail, are mutually exclusive and these are the only events which can happen. The probability distribution is:

$$\begin{aligned} \text{PROB(Head)} &= 1/2 \\ \text{PROB(Tail)} &= 1/2 \end{aligned}$$

We can represent this with a diagram, as in Figure 6.1. Now, let us define a variable, which we will denote by the symbol X, such that $X = 0$ if the coin shows a tail and $X = 1$ if the coin shows a head. X is the number of heads shown on a single toss, which must be 0 or 1. We do not know before the toss what X will be, but do know the probability of it having any possible value. X is a random variable (§6.1) and the probability distribution is also the distribution of X.

What happens if we toss two coins at once? We now have four possible events: a head and a head, a head and a tail, a tail and a head, a tail and a tail. Clearly, these are equally likely and each has probability 1/4. Let Y be the number of heads. Y has three possible values: 0, 1, and 2. $Y = 0$ only when we get a tail and a tail and has probability 1/4. Similarly, $Y = 2$ only when we get a head and a head, so has probability 1/4. However, $Y = 1$ either when we get a head and tail, or when we have a tail and a head, and so has probability $1/4 + 1/4 = 1/2$. We can write this probability distribution

$$\begin{aligned} \text{PROB}(Y = 0) &= 1/4 \\ \text{PROB}(Y = 1) &= 1/2 \\ \text{PROB}(Y = 2) &= 1/4 \end{aligned}$$

as:

The probability distribution of Y is shown in Figure 6.1.

6.4 The Binomial distribution

We have considered the probability distributions of two random variables: X, the number of heads in one toss of a coin, taking values 0 and 1, and Y, the number of heads in two tosses of a coin, taking values 0, 1 or 2. We can increaes the number of two coins; Figure 6.2 shows the distribution of the number of heads obtained when 15 coins are tossed. We do not need the probability of a 'head' to be 0.5: we can count the number of sixes when dice are thrown. Figure 6.2 shows the distribution of the number of sixes obtained from 10 dice. In general, we can think of the coin or the die as trials, which can have outcomes success (head or six) or failure (tail or one to five). The distributions of X and Y, and in Figure 6.2, are all examples of the Binomial distribution, which arises frequently in medical applications. The **Binomial distribution** is the distribution followed by the number of successes in n independent trials when the probability of any single trial being a success is p. The Binomial distribution is in fact a familiy of distributions, the members of which are defined by the values of n and p. The values which define which member of the distribution family we have are called the **parameters** of the distribution.

Simple randomizing devices like coins and dice are of interest in themselves, but not of obvious relevance to medicine. However, suppose we are carrying out a random sample prevalence survey to estimate the unknown prevalence, p, of a disease. Since members of the sample are chosen at random and independently from the population, the probability of any chosen subject having the disease is p. We thus have a series of independent trials, each with probability of success p, and the number of successes, i.e. members of the sample with the disease, will follow a Binomial distribution. As we shall see later, the properties of the Binomial distribution enable us to say how accurate is the estimate of prevalence obtained (§8.4).

We can calculate the probabilities for Binomial distribution by listing all the ways in which, say, 15 coins can fall. However, there are $2^{15} = 32\,768$ combinations of 15 coins, so this is not very practical. Instead, there is a formula for the probability in terms of the number of throws and the probability of a head. This enables us to work these probabilities out for any probability of success and any number of trials. In general, we have n independent trials with the probability that a trial is a success being p. What is the probability of r successes? For any particular series of r successes, each with probability p, and $n-r$ failures, each with probability $1-p$, the probability of the series happening is $p^r(1-p)^{(n-r)}$, since the trials are independent and the multiplicative rule applies. The number of ways in which r things may be chosen from n things is $n!/r!(n-r)!$ (§6A). Only one combination can happen at one time, so we have $n!/r!(n-r)!$ mutually exclusive ways of having r successes, each with probability $p^r(1-p)^{(n-r)}$.

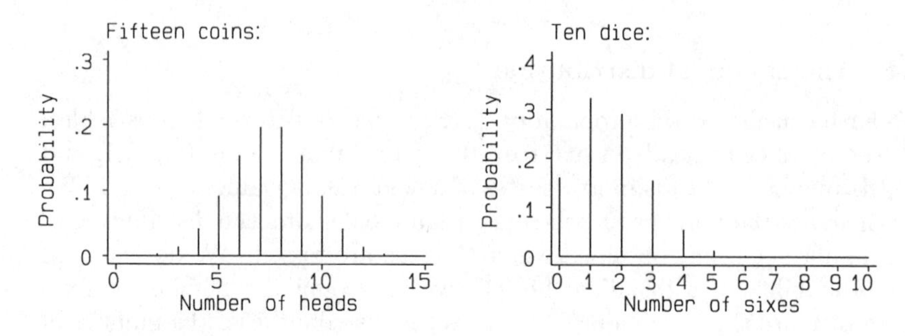

Fig. 6.2. Distribution of the number of heads shown when 15 coins are tossed and of the number of sixes shown when 10 dice are thrown, examples of the Binomial distribution

The probability of having r successes is the sum of these $n!/r!(n-r)!$ probabilities:

$$\text{PROB}(r \text{ successes}) = \frac{n!}{r!(n-r)!}p^r(1-p)^{(n-r)}$$

Those who remember the binomial expansion in mathematics will see that this is one term of it, hence the name Binomial distribution.

We can apply this to the number of heads in two tosses of a coin. The number of heads will be from a Binomial distribution with $p = 0.5$ and $n = 2$. Hence the probability of two heads (r = 2) is:

$$\begin{aligned}\text{PROB}(r=2) &= \frac{n!}{r!(n-r)!}p^r(1-p)^{(n-r)} \\ &= \frac{2!}{2!0!}0.5^2 \times 0.5^0 = \frac{2}{2 \times 1} \times 0.25 \times 1 = 0.25\end{aligned}$$

Note that $0! = 1$ (§6A), and anything to the power 0 is 1. Similary for $r = 1$ and $r = 0$:

$$\text{PROB}(r=1) = \frac{2!}{1!1!}0.5^1 \times 0.5^1 = \frac{2}{1 \times 1} \times 0.5 \times 0.5 = 0.5$$

$$\text{PROB}(r=0) = \frac{2!}{0!2!}0.5^0 \times 0.5^2 = \frac{2}{1 \times 2} \times 1 \times 0.25 = 0.25$$

This is what was found for two coins in §6.3. We can use this distribution as whenever we have a series of trials with two possible outcomes. If we treat a group of patients, the number who recover is from a Binomial distribution. If we measure the blood pressure of a group of people, the number classified as hypertensive is from a Binomial distribution.

Figure 6.3 shows the Binomial distribution for $p = 0.3$ and increasing values of n. The distribution becomes more symmetrical as n increases. It is converging to the Normal distribution, described in the next chapter.

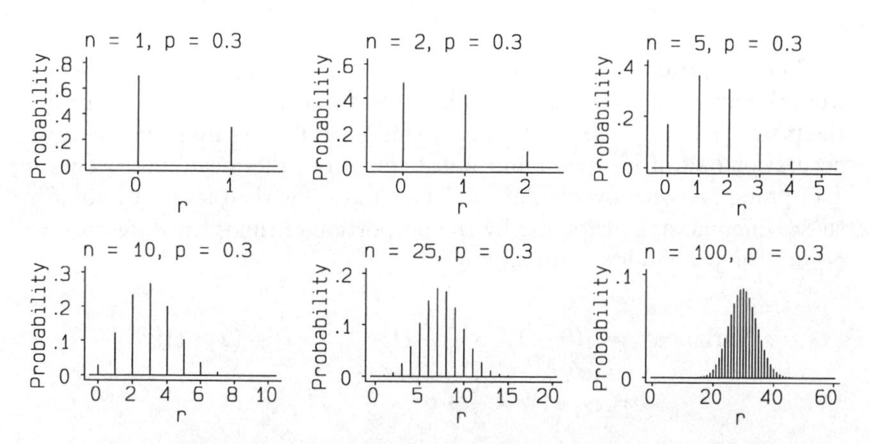

Fig. 6.3. Binomial distributions with different n, $p = 0.3$

6.5 Mean and variance

The number of different probabilities in a Binomial distribution can be very large and unwieldy. When n is large, we usually need to summarize these probabilities in some way. Just as a frequency distribution can be described by its mean and variance, so can a probability distribution and its associated random variable.

The **mean** is the average value of the random variable in the long run. It is also called the **expected value** or **expectation** and the expectation of a random variable X is usually denoted by $\mathrm{E}(X)$. For example, consider the number of heads in tosses of two coins. We get 0 heads in $\frac{1}{4}$ of pairs of coins, i.e. with probability $\frac{1}{4}$. We get 1 head in $\frac{1}{2}$ of pairs of coins, and 2 heads in $\frac{1}{4}$ of pairs. The average value we should get in the long run is found by multiplying each value by the proportion of pairs in which it occurs and adding:

$$0 \times \frac{1}{4} + 1 \times \frac{1}{2} + 2 \times \frac{1}{4} = 0 + \frac{1}{2} + \frac{1}{2} = 1$$

If we kept on tossing pairs of coins, the average number of heads per pair would be 1. Thus for any random variable which takes discrete values the mean, expectation or expected value is found by summing each possible value multiplied by its probability.

Note that the expected value of a random variable does not have to be a value that the random variable can actually take. For example, for the mean number of heads in throws of one coin we have either no heads or 1 head, each with probability half, and the expected value is $0 \times \frac{1}{2} + 1 \times \frac{1}{2} = \frac{1}{2}$. The number of heads must be 0 or 1, but the expected value is half, the average which we would get in the long run.

The **variance** of a random variable is the average squared difference from the mean. For the number of heads in two coin tosses, 0 is 1 unit from the mean and occurs for $\frac{1}{4}$ of pairs of coins, 1 is 0 units from the mean and occurs for half of the pairs and 2 is 1 unit from the mean and occurs for $\frac{1}{4}$ of pairs, i.e. with probability $\frac{1}{4}$. The variance is then found by squaring these differences, multiplying by the proportion of times the difference will occur (the probability) and adding:

$$\begin{aligned}\text{Variance} &= (0-1)^2 \times \frac{1}{4} + (1-1)^2 \times \frac{1}{2} + (2-1)^2 \times \frac{1}{4} \\ &= (-1)^2 \times \frac{1}{4} + 0^2 \times \frac{1}{2} + 1^2 \times \frac{1}{4} \\ &= \frac{1}{2}\end{aligned}$$

We denote the variance of a random variable X by $\text{VAR}(X)$. In mathematical terms,

$$\text{VAR}(X) = \text{E}\left(X^2 - \text{E}(X)^2\right)$$

The square root of the variance is the standard deviation of the random variable or distribution. We often use the Greek letter μ, pronounced 'mu', and σ, 'sigma', for the mean and standard deviation of a probability distribution. The variance is then σ^2.

The mean and variance of the distribution of a continuous variable, of which more in Chapter 7, are defined in a similar way. Calculus is used to define them as integrals, but this need not concern us here. Essentially what happens is that the continuous scale is broken up into many very small intervals and the value of the variable in that very small interval is multiplied by the probability of being in it, then these are added.

6.6 Properties of means and variances

When we use the mean and variance of probability distributions in statistical calculations, it is not the details of their formulae which we need to know, but some of their simple properties. Most of the formulae used in statistical calculations are derived from these. The reasons for these properties are quite easy to see in a non-mathematical way.

If we add a constant to a random variable, the new variable so created has a mean equal to that of the original variable plus the constant. The variance and standard deviation will be unchanged. Suppose our random variable is human height. We can add a constant to the height by measuring the heights of people standing on a box. The mean height of people plus box will now be the mean height of the people plus the constant height of the box. The box will not alter the variability of the heights, however. The difference between the tallest and smallest, for example, will be unchanged. We can subtract a constant by asking the people to stand in a constant hole to be measured. This reduces the mean but leaves the variance unchanged as before.

If we multiply a random variable by a positive constant, the mean and standard deviation are multiplied by the constant, the variance is multiplied by the square of the constant. For example, if we change our units of measurements, say from inches to centimetres, we multiply each measurement by 2.54. This has the effect of multiplying the mean by the constant, 2.54, and multiplying the standard deviation by the constant since it is in the same units as the observations. However, the variance is measured in squared units, and so is multiplied by the square of the constant. Division by a constant works in the same way. If the constant is negative, the mean is multiplied by the constant and so changes sign. The variance is multiplied by the square of the constant, which is positive, so the variance remains positive. The standard deviation, which is the square root of the variance, is always positive. It is multiplied by the absolute value of the constant, i.e. the constant without the negative sign.

If we add two random variables the mean of the sum is the sum of the means, and, *if the two variables are independent*, the variance of the sum is the sum of their variances. We can do this by measuring the height of people standing on boxes of random height. The mean height of people on boxes is the mean height of people + the mean height of the boxes. The variability of the heights is also increased. This is because some short people will find themselves on small boxes, and some tall people will find themselves on large boxes. If the two variables are not independent, something different happens. The mean of the sum remains the sum of the means, but the variance of the sum is not the sum of the variances. Suppose our people have decided to stand on the boxes, not just at a statistician's whim, but for a purpose. They wish to change a light bulb, and so must reach a required height. Now the short people must pick large boxes, whereas tall people can make do with small ones. The result is a reduction in variability to almost nothing. On the other hand, if we told the tallest people to find the largest boxes and the shortest to find the smallest boxes, the variablity would be increased. Independence is an important condition.

If we subtract one random variables from another, the mean of the difference is the difference between the means, and, if the two variables are independent, the variance of the difference is the *sum* of their variances. Suppose we measure the heights above ground level of our people standing in holes of random depth. The mean height above ground is the mean height of the people minus the mean depth of the hole. The variability is increased, because some short people stand in deep holes and some tall people stand in shallow holes. If the variables are not independent, the additivity of the variances breaks down, as it did for the sum of two variables. When the people try to hide in the holes, and so must find a hole deep enough to hold them, the variability is again reduced.

The effects of multiplying two random variables and of dividing one by another are much more complicated. Fortunately we rarely need to do this.

We can now find the mean and variance of the Binomial distribution with parameters n and p. First consider $n = 1$. Then the probability distribution is:

value	probability
0	$1-p$
1	p

The mean is therefore $0 \times (1-p) + 1 \times p = p$. The variance is

$$\begin{aligned}(0-p)^2 \times (1-p) + (1-p)^2 \times p &= p^2(1-p) + p(1-p)^2 \\ &= p(1-p)(p+1-p) \\ &= p(1-p)\end{aligned}$$

Now, a variable from the Binomial distribution with parameters n and p is the sum of n independent variables from the Binomial distribution with parameters 1 and p. So its mean is the sum of n means all equal to p, and its variance is the sum of n variances all equal to p. Hence the Binomial distribution has mean $= np$ and variance $= np(1-p)$. For large sample problems, these are more useful than the Binomial probability formula.

The properties of means and variances of random variables enable us to find a formal solution to the problem of degrees of freedom for the sample variance discussed in Chapter 4. We want an estimate of variance whose expected value is the population variance. The expected value of $\sum(x_i - \bar{x})^2$ can be shown to be $(n-1)\text{VAR}(x)$ (§6B) and hence we divide by $n-1$, not n, to get our estimate of variance.

6.7 The Poisson distribution

The Binomial distribution is one of many probability distributions which are used in statistics. It is a discrete distribution, that is it can take only a finite set of possible values, and is the discrete distribution most commonly encountered in medical applications. One other discrete distribution is worth discussing at this point, the Poisson distribution. Although, like the Binomial, the Poisson distribution arises from a simple probability model, the mathematics involved is more complicated and will be omitted.

Suppose events happen randomly and independently in time at a constant rate. The **Poisson distribution** is the distribution followed by the number of events which happen in a fixed time interval. If events happen

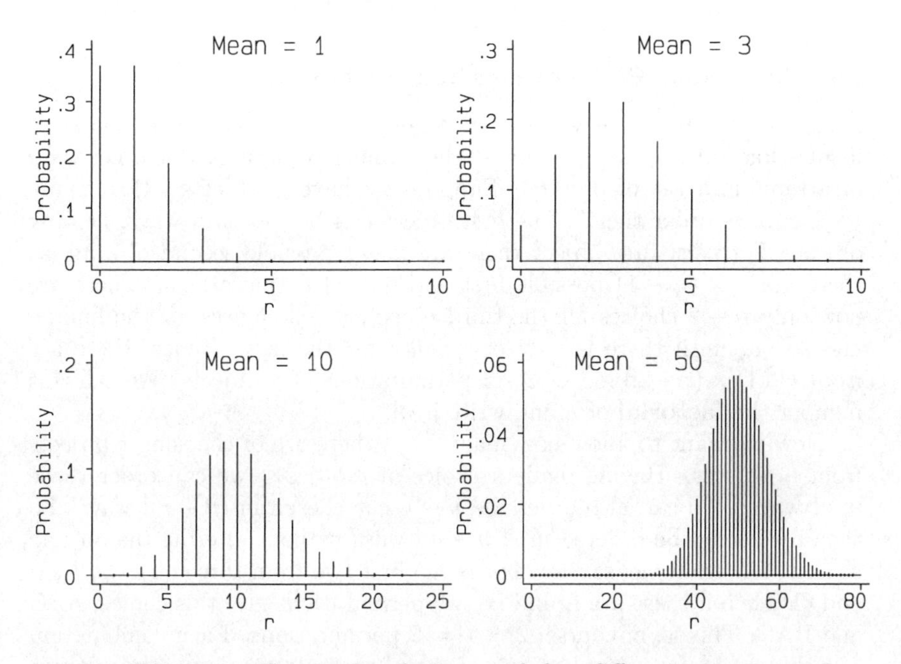

Fig. 6.4. Poisson distributions with four different means

with rate μ events per unit time, the probability of r events happening in unit time is

$$\frac{e^{-\mu}\mu^r}{r!}$$

where e = 2.718..., the mathematical constant. However, there is seldom any need to use individual probabilities of this distribution, as its mean and variance suffice. The mean of the Poisson distribution for the number of events per unit time is simply the rate, μ. The variance of the Poisson distribution is also equal to μ. Thus the Poisson is a family of distributions, like the Binomial, but with only one parameter, μ. This distribution is important, because deaths from many diseases can be treated as occuring randomly and independently in the population. Thus, for example, the number of deaths from lung cancer in one year among people in an occupational group, such as coal miners, will be an observation from a Poisson distribution, and we can use this to make comparisons between mortality rates (§16.3).

Figure 6.4 shows the Poisson distribution for four different means. You will see that as the mean increases the Poisson distribution looks rather like the Binomial distribution in Figure 6.3. We shall discuss this similarity further in the next chapter.

6A Appendix: Permutations and combinations

For those who never knew, or have forgotten, the theory of combinations, it goes like this. First, we look at the number of permutations, i.e. ways of arranging a set of objects. Suppose we have n objects. How many ways can we order them? The first object can be chosen n ways, i.e. any object. For each first object there are $n-1$ possible second objects, so there are $n \times (n-1)$ possible first and second permutations. There are now only $n-2$ choices for the third object, $n-3$ choices for the fourth, and so on, until there is only one choice for the last. Hence, there are $n \times (n-1) \times (n-2) \times \ldots \times 2 \times 1$ permutations of n objects. We call this number the **factorial** of n and write it $n!$.

Now we want to know how many ways there are of choosing r objects from n objects. Having made a choice of r objects, we can order those in $r!$ ways. We can also order the $n-r$ not chosen in $(n-r)!$ ways. So the objects can be ordered in $r!(n-r)!$ ways without altering the objects chosen. For example, say we choose the first two from three objects, A, B and C. Then if these are A and B, two permutations give this choice, ABC and BAC. This is, of course, $2! \times 1! = 2$ permutations. Each combination of r things accounts for $r!(n-r)!$ of the $n!$ permutations possible, so there are

$$\frac{n!}{r!(n-r)!}$$

possible combinations. For example, consider the number of combinations of two objects out of three, say A, B and C. The possible choices are AB, AC and BC. There is no other possibility. Applying the formula, we have $n = 3$ and $r = 2$ so

$$\frac{n!}{r!(n-r)!} = \frac{3!}{2!(3-2)!} = \frac{3 \times 2 \times 1}{2 \times 1 \times 1} = 3$$

Sometimes in using this formula we come across $r = 0$ or $r = n$ leading to $0!$. This cannot be defined in the way we have chosen, but we can calculate its only possible value, $0! = 1$. Because there is only one way of choosing n objects from n, we have

$$1 = \frac{n!}{n!(n-n)!} = \frac{n!}{n! \times 0!} = \frac{1}{0!}$$

so $0! = 1$.

6B Appendix: Expected value of a sum of squares

The properties of means and variances described in §6.6 can be used to answer the question raised in §4.7 and §4A about the divisor in the sample variance. We ask why the variance from a sample is

$$s^2 = \frac{1}{n-1}\sum(x_i - \bar{x})^2$$

and not

$$s_n^2 = \frac{1}{n}\sum(x_i - \bar{x})^2$$

We shall be concerned with the general properties of samples of size n, so we shall treat n as a constant and x_i and $\bar{x}$ as random variables. We shall suppose x_i has mean μ and variance σ^2.

The expected value of the sum of squares is

$$\begin{aligned} \mathrm{E}\left(\sum(x_i - \bar{x})^2\right) &= \mathrm{E}\left(\sum x_i^2 - \frac{1}{n}\left(\sum x_i\right)^2\right) \qquad (\S 4\mathrm{A}) \\ &= \mathrm{E}\left(\sum x_i^2\right) - \frac{1}{n}\mathrm{E}\left(\left(\sum x_i\right)^2\right) \end{aligned}$$

because the expected value of the difference is the difference between the expected values and n is a constant. Now, the population variance σ^2 is the average squared distance from the population mean μ, so

$$\sigma^2 = \mathrm{E}\left((x_i - \mu)^2\right) = \mathrm{E}(x_i^2 - 2\mu x_i + \mu^2) = \mathrm{E}(x_i^2) - 2\mu\mathrm{E}(x_i) + \mu^2$$

because μ is a constant. Because $\mathrm{E}(x_i) = \mu$, we have

$$\sigma^2 = \mathrm{E}(x_i^2) - 2\mu^2 + \mu^2 = \mathrm{E}(x_i^2) - \mu^2$$

and so we find $\mathrm{E}(x_i^2) = \sigma^2 + \mu^2$ and so $\mathrm{E}(\sum x_i^2) = n(\sigma^2 + \mu^2)$, being the sum of n numbers all of which are $\sigma^2 + \mu^2$. We now find the value of $\mathrm{E}\left((\sum x_i)^2\right)$. We need

$$\mathrm{E}\left(\sum x_i\right) = \sum \mathrm{E}(x_i) = \sum \mu = n\mu$$

$$\mathrm{VAR}\left(\sum x_i\right) = \sum \mathrm{VAR}(x_i) = n\sigma^2$$

Just as $\mathrm{E}(x_i^2) = \sigma^2 + \mu^2 = \mathrm{VAR}(x_i) + (\mathrm{E}(x_i))^2$ so

$$\begin{aligned} \mathrm{E}\left(\left(\sum x_i\right)^2\right) &= \mathrm{VAR}\left(\left(\sum x_i\right)^2\right) + \left(\mathrm{E}\left(\left(\sum x_i\right)^2\right)\right)^2 \\ &= n\sigma^2 + (n\mu)^2 \end{aligned}$$

So

$$\mathrm{E}\left(\sum(x_i - \bar{x})^2\right) = \mathrm{E}\left(\sum x_i^2\right) - \frac{1}{n}\mathrm{E}\left(\left(\sum x_i\right)^2\right)$$

$$\begin{aligned} &= n(\sigma^2 + \mu^2) - \frac{1}{n}\left(n\sigma^2 + n^2\mu^2\right) \\ &= n\sigma^2 + n\mu^2 - \sigma^2 - n\mu^2 \\ &= (n-1)\sigma^2 \end{aligned}$$

So the expected value of the sum of squares is $(n-1)\sigma^2$ and we must divide the sum of squares by $n-1$, not n, to obtain the estimate of the variance, σ^2.

We shall find the variance of the sample mean, $\bar{x}$, useful later (§8.2):

$$\mathrm{VAR}(\bar{x}) = \mathrm{VAR}\left(\frac{1}{n}\sum x_i\right) = \frac{n\sigma^2}{n^2} = \frac{\sigma^2}{n}$$

6M Multiple choice questions 25 to 31

(Each branch is either true or false.)

25. The events A and B are mutually exclusive, so:
(a) PROB(A or B) = PROB(A) + PROB(B);
(b) PROB(A and B) = 0;
(c) PROB(A and B) = PROB(A) PROB(B);
(d) PROB(A) = PROB(B);
(e) PROB(A) + PROB(B) = 1.

26. The probability of a woman aged 50 having condition X is 0.20 and the probability of her having condition Y is 0.05. These probabilities are independent:
(a) the probability of her having both conditions is 0.01;
(b) the probability of her having both conditions is 0.25;
(c) the probability of her having either X, or Y, or both is 0.24;
(d) if she has condition X, the probability of her having Y also is 0.01;
(e) if she has condition Y, the probability of her having X also is 0.20.

27. The following variables follow a Binomial distribution:
(a) number of sixes in 20 throws of a die;
(b) human weight;
(c) number of a random sample of patients who respond to a treatment;
(d) number of red cells in 1 ml of blood;
(e) proportion of hypertensives in a random sample of adult men.

28. Two parents each carry the same recessive gene which each transmits to their child with probability 0.5. If their child will develop clinical disease if it inherits the gene from both parents and will be a carrier if it inherits the gene from one parent only then:

(a) the probability that their next child will have clinical disease is 0.25;

(b) the probability that two successive children will both develop clinical disease is 0.25×0.25;

(c) the probability their next child will be a carrier without clinical disease is 0.50;

(d) the probability of a child being a carrier or having clinical disease is 0.75;

(e) if the first child does not have clinical disease, the probability that the second child will not have clinical disease is 0.75^2.

29. If a coin is spun twice in succession:

(a) the expected number of tails is 1.5;

(b) the probability of two tails is 0.25;

(c) the number of tails follows a Binomial distribution;

(d) the probability of at least one tail is 0.5;

(e) the distribution of the number of tails is symmetrical.

30. If X is a random variable, mean μ and variance σ^2:

(a) $\mathrm{E}(X+2) = \mu$;

(b) $\mathrm{VAR}(X+2) = \sigma^2$;

(c) $\mathrm{E}(2X) = 2\mu$;

(d) $\mathrm{VAR}(2X) = 2\sigma^2$;

(e) $\mathrm{VAR}(X/2) = \sigma^2/4$.

31. If X and Y are independent random variables:

(a) $\mathrm{VAR}(X+Y) = \mathrm{VAR}(X) + \mathrm{VAR}(Y)$;

(b) $\mathrm{E}(X+Y) = \mathrm{E}(X) + \mathrm{E}(Y)$;

(c) $\mathrm{E}(X-Y) = \mathrm{E}(X) - \mathrm{E}(Y)$;

(d) $\mathrm{VAR}(X-Y) = \mathrm{VAR}(X) - \mathrm{VAR}(Y)$;

(e) $\mathrm{VAR}(-X) = -\mathrm{VAR}(X)$.

6E Exercise: Probability and the life table

In this exercise we shall apply some of the basic laws of probability to a practical exercise. The data are based on a life table. (I shall say more about these in §16.4.) Table 6.1 shows the number of men, from a group numbering 1000 at birth, who we would expect to be alive at different ages. Thus, for example, after 10 years, we see that 959 survive and so 41 have

Table 6.1. Number of men remaining alive at ten year intervals (from English Life Table No. 11, Males)

Age in years, x	Number surviving, l_x	Age in years, x	Number surviving, l_x
0	1000	60	758
10	959	70	524
20	952	80	211
30	938	90	22
40	920	100	0
50	876		

died, at 20 years 952 survive and so 48 have died, 41 between ages 0 to 9 and 7 between ages 10 to 19.

1. What is the probability that an individual chosen at random will survive to age 10?

2. What is the probability that this individual will die before age 10? Which property of probability does this depend on?

3. What are the probabilities that the individual will survive to ages 10, 20, 30, 40, 50, 60, 70, 80, 90, 100? Is this set of probabilities a probability distribution?

4. What is the probability that an individual aged 60 years survives to age 70?

5. What is the probability that two men aged 60 will both survive to age 70? Which property of probability is used here?

6. If we had 100 individuals aged 60, how many would we expect to attain age 70?

7. What is the probability that a man dies in his second decade? You can use the fact that PROB(death in 2nd) + PROB(survives to 3rd) = PROB(survives to 2nd).

8. For each decade, what is the probability that a given man will die in that decade? This is a probability distribution — why? Sketch the distribution.

9. As an approximation, we can assume that the average number of years lived in the decade of death is 5. Thus, those who die in the 2nd decade will have an average life span of 15 years. The probability of dying in the 2nd decade is 0.007, i.e. a proportion 0.007 of men have a mean lifetime of 15 years. What is the mean lifetime of all men? This is the expectation of life at birth.

7

The Normal distribution

7.1 Probability for continuous variables

When we derived the theory of probability in the discrete case, we were able to say what the probability was of a random variable taking a particular value. As the number of possible values increases, the probability of a particular value decreases. For example, in the Binomial distribution with $p = 0.5$ and $n = 2$, the most likely value, 1, has probability 0.5. In the Binomial distribution with $p = 0.5$ and $n = 100$ the most likely value, 50, has probability 0.08. In such cases we are usually more interested in the probability of a range of values than one particular value.

For a continuous variable, such as height, the set of possible values is infinite and the probability of any particular value is zero (§6.1). We are interested in the probability of the random variable taking values between certain limits rather than taking particular values. If the proportion of individuals in the population whose values are between given limits is p, and we choose an individual at random, the probability of choosing an individual who lies between these limits is equal to p. This comes from our definition of probability, the choice of each individual being equally likely. The problem is finding and giving a value to this probability.

When we find the frequency distribution for a sample of observations, we count the number of values in which fall within certain limits. We can represent this as a histogram such as Figure 7.1 (§4.3). One way of presenting the histogram is as relative frequency density, the proportion of observations in the interval per unit of X (§4.3). Thus, when the interval size is 5, the relative frequency density is the relative frequency divided by 5 (Figure 7.1). The relative frequency in an interval is now represented by the width of the interval multiplied by the density, which gives the area of the rectangle. Thus, the relative frequency between any two points can be found from the area under the histogram between the points. For example, to estimate the relative frequency between 10 and 20 in Figure 7.1 we have the density from 10 to 15 as 0.05 and between 15 and 20 as 0.03. Hence the relative frequency is

$$0.05 \times (15 - 10) + 0.03 \times (20 - 15) = 0.25 + 0.15 = 0.40.$$

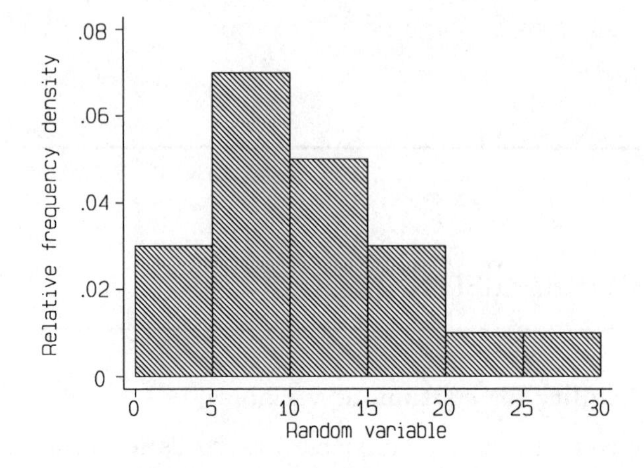

Fig. 7.1. Histogram showing relative frequency density

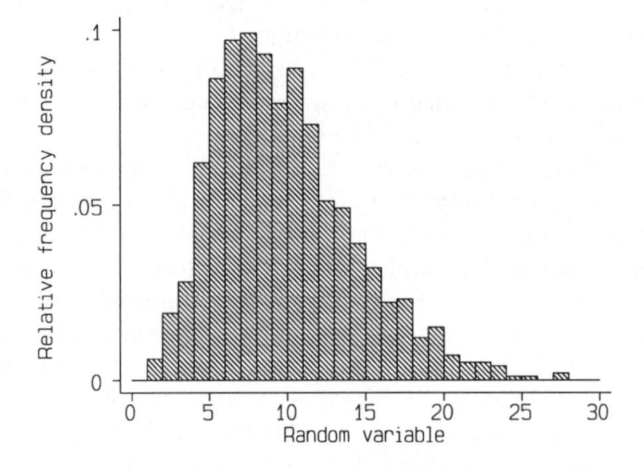

Fig. 7.2. The effect on a frequency distribution of increasing sample size

If we take a larger sample we can use smaller intervals. We get a smoother looking histogram, as in Figure 7.2, and as we take larger and larger samples, and so smaller and smaller intervals, we get a shape very close to a smooth curve (Figure 7.3). As the sample size approaches that of the population, which we can assume to be very large, this curve becomes the relative frequency density of the whole population. Thus we can find the proportion of observations between any two limits by finding the area under the curve, as indicated in Figure 7.3.

If we know the equation of this curve, we can find the area under it. (Mathematically we do this by integration, but we do not need to know how

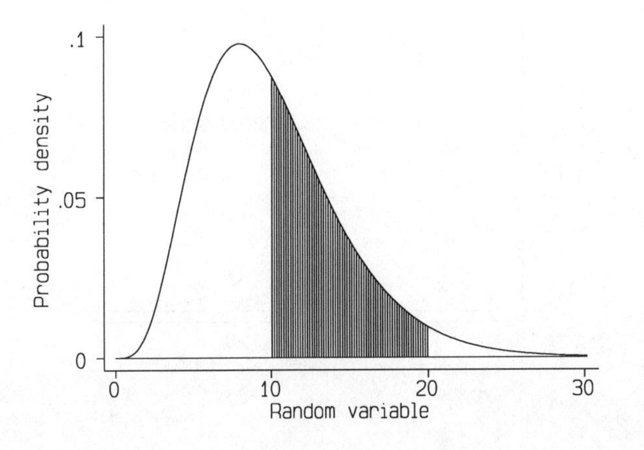

Fig. 7.3. Relative frequency density or probability density function

to integrate to use or to understand practical statistics—all the integrals we need have been done and tabulated.) Now, if we choose an individual at random, the probability that X lies between any given limits is equal to the proportion of individuals who fall between these limits. Hence, the relative frequency distribution for the whole population gives us the probability distribution of the variable. We call this curve the **probability density function.**

Probability density functions have a number of general properties. For example, the total area under the curve must be one, since this is the total probability of all possible events. As was noted in §6.5, continuous random variables have means, variances and standard deviations defined in a similar way to those for discrete random variables and possessing the same properties. The mean will be somewhere near the middle of the curve and most of the area under the curve will be between the mean minus two standard deviations and the mean plus two standard deviations (Figure 7.4).

The precise shape of the curve is more difficult to ascertain. There are many possible probability density functions and some of these can be shown to arise from simple probability situations, as were the Binomial and Poisson distributions. However, most continuous variables with which we have to deal, such as height, blood pressure, serum cholesterol, etc., do not arise from simple probability situations. As a result, we do not know the probability distribution for these measurements on theoretical grounds. As we shall see, we can often find a standard distribution whose mathematical properties are known, which fits observed data well and which enables us to draw conclusions about them. Further, as sample size increases the

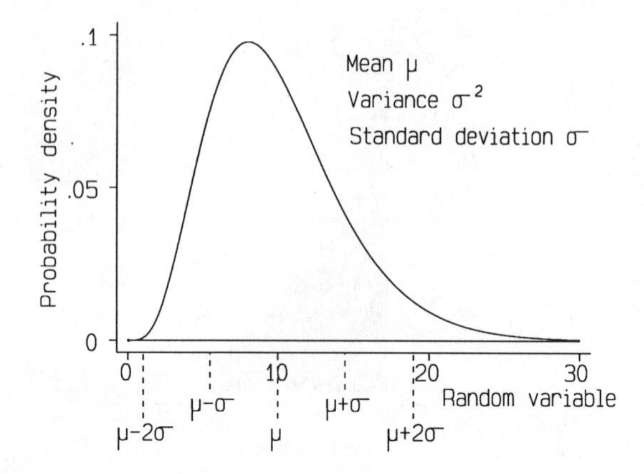

Fig. 7.4. Mean, μ, standard deviation, σ, and a probability density function

distribution of certain statistics calculated from the data, such as the mean, become independent of the distribution of the observations themselves and follow one particular distribution form, the Normal distribution. We shall devote the remainder of this chapter to a study of this distribution.

7.2 The Normal distribution

The Normal distribution, also known as the Gaussian distribution, may be regarded as the fundamental probability distribution of statistics. The word 'normal' here is not used in its common meaning of 'ordinary or common', or its medical meaning of 'not diseased'. The usage relates to its older meaning of 'conforming to a rule or pattern', and as we shall see, the Normal distribution is the form to which the Binomial distribution tends as its parameter n increases. There is no implication that most variables follow a Normal distribution.

We shall start by considering the Binomial distribution as n increases. We saw in §6.4 that, as n increases, the shape of the distribution changes. The most extreme possible values become less likely and the distribution becomes more symmetrical. This happens whatever the value of p. The position of the distribution along the horizontal axis, and its spread, are still determined by p, but the shape is not. A smooth curve can be drawn which goes very close to these points. This is the Normal distribution curve, the curve of the continuous distribution which the Binomial distribution approaches as n increases. Any Binomial distribution may be approximated by the Normal distribution of the same mean and variance provided n is large enough. Figure 7.5 shows the Binomial distributions of Figure 6.3 with the corresponding Normal distribution curves. From $n = 10$ onwards

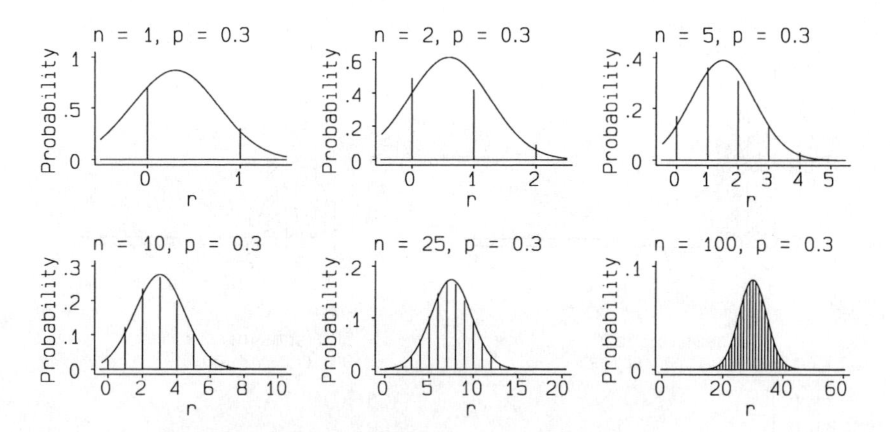

Fig. 7.5. Binomial distributions for $p = 0.3$ and six different values of n, with corresponding Normal distribution curves

the two distributions are very close. Generally, if both np and $n(1 - p)$ exceed 5 the approximation of the Binomial to the Normal distribution is quite good enough for most practical purposes. See §8.4 for an application. The Poisson distribution has the same property, as Figure 6.4 suggests.

The Binomial variable may be regarded as the sum of n independent identically distributed random variables, each being the outcome of one trial taking value 1 with probability p. In general, if we have any series of independent, identically distributed random variables, then their sum tends to a Normal distribution as the number of variables increases. This is known as the **central limit theorem**. As most sets of measurements are observations of such a series of random variables, this is a very important property. From it, we can deduce that the sum or mean of any large series of independent observations follows a Normal distribution.

For example, consider the **Uniform** or **Rectangular distribution**. This is the distribution where all values between two limits, say 0 and 1, are equally likely and no other values are possible. Observations from this arise if we take random digits from a table of random numbers such as Table 2.3. Each observation of the Uniform variable is formed by a series of such digits placed after a decimal point. On a microcomputer, this is usually the distribution produced by the RND(X) function in the BASIC language. Figure 7.6 shows the histogram for the frequency distribution of 500 observations from the Uniform distribution between 0 and 1. It is quite different from the Normal distribution. Now suppose we create a new variable by taking two Uniform variables and adding them (Figure 7.6). The shape of the distribution of the sum of two is quite different from the shape of the Uniform distribution. The sum is unlikely to be close to either

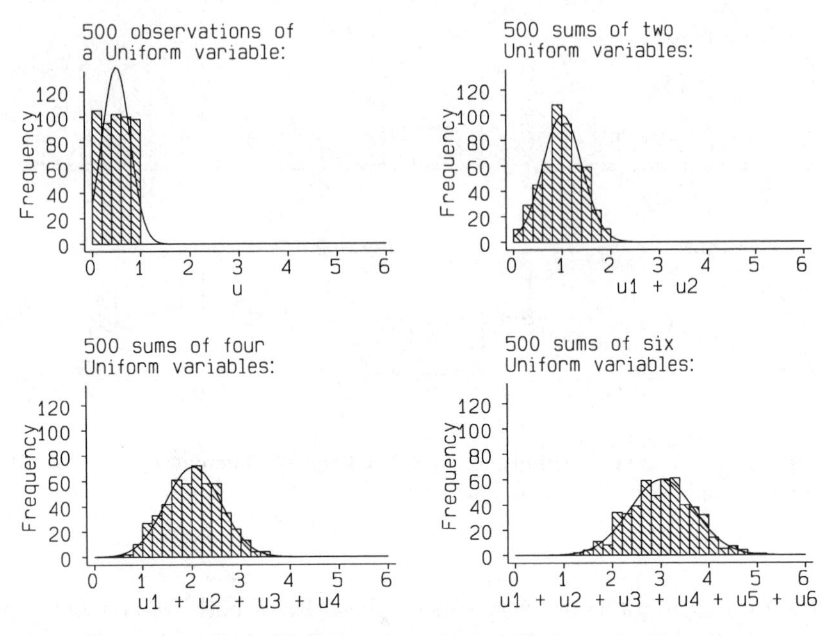

Fig. 7.6. Sums of observations from a Uniform distribution

extreme, here 0 or 2, and observations are concentrated in the middle near the expected value. The reason for this is that to obtain a low sum, both the Uniform variables forming it must be low; to make a high sum both must be high. But we get a sum near the middle if the first is high and the second low, or the first is low and second high, or both first and second are moderate. The distribution of the sum of two is much closer to the Normal than is the Uniform distribution itself. However, the abrupt cut-off at 0 and at 2 is unlike the corresponding Normal distribution. Figure 7.6 also shows the result of adding four Uniform variables and six Uniform variables. The similarity to the Normal distribution increases as the number added increases and for the sum of six the correspondence is so close that the distributions could not easily be told apart.

The approximation of the Binomial to the Normal distribution is a special case of the central limit theorem. The Poisson distribution is another. If we take a set of Poisson variables with the same rate and add them, we will get a variable which is the number of random events in a longer time interval (the sum of the intervals for the individual variables) and which is therefore a Poisson distribution with increased mean. As it is the sum of a set of independent, identically distributed random variables it will tend towards the Normal as the mean increases. Hence as the mean increases the Poisson distribution becomes approximately Normal. For most prac-

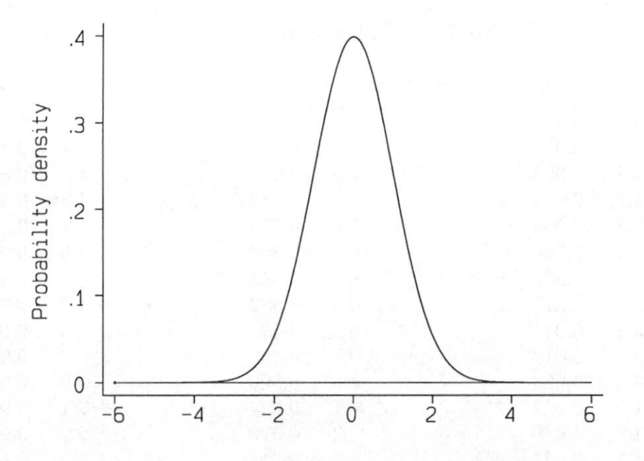

Fig. 7.7. The Standard Normal distribution

tical purposes this is when the mean exceeds 10. The similarity between the Poisson and the Binomial noted in §6.7 is a part of a more general convergence shown by many other distributions.

7.3 Properties of the Normal distribution

In its simplest form the equation of the Normal distribution curve, called the **Standard Normal distribution**, is usually denoted by $\phi(x)$, where ϕ is the Greek letter 'phi':

$$\phi(x) = \frac{1}{\sqrt{2\pi}} \exp\left(-\frac{x^2}{2}\right)$$

where π is the usual mathematical constant. The medical reader can be reassured that we do not need to use this forbidding formula in practice. The Standard Normal distribution has a mean of 0, a standard deviation of 1 and a shape as shown in Figure 7.7. The curve is symmetrical about the mean and often described as 'bell-shaped' (though I've never seen a bell like it). We can note that most of the area, i.e. the probability, is between –1 and +1, the large majority between –2 and +2, and almost all between –3 and +3.

Although the Normal distribution curve has many remarkable properties, it has one rather awkward one: it cannot be integrated. In other words, there is no simple formula for the probability of a random variable from a Normal distribution lying between given limits. The areas under the curve can be found numerically, however, and these have been calculated and tabulated. Table 7.1 shows the area under the Normal distribution curve for different values of the Normal distribution. To be more precise,

Table 7.1. The Normal distribution

x	$\Phi(x)$	x	$\Phi(x)$	x	$\Phi(x)$
–3.0	0.001	–1.0	0.159	1.0	0.841
–2.9	0.002	–0.9	0.184	1.1	0.864
–2.8	0.003	–0.8	0.212	1.2	0.885
–2.7	0.003	–0.7	0.242	1.3	0.903
–2.6	0.005	–0.6	0.274	1.4	0.919
–2.5	0.006	–0.5	0.309	1.5	0.933
–2.4	0.008	–0.4	0.345	1.6	0.945
–2.3	0.011	–0.3	0.382	1.7	0.955
–2.2	0.014	–0.2	0.421	1.8	0.964
–2.1	0.018	–0.1	0.460	1.9	0.971
–2.0	0.023	0.0	0.500	2.0	0.977
–1.9	0.029	0.1	0.540	2.1	0.982
–1.8	0.036	0.2	0.579	2.2	0.986
–1.7	0.045	0.3	0.618	2.3	0.989
–1.6	0.055	0.4	0.655	2.4	0.992
–1.5	0.067	0.5	0.691	2.5	0.994
–1.4	0.081	0.6	0.726	2.6	0.995
–1.3	0.097	0.7	0.758	2.7	0.997
–1.2	0.115	0.8	0.788	2.8	0.997
–1.1	0.136	0.9	0.816	2.9	0.998
–1.0	0.159	1.0	0.841	3.0	0.999

for a value x the table shows the area under the curve to the left of x, i.e. from minus infinity to x (Figure 7.8). Thus $\Phi(x)$ is the probability that a value chosen at random from the Standard Normal distribution will be less than x. Φ is the Greek capital 'phi'. Note that half this table is not strictly necessary. We need only the half for positive x as $\Phi(-x) + \Phi(x) = 1$. This arises from the symmetry of the distribution. To find the probability of x lying between two values a and b, where $b > a$, we find $\Phi(b) - \Phi(a)$. To find the probability of x being greater than a we find $1 - \Phi(a)$. These formulae are all examples of the additive law of probability. Table 7.1 gives only a few values of x, and much more extensive ones are available (Lindley and Miller 1955; Pearson and Hartley 1970). Good statistical computer programs will calculate these values when they are needed.

There is another way of tabulating a distribution, using what are called percentage points. The **one sided p percentage point** of a distribution is the value x such that there is a probability $p\%$ of an observation from that distribution being greater than or equal to x (Figure 7.8). The **two sided p percentage point** is the value x such that there is a probability $p\%$ of an observation being greater than or equal to x or less than or equal to $-x$ (Figure 7.8). Table 7.2 shows both one sided and two sided percentage points for the Normal distribution. The probability is quoted as a percentage because when we use percentage points we are usually concerned with rather small probabilities, such as 0.05 or 0.01, and use of

Table 7.2. Percentage points of the Normal distribution

One sided		Two sided	
$P_1(x)$	x	$P_2(x)$	x
50	0.00		
25	0.67	50	0.67
10	1.28	20	1.28
5	1.64	10	1.64
2.5	1.96	5	1.96
1	2.33	2	2.33
0.5	2.58	1	2.58
0.1	3.09	0.2	3.09
0.05	3.29	0.1	3.29

The table shows the probability $P_1(x)$ of a Normal variable with mean 0 and variance 1 being greater than x, and the probability $P_2(x)$ of a Normal variable with mean 0 and variance 1 being less than $-x$ or greater than x.

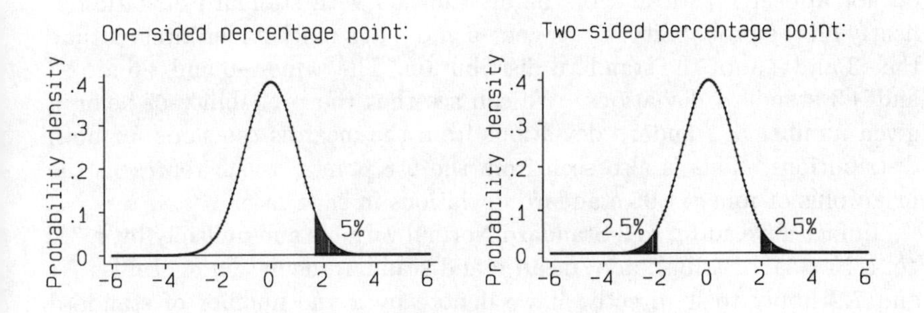

Fig. 7.8. One- and two-sided percentage points (5%) of the Standard Normal distribution

the percentage form, making them 5% and 1%, cuts out the leading zero.

So far we have examined the Normal distribution with mean 0 and standard deviation 1. If we add a constant μ to a Standard Normal variable, we get a new variable which has mean μ (see §6.6). Figure 7.9 shows the Normal distribution with mean 0 and the distribution obtained by adding 1 to it together with their two sided 5% points. The curves are identical apart from a shift along the axis. On the curve with mean 0 nearly all the probability is between -3 and $+3$. For the curve with mean 1 it is between -2 and $+4$, i.e. between the mean-3 and the mean$+3$. The probability of being a given number of units from the mean is the same for both distributions, as is also shown by the 5% points.

If we take a Standard Normal variable, with standard deviation 1, and

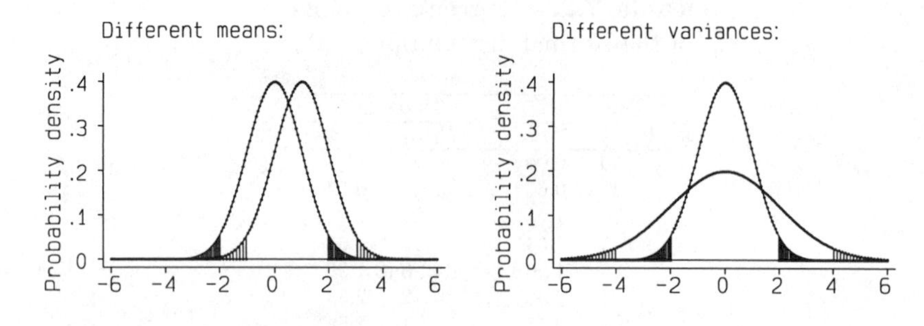

Fig. 7.9. Normal distributions with different means and with different variances, showing two sided 5% points

multiply by a constant σ we get a new variable which has standard deviation σ. Figure 7.9 shows the Normal distribution with mean 0 and standard deviation 1 and the distribution obtained by multiplying by 2. The curves do not appear identical. For the distribution with standard deviation 2, nearly all the probability is between –6 and +6, a much wider interval than the –3 and +3 for the standard distribution. The values –6 and +6 are –3 and +3 standard deviations. We can see that the probability of being a given number of standard deviations from the mean is the same for both distributions. This is also seen from the 5% points, which represent the mean plus or minus 1.96 standard deviations in each case.

In fact if we add μ to a Standard Normal variable and multiply by σ, we get a Normal distribution of mean μ and standard deviation σ. Tables 7.1 and 7.2 apply to it directly, if we denote by x the number of standard deviations above the mean, rather than the numerical value of the variable. Thus, for example, the two sided 5% points of a Normal distribution with mean 10 and standard deviation 5 are found by $10 + 1.96 \times 5 = 19.8$ and $10 - 1.96 \times 5 = 0.2$, the value 1.96 being found from Table 7.2.

This property of the Normal distribution, that multiplying or adding constants still gives a Normal distribution, is not as obvious as it might seem. The Binomial does not have it, for example. Take a Binomial variable with $n = 3$, possible values 0, 1, 2, and 3, and multiply by 2. The possible values are now 0, 2, 4, and 6. The Binomial distribution with $n = 6$ has also possible values 1, 3, and 5, so the distributions are different and the one which we have derived is not a member of the Binomial family.

We have seen that adding a constant to a variable from a Normal distribution gives another variable which follows a Normal distribution. If we add two variables from Normal distributions together, even with different means and variances, the sum follows a Normal distribution. The difference between two variables from Normal distributions also follows a Normal distribution.

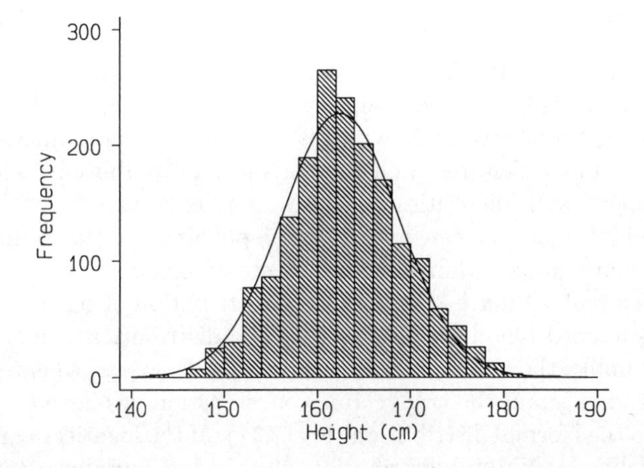

Fig. 7.10. Distribution of height in a sample of 1794 pregnant women (data of Brooke *et al.* 1989)

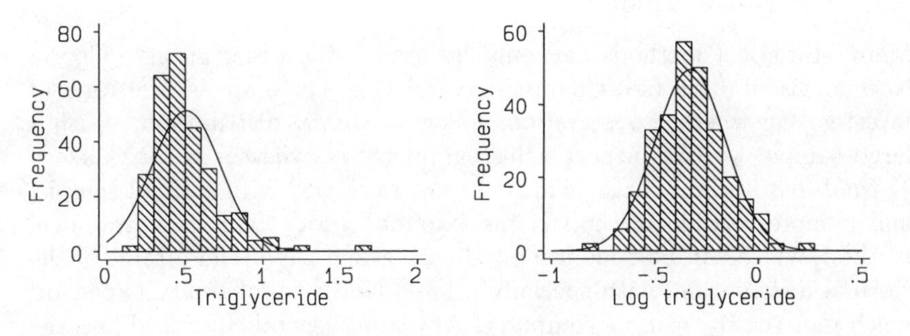

Fig. 7.11. Distribution of serum triglyceride and $\log_{10}$ trigliceride in cord blood for 282 babies, with corresponding Normal distribution curves

7.4 Variables which follow a Normal distribution

So far we have discussed the Normal distribution as it arises from sampling as the sum or limit of other distributions. However, many naturally occurring variables, such as human height, appear to follow a Normal distribution very closely. We might expect this to happen if the variable were the result of adding variation from a number of different sources. The process shown by the central limit theorem may well produce a result close to Normal. Figure 7.10 shows the distribution of height in a sample of pregnant women, and the corresponding Normal distribution curve. The fit to the Normal distribution is very good.

If the variable we measure is the result of multiplying several different sources of variation we would not expect the result to be Normal from the properties discussed in §7.2, which were all based on addition of variables. However, if we take the log transformation of such a variable (§5A) we would then get a new variable which is the sum of several different sources of variation and which may well have a Normal distribution. This process often happens with quantities which are part of metabolic pathways, the rate at which reaction can take place depending on the concentrations of other compounds. Many measurements of blood constituents exhibit this, for example. Figure 7.11 shows the distribution of serum triglyceride measured in cord blood for 282 babies. The distribution is highly skewed and quite unlike the Normal distribution curve. However, when we take the log transformation of the triglyceride concentration, we have a remarkably good fit to the Normal distribution (Fig. 7.11). If the logarithm of a random variable follows a Normal distribution, the random variable itself follows a **Lognormal distribution**.

7.5 The Normal plot

Many statistical methods can only be used if the observations follow a Normal distribution (see Chapters 10 and 11). There are several ways of investigating whether observations follow a Normal distribution. With a large sample we can inspect a histogram to see whether it looks like a Normal distribution curve. This does not work well with a small sample, and a more reliable method is the **Normal plot**. This is a graphical method, which can be done using ordinary graph paper and a table of the Normal distribution, with specially printed Normal probability paper, or, much more easily, using a computer. Any good general statistical package will give Normal plots; if it doesn't then it isn't a good package.

The Normal plot is a plot of the cumulative frequency distribution for the data against the cumulative frequency distribution for the Normal distribution. To construct a Normal plot we order the data from lowest to highest. For each ordered observation we find the expected value of the observation if the data followed a Standard Normal distribution. There are several approximate formulae for this. I shall follow Armitage and Berry (1987) and use for the ith observation x where $\Phi(x) = (i - 0.5)/n$. Exact tables are given by Pearson and Hartley (1972), but Normal plots are best done using computer programs. Having ordered the data, we find from a table of the Normal distribution the values of x which correspond to $\Phi(x) = 0.5/n$, $1.5/n$, etc. For 5 points, for example, we have $\Phi(x) = 0.1$, 0.3, 0.5, 0.7, and 0.9, and $x = -1.3$, -0.5, 0, 0.5, and 1.3. These are the points of the Standard Normal distribution which correspond to the observed data. Now, if the observed data come from a Normal distribution of mean μ and variance σ^2, the observed point should equal $\sigma x + \mu$, where x is the corresponding point of the Standard Normal distribution. If we

Table 7.3. Vitamin D levels measured in the blood of 26 healthy men, data of Hickish *et al.* (1989)

14	25	30	42	54
17	26	31	43	54
20	26	31	46	63
21	26	32	48	67
22	27	35	52	83
24				

Table 7.4. Calculation of the Normal plot for the vitamin D data

i	Vit D	$\Phi(x)$	x	i	Vit D	$\Phi(x)$	x
1	14	0.019	−2.07	14	31	0.519	0.05
2	17	0.058	−1.57	15	32	0.558	0.15
3	20	0.096	−1.30	16	35	0.596	0.24
4	21	0.135	−1.10	17	42	0.635	0.34
5	22	0.173	−0.94	18	43	0.673	0.45
6	24	0.212	−0.80	19	46	0.712	0.56
7	25	0.250	−0.67	20	48	0.750	0.67
8	26	0.288	−0.56	21	52	0.788	0.80
9	26	0.327	−0.45	22	54	0.827	0.94
10	26	0.365	−0.34	23	54	0.865	1.10
11	27	0.404	−0.24	24	63	0.904	1.30
12	30	0.442	−0.15	25	67	0.942	1.57
13	31	0.481	−0.05	26	83	0.981	2.07

$\Phi(x) = (i - 0.5)/26$

plot the Standard Normal points against the observed values we should get something close to a straight line. We can write the equation of this line as $\sigma x_{SND} + \mu = x_{obs}$, where x_{obs} is the observed variable and x_{SND} the corresponding quantile of the Standard Normal distribution. We can rewrite this as

$$x_{SND} = \frac{x_{obs}}{\sigma} - \frac{\mu}{\sigma}$$

which goes through the point defined by $(\mu, 0)$ and has slope $1/\sigma$ (see §11.1). If the data are not from a Normal distribution we will get a curve of some sort. Because we plot the quantiles of the observed frequency distribution against the corresponding quantiles of the theoretical (here Normal) distribution, this is also referred to as a **quantile–quantile plot** or **q–q plot.**

Table 7.3 shows vitamin levels measured in the blood of 26 healthy men. The calculation of the Normal plot is shown in Table 7.4. Note that the $\Phi(x) = (i - 0.5)/26$ and x are symmetrical, the second half being the first half inverted. The value of the Standard Normal deviate, x, can be found by interpolation in Table 7.1, by using a fuller table, or by computer.

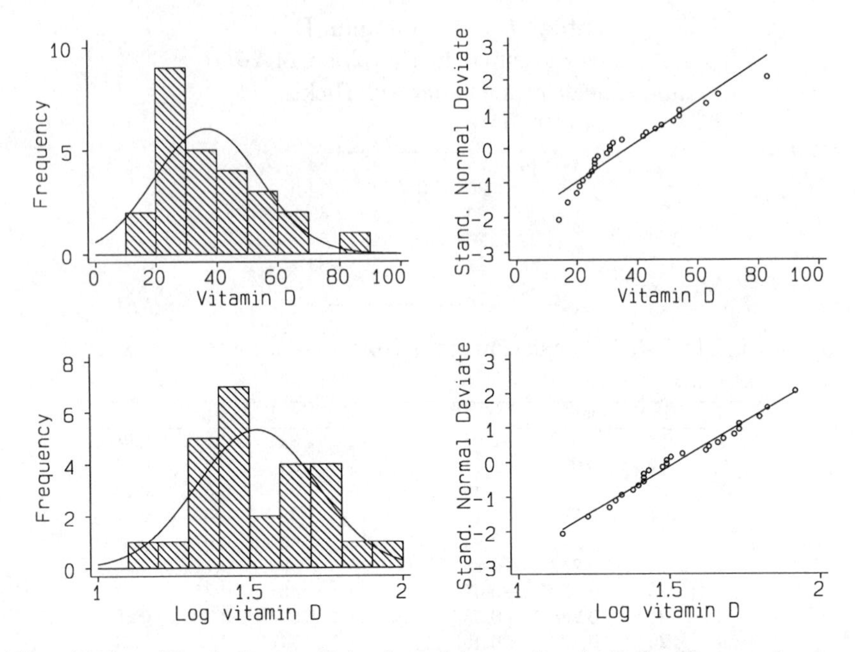

Fig. 7.12. Blood vitamin D levels and $\log_{10}$ vitamin D for 26 normal men, with Normal plots

Figure 7.12 shows the histogram and the Normal plot for these data. The distribution is skew and the Normal plot shows a pronounced curve. Figures 7.12 also shows the vitamin D data after log transformation. It is quite easy to produce the Normal plot, as the corresponding Standard Normal deviate, x, is unchanged. We only need to log the observations and plot again. The Normal plot for the transformed data conforms very well to the theoretical line, suggesting that the distribution of log vitamin D level is close to the Normal.

The Normal plot method can be used to investigate the Normal assumption in samples of any size, and is a very useful check when using methods such as the t distribution methods described in Chapter 10. There are several different formulae in use for calculating the centiles, but the differences are not important.

7A Appendix: Chi-squared, t, and F

Less mathematically inclined readers can skip this section, but those who persevere should find that applications like chi-squared tests (Chapter 13) appear much more logical.

Many probability distributions can be derived for functions of Normal variables which arise in statistical analysis. Three of these are particu-

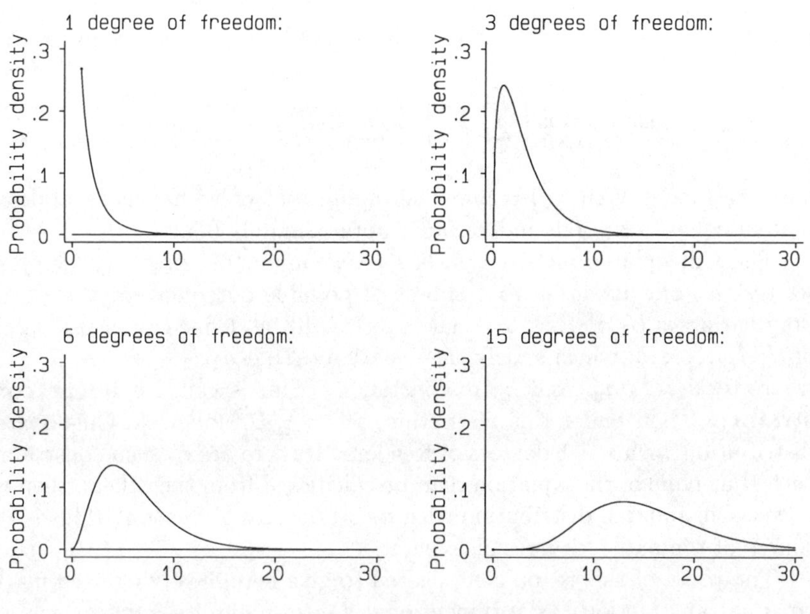

Fig. 7.13. Some Chi-squared distributions

larly important: the Chi-squared, t and F distributions. These have many applications, some of which we shall discuss in later chapters.

The Chi-squared distribution is defined as follows. Suppose U is a Standard Normal variable, so having mean 0 and variance 1. Then the variable formed by U^2 follows the Chi-squared distribution with 1 degree of freedom. If we have n such independent Standard Normal variables, U_1, U_2 , ..., U_n then the variable defined by

$$\chi^2 = U_1 + U_2 + \cdots + U_n$$

is defined to be the **Chi-squared distribution with n degrees of freedom.** χ is the Greek letter 'chi', pronounced 'ki' as in 'kite'. The distribution curves for several different numbers of degrees of freedom are shown in Figure 7.13. The mathematical description of this curve is rather complicated, but we do not need to go into this.

Some properties of the Chi-squared distribution are easy to deduce. As the distribution is the sum of n independent identically distributed random variables it tends to the Normal as n increases, from the central limit theorem. The convergence is slow, however, (Figure 7.13) and the square root of chi-squared converges much more quickly. The expected value of U^2 is the variance of U, the expected value of U being 0, and so

$\mathrm{E}(U^2) = 1$. The expected value of chi-squared with n degrees of freedom is thus n:

$$\mathrm{E}(\chi^2) = \mathrm{E}\left(\sum_{i=1}^{n} U_i^2\right) = \sum_{i=1}^{n} \mathrm{E}(U_i^2) = \sum_{i=1}^{n} 1 = n$$

The variance is $\mathrm{VAR}(\chi^2) = 2n$. The square root of χ^2 has mean approximately equal to $n - 0.5$ and variance approximately 0.5.

The Chi-squared distribution has a very important property. Suppose we restrict our attention to a subset of possible outcomes for the n random variables $U_1, U_2, \ldots, U_n$. The subset will be defined by those values of $U_1, U_2, \ldots, U_n$ which satisfy the equation $a_1U_1 + a_2U_2 + \cdots + a_nU_n = k$, where $a_1, a_2, \ldots, a_n$, and k are constants. (This is called a **linear constraint**). Then under this restriction, $\chi^2 = \sum U_i^2$ follows a Chi-squared distribution with $n - 1$ degrees of freedom. If there are m such constraints such that none of the equations can be calculated from the others, then we have a Chi-squared distribution with $n - m$ degrees of freedom. This is the source of the name 'degrees of freedom'.

The proof of this is too complicated to give here, involving such mathematical abstractions as n dimensional spheres, but its implications are very important. First, consider the sum of squares about the population mean μ of a sample of size n from a Normal distribution, divided by σ^2. $\sum(x_i - \mu)^2/\sigma^2$ will follow a Chi-squared distribution with n degrees of freedom, as the $(x_i - \mu)/\sigma$ have mean 0 and variance 1 and they are independent. Now suppose we replace μ by an estimate calculated from the data, $\bar{x}$. The variables are no longer independent, they must satisfy the relationship $\sum(x_i - \bar{x}) = 0$ and we now have $n - 1$ degrees of freedom. Hence $\sum(x_i - \bar{x})^2/\sigma^2$ follows a Chi-squared distribution with $n - 1$ degrees of freedom. The sum of squares about the mean of any Normal sample with variance σ^2 follows the distribution of a Chi-squared variable multiplied by σ^2. It therefore has expected value $(n - 1)\sigma^2$ and we divide by $n - 1$ to give the estimate of σ^2.

Thus, provided the data are from a Normal distribution, not only does the sample mean follow a Normal distribution, but the sample variance is from a Chi-squared distribution times σ^2. Because the square root of the Chi-squared distribution converges quite rapidly to the Normal, the distribution of the sample standard deviation is approximately Normal for $n > 20$, provided the data themselves are from a Normal distribution. Another important property of the variances of Normal samples is that the sample variance and sample mean are independent if, and only if, the data are from a Normal distribution.

Student's t distribution with n degrees of freedom is the distribution of $U/\sqrt{\chi^2/n}$, where U is a Standard Normal variable and χ^2 is independent of it and has n degrees of freedom. It is the distribution followed by the ratio of the mean to its standard error (§10A). The combined variance in the two sample t method (§10.3) is to give the single sum of squares on the bottom row.

The F distribution with m and n degrees of freedom is the distribution of $(\chi^2_m/m)/(\chi^2_n/n)$, the ratio of two independent χ^2 variables each divided by its degrees of freedom. This distribution is used for comparing variances. If we have two independent estimates of the same variance calculated from Normal data, the variance ratio will follow the F distribution. We can use this for comparing two estimates of variance (§10.8), but it main uses are in comparing groups of means (§10.9) and in examining the effects of several factors together (§17.2).

7M Multiple choice questions 32 to 37

(Each branch is either true or false)

32. The Normal distribution:

(a) is also called the Gaussian distribution;

(b) is followed by many variables;

(c) is a family of distributions with two parameters;

(d) is followed by all measurements made in healthy people;

(e) is the distribution towards which the Poisson distribution tends as its mean increases.

33. The Standard Normal distribution:

(a) is skew to the left;

(b) has mean = 1.0;

(c) has standard deviation = 0.0;

(d) has variance = 1.0;

(e) has the median equal to the mean.

34. The PEFRs of a group of 11 year old girls follow a Normal distribution with mean 300 l/min and a standard deviation 20 l/min:

(a) about 95% of the girls have PEFR between 260 and 340 l/min;

(b) 50% of the girls have PEFR above 300 l/min;

(c) the girls have healthy lungs;

(d) about 5% of girls have PEFR below 260 l/min;

(e) all the PEFRs must be less than 340 l/min.

35. The mean of a large sample:

(a) is always greater than the median;

(b) is calculated from the formula $\sum x_i/n$;

(c) is from an approximately Normal distribution;

(d) increases as the sample size increases;

(e) is always greater than the standard deviation.

36. If X and Y are independent variables which follow Standard Normal distributions, a Normal distribution is also followed by:

(a) $5X$;

(b) X^2;

(c) $X + 5$;

(d) $X - Y$;

(e) X/Y.

37. When a Normal plot is drawn with the Standard Normal deviate on the y axis:

(a) a straight line indicates that observations are from a Normal Distribution;

(b) a curve with decreasing slope indicates positive skewness;

(c) an 'S' shaped curve (or ogive) indicates long tails;

(d) a vertical line will occur if all observations are equal;

(e) if there is a straight line its slope depends on the standard deviation.

7E Exercise: A Normal plot

In this exercise we shall return to the blood glucose data of §4E and try to decide how well they conform to a Normal distribution.

1. From the box and whisker plot and the histogram found in exercise §4E (if you have not tried exercise §4E see the solution in Chapter 19), do the blood glucose levels look like a Normal distribution?

2. Construct a Normal plot for the data. This is quite easy as they are ordered already. Find $(i - 0.5)/n$ for $i = 1$ to 40 and obtain the corresponding cumulative Normal probabilities from Table 7.1. Now plot these probabilities against the corresponding blood glucose.

3. Does the plot appear to give a straight line? Do the data follow a Normal distribution?

8 Estimation

8.1 Sampling distributions

We have seen in Chapter 3 how samples are drawn from much larger populations. Data are collected about the sample so that we can find out something about the population. We use samples to estimate quantities such as disease prevalence, mean blood pressure, mean exposure to a carcinogen, etc. We also want to know by how much these estimates might vary from sample to sample.

In Chapters 6 and 7 we saw how the theory of probability enables us to link random samples with the populations from which they are drawn. In this chapter we shall see how probability theory enables us to use samples to estimate quantities in populations, and to determine the precision of these estimates. First we shall consider what happens when we draw repeat samples from the same population. Table 8.1 shows a set of 100 random digits which we can use as the population for a sampling experiment. The distribution of the numbers in this population is shown in Figure 8.1. The population mean is 4.7 and the standard deviation is 2.9.

The sampling experiment is done by using a suitable random sampling method to draw repeated samples from the population. In this case decimal dice were a convenient method. A sample size four was chosen: 6, 4, 6 and 1. The mean was calculated: $17/4 = 4.25$. This was repeated to draw a second sample of 4 numbers: 7, 8, 1, 8. Their mean is 6.00. This sampling procedure was done 20 times altogether, to give the samples and their means shown in Table 8.2.

These sample means are not all the same. They show random variation. If we were able to draw all of the 3 921 225 possible samples of size 4 and

Table 8.1. Population of 100 random digits for a sampling experiment

9	1	0	7	5	6	9	5	8	8	1	0	5	7	6	5	0	2	1	2
1	8	8	8	5	2	4	8	3	1	6	5	5	7	4	1	7	3	3	3
2	8	1	8	5	8	4	0	1	9	2	1	6	9	4	4	7	6	1	7
1	9	7	9	7	2	7	7	0	8	1	6	3	8	0	5	7	4	8	6
7	0	2	8	8	7	2	5	4	1	8	6	8	3	5	8	2	7	2	4

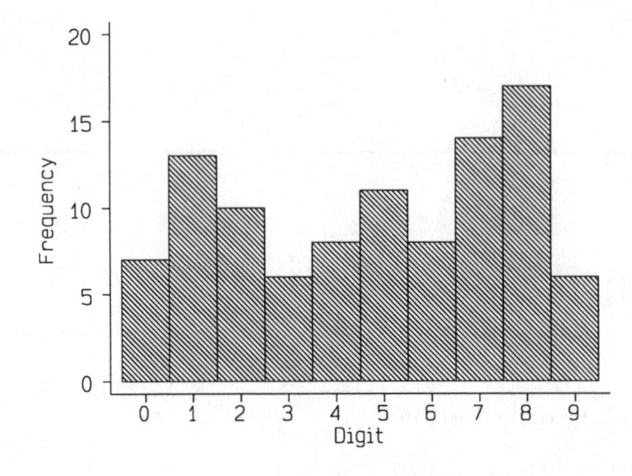

Fig. 8.1. Distribution of the population of Table 8.1

calculate their means, these means themselves would form a distribution. Our 20 sample means are themselves a sample from this distribution. The distribution of all possible sample means is called the **sampling distribution** of the mean. In general, the sampling distribution of any statistic is the distribution of the values of the statistic which would arise from all possible samples.

8.2 Standard error of a sample mean

For the moment we shall consider the sampling distribution of the mean only. As our sample of 20 means is a random sample from it, we can use this to estimate some of the parameters of the distribution. The twenty means have their own mean and standard deviation. The mean is 5.1 and the standard deviation is 1.1. Now the mean of the whole population is 4.7, which is close to the mean of the samples. But the standard deviation of the population is 2.9, which is considerably greater than that of the sample

Table 8.2. Random samples drawn in a sampling experiment

Sample	6	7	7	1	5	5	4	7	2	8
	4	8	9	8	2	5	2	4	8	1
	6	1	2	8	9	7	7	0	7	2
	1	8	7	4	5	8	6	1	7	0
Mean	4.25	6.00	6.25	5.25	5.25	6.25	4.75	3.00	6.00	2.75
Sample	7	7	2	8	3	4	5	4	4	7
	8	3	5	0	7	8	5	3	5	4
	7	8	0	7	4	7	8	1	8	6
	2	7	8	7	8	7	3	6	2	3
Mean	6.00	6.25	3.75	5.50	5.50	6.50	5.25	3.50	4.75	5.00

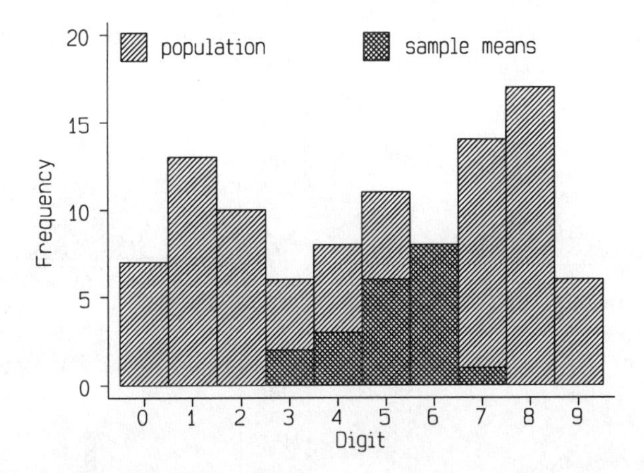

Fig. 8.2. Distribution of the population of Table 8.1 and of the sample of the means of Table 8.2

means. If we plot a histogram for the sample of means (Figure 8.2) we see that the centre of the sampling distribution and the parent population distribution are the same, but the scatter of the sampling distribution is much less.

Another sampling experiment, on a larger scale, will illustrate this further. This time our parent distribution will be the Normal distribution with mean 0 and standard deviation 1. Figure 8.3 shows the distribution of a random sample of 500 observations from this distribution. Figure 8.3 also shows the distribution of means from 500 random samples of size 4 from this population, the same sample size as in Figure 8.2. Figure 8.3 also shows the distributions of 500 means of samples of size 9 and of size 16. In all four distributions the means are close to zero, the mean of the parent distribution. But the standard deviations are not the same. They are, in fact, approximately 1 (parent distribution); 1/2 (means of 4), 1/3 (means of 9) and 1/4 (means of 16). In fact, the sampling distribution of the mean has standard deviation $\sigma/\sqrt{n}$ or $\sqrt{\sigma^2/n}$, where σ is the standard deviation of the parent distribution and n is the sample size (Appendix §6B). The mean of the sampling distribution is equal to the mean of the parent distribution. The actual, as opposed to simulated, distribution of the mean of four observations from a Normal distribution is shown in Figure 8.4.

The sample mean is an estimate of the population mean. The standard deviation of its sampling distribution is called the **standard error** of the estimate. It provides a measure of how far from the true value the estimate is likely to be. In most estimation, the estimate is likely to be within one standard error of the true mean and unlikely to be more than two standard errors from it. We shall look at this more precisely in §8.3.

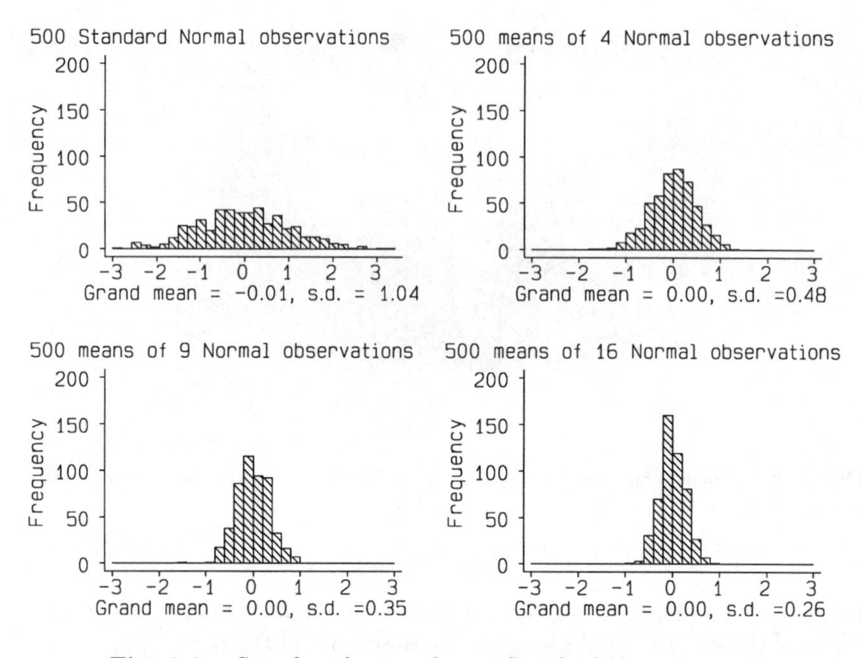

Fig. 8.3. Samples of means from a Standard Normal variable

In almost all practical situations we do not know the true value of the population variance σ^2 but only its estimate s^2 (§4.7). We can use this to estimate the standard error by $s/\sqrt{n}$. This estimate is also referred to as the standard error of the mean. It is usually clear from the context whether the standard error is the true value or that estimated from the data.

When the sample size n is large, the sampling distribution of $\bar{x}$ tends to a Normal distribution. Also, we can assume that s^2 is a good estimate of σ^2. So for large n, $\bar{x}$ is, in effect, an observation from a Normal distribution with mean μ and standard deviation estimated by $s/\sqrt{n}$. So with probability 0.95, $\bar{x}$ is within two, or more precisely is within 1.96 standard errors of μ. With small samples we cannot assume either a Normal distribution or, more importantly, that s^2 is a good estimate of σ^2. We shall discuss this in Chapter 10.

For an example, consider the 57 FEV1 measurements of Table 4.4. We have $\bar{x} = 4.062$ litres, $s^2 = 0.449\,174$, $s = 0.67$ litres. Then the standard error of $\bar{x}$ is $\sqrt{s^2/n} = \sqrt{0.449\,174/57} = 0.007\,880 = 0.089$. The best estimate of the mean FEV1 in the population is then 4.06 litres with standard error 0.089 litres.

The mean and standard error are often written as 4.062 ± 0.089. This is rather misleading, as the true value may be up to two standard errors from the mean with a reasonable probability. This practice is not recommended.

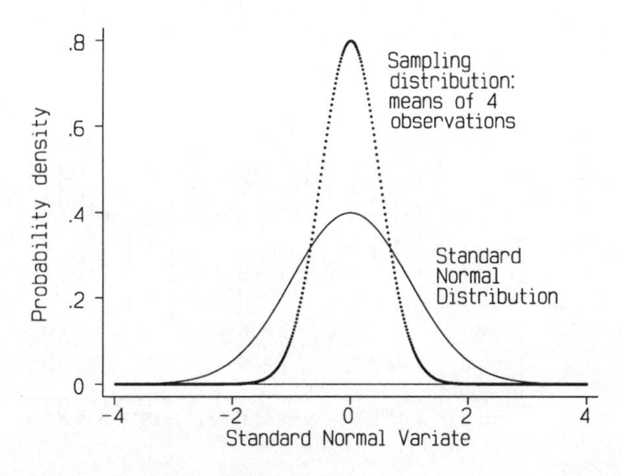

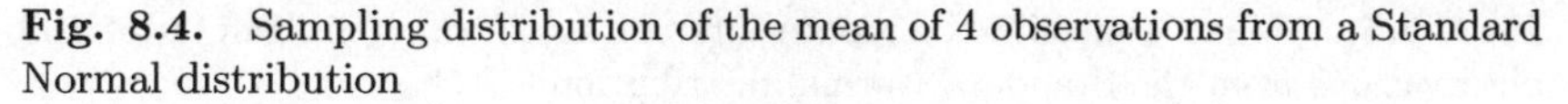
Fig. 8.4. Sampling distribution of the mean of 4 observations from a Standard Normal distribution

There is often confusion between the terms 'standard error' and 'standard deviation'. This is understandable, as the standard error is a standard deviation (of the sampling distribution) and the terms are often interchanged in this context. The convention is this: we use the term 'standard error' when we measure the precision of estimates, and the term 'standard deviation' when we are concerned with the variability of samples, populations or distributions. If we want to say how good our estimate of the mean FEV1 measurement is, we quote the standard error of the mean. If we want to say how widely scattered the FEV1 measurements are, we quote the standard deviation, s.

8.3 Confidence intervals

The estimate of mean FEV1 is a single value and so is called a **point estimate.** There is no reason to suppose that the population mean will be exactly equal to the point estimate, the sample mean. It is likely to be close to it, however, and the amount by which it is likely to differ from the estimate can be found from the standard error. What we do is find limits which are likely to include the population mean, and say that we estimate the population mean to lie somewhere in the interval (the set of all possible values) between these limits. This is called an **interval estimate.**

For instance, if we regard the 57 FEV measurements as being a large sample we can assume that the sampling distribution of the mean is Normal, and that the standard error is a good estimate of its standard deviation (see §10.6 for a discussion of how large is large). We therefore expect about 95% of such means to be within 1.96 standard errors of the

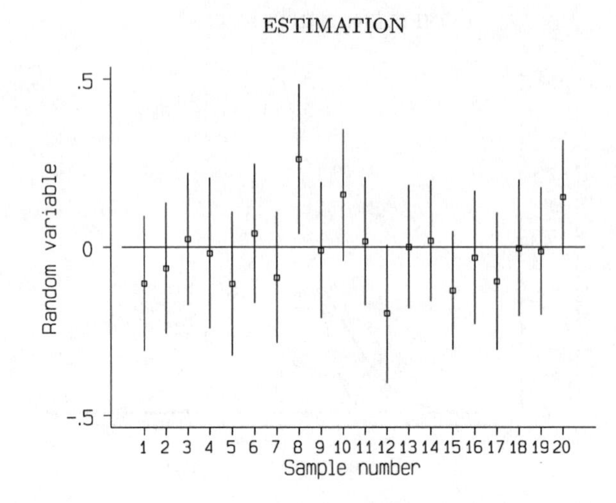

Fig. 8.5. Mean and 95% confidence interval for 20 random samples of 100 observations from the Standard Normal distribution

population mean, μ. Hence, for about 95% of all possible samples, the population mean must be greater than the sample mean minus 1.96 standard errors and less than the sample mean plus 1.96 standard errors. If we calculated $\bar{x} - 1.96se$ and $\bar{x} + 1.96se$ for all possible samples, 95% of such intervals would contain the population mean. In this case these limits are $4.062 - 1.96 \times 0.089$ to $4.062 + 1.96 \times 0.089$ which gives 3.89 to 4.24, or 3.9 to 4.2 litres, rounding to 2 significant figures. 3.9 and 4.2 are called the **95% confidence limits** for the estimate, and the set of values between 3.9 and 4.2 is called the **95% confidence interval.** The confidence limits are the values at the ends of the confidence interval.

Strictly speaking, it is incorrect to say that there is a probability of 0.95 that the population mean lies between 3.9 and 4.2, though it is often put that way (even by me). The population mean is a number, not a random variable, and has no probability. It is the probability that limits calculated from a random sample will include the population value which is 95%. Figure 8.5 shows confidence intervals for the mean for 20 random samples of 100 observations from the Standard Normal distribution. The population mean is, of course, 0.0, shown by the horizontal line. Some sample means are close to 0.0, some further away, some above and some below. The population mean is contained by 19 of the 20 confidence intervals. In general, for 95% of confidence intervals it will be true to say that the population value lies within the interval. We just don't know which 95%. We express this by saying that we are 95% confident that the mean lies between these limits.

In the FEV1 example, the sampling distribution of the mean is Normal and its standard deviation is well estimated because the sample is large.

This is not always true and although it is usually possible to calculate confidence intervals for an estimate they are not all quite as simple as that for the mean estimated from a large sample. We shall look at the mean estimated from a small sample in Chapter 10.

There is no necessity for the confidence interval to have a probability of 95%. For example, we can also calculate 99% confidence limits. The upper 0.5% point of the Standard Normal distribution is 2.58 (Table 7.2), so the probability of a Standard Normal deviate being above 2.58 or below –2.58 is 1% and the probability of being within these limits is 99%. The 99% confidence limits for the mean FEV1 are therefore, $4.062 - 2.58 \times 0.089$ and $4.062 + 2.58 \times 0.089$, i.e. 3.8 and 4.3 litres. These give a wider interval than the 95% limits, as we would expect since we are more confident that the mean will be included. The probability we choose for a confidence interval is thus a compromise between the desire to include the estimated population value and the desire to avoid parts of scale where there is a low probability that the mean will be found. For most purposes, 95% confidence intervals have been found to be satisfactory.

8.4 Standard error of a proportion

The standard error of a proportion estimate can be calculated in the same way. Suppose the proportion of individuals who have a particular condition in a given population is p, and we take a random sample of size n, the number observed with the condition being r. Then the estimated proportion is r/n. We have seen (§6.4) that r comes from a Binomial distribution with mean np and variance $np(1-p)$. Provided n is large, this distribution is approximately Normal. So r/n, the estimated proportion, is Normally distributed with mean given by $np/n = p$, and variance given by

$$\mathrm{VAR}\left(\frac{r}{n}\right) = \frac{1}{n^2}\mathrm{VAR}(r) = \frac{1}{n^2}np(1-p) = \frac{p(1-p)}{n}$$

since n is constant, and the standard error is

$$\sqrt{\frac{p(1-p)}{n}}$$

We can estimate this by replacing p by r/n.

For example, in a random sample of first year secondary school-children in Derbyshire (Banks *et al.* 1978), 118 out of 2 837 boys said that they usually coughed first thing in the morning. This gave a prevalence estimate of $118/2\,837 = 0.041\,6$, with standard error $\sqrt{0.041\,6 \times (1 - 0.041\,6)/2\,837} = 0.003\,7$. The sample is large so we can assume that the estimate is from a Normal distribution and that the standard error is well estimated. The

95% confidence interval for the prevalence is thus $0.0416 - 1.96 \times 0.0037$ to $0.0416 + 1.96 \times 0.0037 = 0.034$ to 0.049. Even with this fairly large sample the estimate is not very precise.

The standard error of the proportion is only of use if the sample is large enough for the Normal approximation to apply. A rough guide to this is that np and $n(1-p)$ should both exceed 5. This is usually the case when we are concerned with straightforward estimation. If we try to use the method for smaller samples, we may get absurd results. For example, in a study of the prevalence of HIV in ex-prisoners (Turnbull *et al.* 1992), of 29 women who did not inject drugs one was HIV positive. The authors reported this to be 3.4%, with a 95% confidence interval –3.1% to 9.9%. The lower limit of –3.1%, obtained from the observed proportion minus 1.96 standard errors, is impossible. As Newcombe (1992) pointed out, the correct 95% confidence interval can be obtained from the exact probabilities of the Binomial distribution and is 0.1% to 17.8% (see Pearson and Hartley 1970).

8.5 The difference between two means

In many studies we are more interested in the difference between two parameters than in their absolute value. These could be means, proportions, the slopes of lines, and many other statistics. This is usually straightforward if the parameters are estimated from two independent samples. It can be more difficult if the samples are matched or are two observations on the same sample (§13.9).

When samples are large we can assume that sample means and proportions are observations from a Normal distribution, and that the calculated standard errors are good estimates of the standard deviations of these Normal distributions. We can use this to find confidence intervals.

For example, suppose we wish to compare the means, $\bar{x}_1$ and $\bar{x}_2$, of two large samples, sizes n_1 and n_2. The expected difference between the sample means is equal to the difference between the population means, i.e. $E(\bar{x}_1 - \bar{x}_2) = \mu_1 - \mu_2$. What is the standard error of the difference? The variance of the difference between two independent random variables is the sum of their variances (§6.6). Hence, the standard error of the difference between two independent estimates is the square root of the sum of the squares of their standard errors. The standard error of a mean is $\sqrt{s^2/n}$, so the standard error of the difference between two independent means is

$$\sqrt{\frac{s_1^2}{n_1} + \frac{s_2^2}{n_2}}$$

For an example, in a study of respiratory symptoms in school children (Bland *et al.* 1974), we wanted to know whether children reported by their parents to have respiratory symptoms had worse lung function than children who were not reported to have symptoms. Ninety two children were reported to have cough during the day or at night, and their mean PEFR was 294.8 litre/min with standard deviation 57.1 litre/min, and 1643 children were not reported to have this symptom, their mean PEFR being 313.6 litre/min with standard deviation 55.2 litre/min. We thus have two large samples, and can apply the Normal distribution. We have

$$n_1 = 92, \ \bar{x}_1 = 294.8, \ s_1 = 57.1, \ n_2 = 1643, \ \bar{x}_2 = 313.6, \ s_2 = 55.2$$

The difference between the two groups is $\bar{x}_1 - \bar{x}_2 = 294.3 - 313.6 = -18.8$. The standard error of the difference is

$$\sqrt{se_1^2 + se_2^2} = \sqrt{\frac{s_1^2}{n_1} + \frac{s_2^2}{n_2}} = \sqrt{\frac{57.1^2}{92} + \frac{55.2^2}{1643}} = 6.11$$

We shall treat the sample as being large, so the difference between the means can be assumed to come from a Normal distribution and the estimated standard error to be a good estimate of the standard deviation of this distribution. (For small samples see §10.3 and §10.6.) The 95% confidence limits for the difference are thus $-18.8 - 1.96 \times 6.11$ and $-18.8 + 1.96 \times 6.11$, i.e. -6.8 and -30.8 l/min. The confidence interval does not include zero, so we have good evidence that, in this population, children reported to have day or night cough have lower mean PEFR than others. The difference is estimated to be between 7 and 31 litre/min lower in children with the symptom, so it may be quite small.

8.6 Comparison of two proportions

We can apply the method of §8.5 to two proportions. The standard error of a proportion p is $\sqrt{p_1(1-p_1)/n_1}$. For two independent proportions, p_1 and p_2, the standard error of the difference between them is

$$\sqrt{\frac{p_1(1-p_1)}{n_1} + \frac{p_2(1-p_2)}{n_2}}$$

Provided the conditions of Normal approximation are met (see §8.4) we can find a confidence interval for the difference in the usual way.

For example, consider Table 8.3. The researchers wanted to know to what extent children with bronchitis in infancy get more respiratory symptoms in later life than others. We can estimate the difference between the proportions reported to cough during the day or at night among children with and children without a history of bronchitis before age 5

Table 8.3. Cough during the day or at night at age 14 and bronchitis before age 5 (Holland *et al.* 1978)

Cough at 14	Bronchitis at 5		
	Yes	No	Total
Yes	26	44	70
No	247	1002	1249
Total	273	1046	1319

years. We have estimates of two proportions, $p_1 = 26/273 = 0.095\,24$ and $p_2 = 44/1\,046 = 0.042\,07$. The difference between them is $p_1 - p_2 = 0.095\,24 - 0.042\,07 = 0.053\,17$. The standard error of the difference is

$$\begin{aligned} & \sqrt{\frac{p_1(1-p_1)}{n_1} + \frac{p_2(1-p_2)}{n_2}} \\ & \qquad = \sqrt{\frac{0.095\,24 \times (1-0.095\,24)}{273} + \frac{0.042\,07 \times (1-0.042\,07)}{1\,046}} \\ & \qquad = \sqrt{0.000\,315\,639 + 0.000\,038\,528} \\ & \qquad = \sqrt{0.000\,354\,167} \\ & \qquad = 0.018\,8 \end{aligned}$$

The 95% confidence interval for the difference is $0.053\,17 - 1.96 \times 0.018\,8$ to $0.053\,17 + 1.96 \times 0.018\,8 = 0.016$ to 0.090. Although the difference is not very precisely estimated, the confidence interval does not include zero and gives us clear evidence that children with bronchitis reported in infancy are more likely than others to be reported to have respiratory symptoms in later life. The data on lung function in §8.5 gives us some reason to suppose that this is not entirely due to response bias (§3.9). As in §8.4, the confidence interval must be estimated differently for small samples.

This difference in proportions may not very easy to interpret. The ratio of two proportions is often more useful. Another method, the odds ratio, is described in §13.7. The ratio of the proportion with cough at age 14 for bronchitis before 5 to the proportion with cough at age 14 for those without bronchitis before 5 is $p_1/p_2 = 0.09524/0.04207 = 2.26$. Children with bronchitis before 5 are more than twice as likely to cough during the day or at night at age 14 than children with no such history.

The standard error for this ratio is complex, and as it is a ratio rather than a difference it does not approximate well to a Normal distribution. If we take the logarithm of the ratio, however, we get the difference between two logarithms, because $\log(p_1/p_2) = \log(p_1) - \log(p_2)$ (§5A). We can find the standard error for the log ratio quite easily. We use the result that, for any random variable X with mean μ and variance σ^2, the approximate variance of $\log(X)$ is given by $\text{VAR}\,(log_e(X)) = \sigma^2/\mu^2$ (see Kendall and Stuart 1969). Hence, the variance of $\log(p)$ is

$$\text{VAR}\,(\log(p)) = \frac{p(1-p)/n}{p^2} = \frac{1-p}{np}$$

For the difference between the two logarithms we get

$$\begin{aligned}\text{VAR}(log_e(p_1/p_2)) &= \text{VAR}(log_e(p_1)) + \text{VAR}(log_e(p_2)) \\ &= \frac{1-p_1}{n_1p_1} + \frac{1-p_2}{n_2p_2}\end{aligned}$$

The standard error is the square root of this. For the example the log ratio is $log_e(2.263\,85) = 0.817\,07$ and the standard error is

$$\begin{aligned}\sqrt{\frac{1-p_1}{n_1p_1} + \frac{1-p_2}{n_2p_2}} &= \sqrt{\frac{1-0.095\,24}{273\times 0.095\,24} + \frac{1-0.042\,07}{1\,046\times 0.042\,07}} \\ &= \sqrt{\frac{0.904\,76}{26} + \frac{0.957\,93}{44}} \\ &= \sqrt{0.056\,57} \\ &= 0.237\,84\end{aligned}$$

The 95% confidence interval for the log ratio is therefore $0.817\,07 - 1.96 \times 0.237\,84$ to $0.817\,07 + 1.96 \times 0.237\,84 = 0.350\,89$ to $1.283\,24$. The 95% confidence interval for the ratio of proportions itself is the antilog of this: $e^{0.35089}$ to $e^{1.28324} = 1.42$ to 3.61. Thus we estimate that the proportion of children reported to cough during the day or at night among those with a history of bronchitis is between 1.4 to 3.6 times the proportion among those without a history of bronchitis.

The proportion of individuals in a population who develop a disease or symptom is equal to the probability that any given individual will develop the disease, called the **risk** of an individual developing a disease. Thus in Table 8.3 the risk that a child with bronchitis before age 5 will cough at age 14 is $26/273 = 0.09524$, and the risk for a child without bronchitis before age 5 is $44/1046 = 0.04207$. To compare risks for people with and without a particular risk factor, we look at the ratio of the risk with the factor to the risk without the factor, the **relative risk**. The relative risk of cough at age 14 for bronchitis before 5 is thus 2.26. To estimate the relative risk directly, we need a cohort study (§3.7) as in Table 8.3. We estimate relative risk for a case-control study in a different way (§13.7).

8.7 Standard error of a sample standard deviation

We can find a standard error and confidence interval for almost any estimate we make from a sample, but sometimes this depends on the distribution of the observations themselves. The sample standard deviation, s, is one such statistic. Provided the observations come from a Normal distribution, $(n-1)s^2/\sigma^2$ is from a Chi-squared distribution with $n-1$ degrees of freedom (§7A). The square root of this Chi-squared distribution is approximately Normal with variance 1/2 if n is large enough, so $\sqrt{(n-1)s^2/\sigma^2}$ is approximately Normally distributed with variance 1/2. Hence s is approximately Normally distributed with variance $\sigma^2/2(n-1)$. The standard error of s is thus $\sqrt{\sigma^2/2(n-1)}$, estimated by $s/\sqrt{2(n-1)}$. This is only true when the observations themselves are from a Normal distribution.

8M Multiple choice questions 38 to 43

(Each branch is either true or false)

38. The standard error of the mean of a sample:
(a) measures the variability of the observations;
(b) is the accuracy with which each observation is measured;
(c) is a measure of how far the sample mean is likely to be from the population mean;
(d) is proportional to the number of observations;
(e) is greater than the estimated standard deviation of the population.

39. 95% confidence limits for the mean estimated from a set of observations
(a) are limits between which, in the long run, 95% of observations fall;
(b) are a way of measuring the precision of the estimate of the mean;
(c) are limits within which the sample mean falls with probability 0.95;
(d) are limits which would include the population mean for 95% of possible samples;
(e) are a way of measuring the variability of a set of observations.

40. If the size of a random sample were increased, we would expect:
(a) the mean to decrease;
(b) the standard error of the mean to decrease;
(c) the standard deviation to decrease;
(d) the sample variance to increase;
(e) the degrees of freedom for the estimated variance to increase.

41. The prevalence of a condition in a population is 0.1. If the prevalence is estimated repeatedly from samples of size 100, these estimates will form a distribution which:

(a) is a sampling distribution;

(b) is approximately Normal;

(c) has mean = 0.1;

(d) have variance = 9;

(e) is Binomial.

42. It is necessary to estimate the mean FEV1 by drawing a sample from a large population. The accuracy of the estimate will depend on:

(a) the mean FEV1 in the population;

(b) the number in the population;

(c) the number in the sample;

(d) the way the sample is selected;

(e) the variance of FEV1 in the population.

43. In a study of 88 births to women with a history of thrombocytopenia (Samuels *et al.* 1990), the same condition was recorded in 20% of babies (95% confidence interval 13% to 30%):

(a) Another sample of the same size will show a rate of thrombocytopenia between 13% and 30%;

(b) 95% of such women have a probability of between 13% and 30% of having a baby with thrombocytopenia;

(c) It is likely that between 13% and 30% of births to such women in the area would show thrombocytopenia;

(d) If the sample were increased to 880 births, the 95% confidence interval would be narrower;

(e) It would be impossible to get these data if the rate for all women was 10%.

8E Exercise: Means of large samples

Table 8.4 summarizes data collected in a study of plasma magnesium in diabetics. The diabetic subjects were all insulin dependent subjects attending a diabetic clinic over a 5 month period. The non-diabetic controls were a mixture of blood donors and people attending day centres for the elderly, to give a wide age distribution. Plasma magnesium follows a Normal distribution very closely.

1. Calculate an interval which would include 95% of plasma magnesium measurements from the control population. This is what we call the 95% reference interval, described in detail in §15.5. It tells us something about the distribution of plasma magnesium in the population.

2. What proportion of insulin dependent diabetics would lie within this

Table 8.4. Plasma magnesium in insulin-dependent diabetics and healthy controls

	Number	Mean	Standard deviation
Insulin-dependent diabetics	227	0.719	0.068
Non-diabetic controls	140	0.810	0.057

95% reference interval? (Hint: find how many standard deviations from the diabetic mean the lower limit is, then use the table of the Normal distribution, Table 7.1, to find the probability of exceeding this.)
3. Find the standard error of the mean plasma magnesium for each group.
4. Find a 95% confidence interval for the mean plasma magnesium in the healthy population. How does the confidence interval differ from the 95% reference interval? Why are they different?
5. Find the standard error of the difference in mean plasma magnesium between insulin dependent diabetics and healthy people.
6. Find a 95% confidence interval for the difference in mean plasma magnesium between insulin dependent diabetics and healthy people. Is there any evidence that diabetics have lower plasma magnesium than non-diabetics in the population from which these data come?
7. Would plasma magnesium be a good diagnostic test for diabetes?

9

Significance tests

9.1 Testing a hypothesis

In Chapter 8 I dealt with estimation and the precision of estimates. This is one form of statistical inference, the process by which we use samples to draw conclusions about the populations from which they are taken. In this chapter I shall introduce a different form of inference, the significance test or hypothesis test.

A significance test enables us to measure the strength of evidence which the data supply concerning some proposition of interest. For example, consider the cross-over trial of pronethalol for the treatment of angina (§2.6). Table 9.1 shows the number of attacks over four weeks on each treatment. These 12 patients are a sample from the population of all patients. Would the other members of this population experience fewer attacks while using pronethalol? We can see that the number of attacks is highly variable from one patient to another, and it is quite possible that this is true from one period of time to another as well. So it could be that some patients would have fewer attacks while on pronethalol than while on placebo quite by chance. In a significance test, we ask whether the difference observed was small enough to have occurred by chance if there were really no difference in the population. If it were so, then the evidence in favour of there being a difference between the treatment periods would be weak. On the other hand, if the difference were much larger than we would expect due to chance if there were no real population difference, then the evidence in favour of a real difference would be strong.

To carry out the test of significance we suppose that, in the population, there is no difference between the two treatments. The hypothesis of 'no difference' or 'no effect' in the population is called the **null hypothesis**. If this is not true, then the **alternative hypothesis** must be true, that there is a difference between the treatments in one direction or the other. We then find the probability of getting data as different from what would be expected, if the null hypothesis were true, as are those data actually observed. If this probability is large the data are consistent with the null hypothesis; if it is small the data are unlikely to have arisen if the

Table 9.1. Trial of pronethalol for the prevention of angina pectoris

Number of attacks while on:		Difference	Sign of
placebo	pronethalol	placebo − pronethalol	difference
71	29	42	+
323	348	−25	−
8	1	7	+
14	7	7	+
23	16	7	+
34	25	9	+
79	65	14	+
60	41	19	+
2	0	2	+
3	0	3	+
17	15	2	+
7	2	5	+

null hypothesis were true and the evidence is in favour of the alternative hypothesis.

9.2 An example: the sign test

I shall now describe a particular test of significance, the **sign test**, to test the null hypothesis that placebo and pronethalol have the same effect on angina. Consider the differences between the number of attacks on the two treatments for each patient, as in Table 9.1. If the null hypothesis were true, then differences in number of attacks would be just as likely to be positive as negative, they would be random. The probability of a change being negative would be equal to the probability of it being positive, so both probabilities would be 0.5. Then the number of negatives would be an observation from a Binomial distribution (§6.4) with $n = 12$ and $p = 0.5$. (If there were any subjects who had the same number of attacks on both regimes we would omit them, as they provide no information about the direction of any difference between the treatments. In this test, n is the number of subjects for whom there is a difference, one way or the other.)

If the null hypothesis were true, what would be the probability of getting an observation from this distribution as extreme as the value we have actually observed? The expected number of negatives would be $np = 6$. What is the probability of getting a value as far from this as is that observed? The number of negative differences is 1. The probability of getting 1 negative change is

$$\frac{n!}{r!(n-r)!}p^r(1-p)^{n-r} = \frac{12!}{1!11!} \times 0.5^1 \times 0.5^{11} = 12 \times 0.5^{12} = 0.002\,93$$

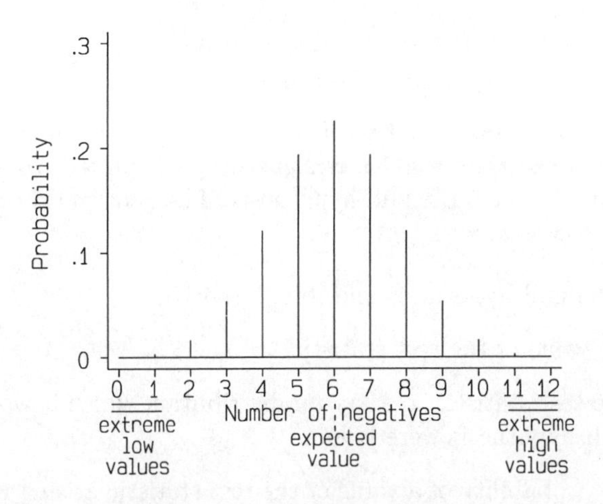

Fig. 9.1. Extremes of the Binomial distribution for the sign test

This is not a likely event in itself. However, we are interested in the probability of getting a value as far or further from the expected value, 6, as is 1, and clearly 0 is further and must be included. The probability of no negative changes is

$$\frac{12!}{0!12!} \times 0.5^0 \times 0.5^{12} = 0.000\,24$$

So the probability of one or fewer negative changes is $0.002\,93 + 0.000\,24 = 0.003\,17$. The null hypothesis is that there is no difference, so the alternative hypothesis is that there is a difference in one direction or the other. We must, therefore, consider the probability of getting a value as extreme on the other side of the mean, that is 11 or 12 negatives (Figure 9.1). The probability of 11 or 12 negatives is also 0.003 17, because the distribution is symmetrical. Hence, the probability of getting as extreme a value as that observed, in either direction, is $0.003\,17 + 0.003\,17 = 0.006\,34$. This means that if the null hypothesis were true we would have a sample which is so extreme that the probability of it arising by chance is 0.006, less than one in a hundred.

Thus, we would have observed a very unlikely event if the null hypothesis were true. This means that the data are not consistent with null hypothesis, so we can conclude that there is strong evidence in favour of a difference between the treatments. (Since this was a double blind randomized trial, it is reasonable to suppose that this was caused by the activity of the drug.)

9.3 Principles of significance tests

The sign test is an example of a test of significance. The number of negative changes is called the **test statistic**, something calculated from the data which can be used to test the null hypothesis. The general procedure for a significance test is as follows:

1. Set up the null hypothesis and its alternative.
2. Find the value of the test statistic.
3. Refer the test statistic to a known distribution which it would follow if the null hypothesis were true.
4. Find the probability of a value of the test statistic arising which is as or more extreme than that observed, if the null hypothesis were true.
5. Conclude that the data are consistent or inconsistent with the null hypothesis.

We shall deal with several different significance tests in this and subsequent chapters. We shall see that they all follow this pattern.

If the data are not consistent with the null hypothesis, the difference is said to be **statistically significant**. If the data do not support the null hypothesis, it is sometimes said that we reject the null hypothesis, and if the data are consistent with the null hypothesis it is said that we accept it. Such an 'all or nothing' decision making approach is seldom appropriate in medical research. It is preferable to think of the significance test probability as an index of the strength of evidence against the null hypothesis. The term 'accept the null hypothesis' is also misleading because it implies that we have concluded that the null hypothesis is true, which we should not do. We cannot prove statistically that something, such as a treatment effect, does not exist. It is better to say that we have not rejected or have failed to reject the null hypothesis.

The probability of such an extreme value of the test statistic occurring if the null hypothesis were true is often called the **P value**. It is *not* the probability that the null hypothesis is true. This is a common misconception. The null hypothesis is either true or it is not; it is not random and has no probability. I suspect that many researchers have managed to use significance tests quite effectively despite holding this incorrect view.

9.4 Significance levels and types of error

We must still consider the question of how small is small. A probability of 0.006, as in the example above, is clearly small and we have a quite unlikely event. But what about 0.06, or 0.1? Suppose we take a probability of 0.01

or less as constituting reasonable evidence against the null hypothesis. If the null hypothesis is true, we shall make a wrong decision one in a hundred times. Deciding against a true null hypothesis is called an **error of the first kind, type I error,** or **α error**. We get an **error of the second kind, type II error,** or **β error** if we do not reject a null hypothesis which is in fact false. (α and β are the Greek letters 'alpha' and 'beta'.) Now the smaller we demand the probability be before we decide against the null hypothesis, the larger the observed difference must be, and so the more likely we are to miss real differences. By reducing the risk of an error of the first kind we increase the risk of an error of the second kind.

The conventional compromise is to say that differences are significant if the probability is less than 0.05. This is a reasonable guide-line, but should not be taken as some kind of absolute demarcation. There is not a great difference between probabilities of 0.06 and 0.04, and they surely indicate similar strength of evidence. It is better to regard probabilities around 0.05 as providing some evidence against the null hypothesis, which increases in strength as the probability falls. If we decide that the difference is significant, the probability is sometimes referred to as the **significance level**. We say that the significance level is high if the P value is low.

9.5 One and two sided tests of significance

In the above example, the alternative hypothesis was that there was a difference in one direction or the other. This is called a **two sided** or **two tailed** test, because we used the probabilities of extreme values in both directions. It would have been possible to have the alternative hypothesis that there was a decrease in the pronethalol direction, in which case the null hypothesis would be that the number of attacks on the placebo was less than or equal to the number on pronethalol. This would give P = 0.00317, and of course, a higher significance level than the two sided test. This would be a **one sided** or **one tailed** test (Figure 9.2). The logic of this is that we should ignore any signs that the active drug is harmful to the patients. If what we were saying was 'if this trial does not give a significant reduction in angina using pronethalol we will not use it again', this might be reasonable, but the medical research process does not work like that. This is one of several pieces of evidence and so we should certainly use a method of inference which would enable us to detect effects in either direction.

The question of whether one or two sided tests should be the norm has been the subject of considerable debate among practitioners of statistical methods. Perhaps the position taken depends on the field in which the testing is usually done. In biological science, treatments seldom have only one effect and relationships between variables are usually complex. Two

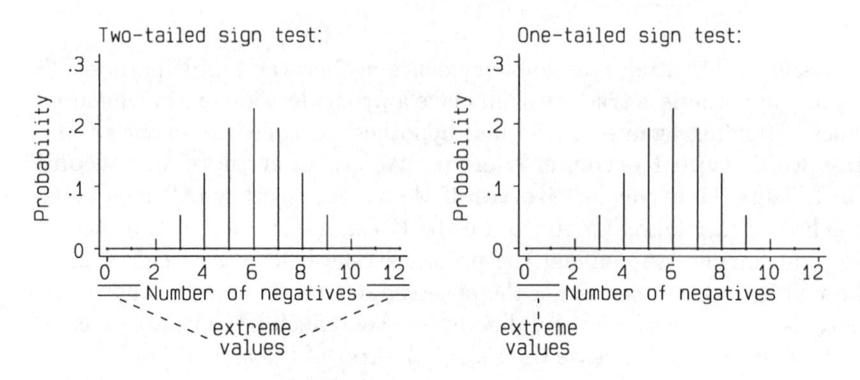

Fig. 9.2. One and two sided tests

sided tests are almost always preferable.

There are circumstances in which a one sided test is appropriate. In a study of the effects of an investigative procedure, laparoscopy and hydrotubation, on the fertility of sub-fertile women (Luthra *et al.* 1982), we studied women presenting at an infertility clinic. These women were observed for several months, during which some conceived, before laparoscopy was carried out on those still infertile. These were then observed for several months afterwards and some of these women also conceived. We compared the conception rate in the period before laparoscopy with that afterwards. Of course, women who conceived during the first period did not have a laparoscopy. We argued that the less fertile a woman was, the longer it was likely to take her to conceive. Hence, the women who had the laparoscopy should have a lower conception rate (by an unknown amount) than the larger group who entered the study, because the more fertile women had conceived before their turn for laparoscopy came. To see whether laparoscopy increased fertility, we could test the null hypothesis that the conception rate after laparoscopy was less than or equal to that before. The alternative hypothesis was that the conception rate after laparoscopy was higher than that before. A two sided test was inappropriate because if the laparoscopy had no effect on fertility the post laparoscopy rate was expected to be lower; chance did not come into it. In fact the post laparoscopy conception rate was very high and the difference clearly significant.

9.6 Significant, real and important

If a difference is statistically significant, then it may well be real, but not necessarily important. For example, we may look at the effect of a drug, given for some other purpose, on blood pressure. Suppose we find that the drug raises blood pressure by an average of 1 mm Hg, and that this is significant. A rise in blood pressure of 1 mm Hg is not clinically important, so, although it may be there, it does not matter. It is (statistically) significant, and real, but not important.

On the other hand, if a difference is not statistically significant, it could still be real. We may simply have too small a sample to show that a difference exists. Furthermore, the difference may still be important. The difference in mortality in the anticoagulant trial of Carleton *et al.* (1960), described in Chapter 2, was not significant, the difference in percentage survival being 5.5 in favour of the active treatment. However, the authors also quote a confidence interval for the difference in percentage survival of 24.2 percentage points in favour of heparin to 13.3 percentage points in favour of the control treatment. A difference in survival of 24 percentage points in favour of the treatment would certainly be important if it turned out to be the case. 'Not significant' does not imply that there is no effect. It means that we have failed to demonstrate the existence of one.

9.7 Comparing the means of large samples

We have already seen in §8.5 that if we have two samples of size n_1 and n_2, with sample means $\bar{x}_1$ and $\bar{x}_2$ and with standard errors se_1 and se_2, the standard error of the difference estimate $\bar{x}_1 - \bar{x}_2$ is $\sqrt{se_1^2 + se_2^2}$. Furthermore, if n_1 and n_2 are large, $\bar{x}_1 - \bar{x}_2$ will be from a Normal distribution with mean $\mu_1 - \mu_2$, the population difference, and its standard deviation well estimated by the standard error estimate. We can use this to find a confidence interval for the difference between the means:

$$\bar{x}_1 - \bar{x}_2 - 1.96\sqrt{se_1^2 + se_2^2} \text{ to } \bar{x}_1 - \bar{x}_2 + 1.96\sqrt{se_1^2 + se_2^2}$$

We can use this confidence interval to carry out a significance test of the null hypothesis that the difference between the means is zero, i.e. the alternative hypothesis is that μ_1 and μ_2 are not equal. If the confidence interval includes zero, then the probability of getting such extreme data if the null hypothesis were true is greater than 0.05 (i.e. 1 – 0.95). If the confidence interval excludes zero, then the probability of such extreme data under the null hypothesis is less than 0.05 and the difference is significant. Another way of doing the same thing is to note that

$$z = \frac{\bar{x}_1 - \bar{x}_2 - (\mu_1 - \mu_2)}{\sqrt{se_1^2 + se_2^2}}$$

is from a Standard Normal distribution, i.e. mean 0 and variance 1. Under the null hypothesis that $\mu_1 = \mu_2$ or $\mu_1 - \mu_2 = 0$, this is

$$z = \frac{\bar{x}_1 - \bar{x}_2}{\sqrt{se_1^2 + se_2^2}}$$

This is the test statistic, and if it lies between −1.96 and +1.96 then the probability of such an extreme value is greater than 0.05 and the difference is not significant. If the test statistic is greater than 1.96 or less than −1.96, there is a less than 0.05 probability of such data arising if the null hypothesis were true, and the data are not consistent with null hypothesis; the difference is significant at the 0.05 or 5% level.

For an example, in a study of respiratory symptoms in school-children (§8.5), we wanted to know whether children reported by their parents to have respiratory symptoms had worse lung function than children who were not reported to have symptoms. 92 children were reported to have cough during the day or at night, and their mean PEFR was 294.8 litre/min with standard deviation 57.1 litre/min. 1643 children were reported not to have the symptom, and their mean PEFR was 313.6 litre/min with standard deviation 55.2 litre/min. We thus have two large samples, and can apply the Normal test. We have

$$se_1 = \sqrt{\frac{s_1^2}{n_1}} = \sqrt{\frac{57.1^2}{92}} \qquad se_2 = \sqrt{\frac{s_2^2}{n_2}} = \sqrt{\frac{55.2^2}{1643}}$$

The difference between the two groups is $\bar{x}_1 - \bar{x}_2 = 294.8 - 313.6 = -18.8$. The standard error of the difference is

$$\text{SE}(\bar{x}_1 - \bar{x}_2) = \sqrt{\frac{s_1^2}{n_1} + \frac{s_2^2}{n_2}} = \sqrt{\frac{57.1^2}{92} + \frac{55.2^2}{1643}} = 6.11$$

The test statistic is

$$\frac{\bar{x}_1 - \bar{x}_2}{\text{SE}(\bar{x}_1 - \bar{x}_2)} = \frac{-18.8}{6.11} = -3.1$$

Under the null hypothesis this is an observation from a Standard Normal distribution, and so $P < 0.01$ (Table 7.2). If the null hypothesis were true, the data which we have observed would be unlikely. We can conclude that there is good evidence that children reported to have cough during the day or at night have lower PEFR than other children.

In this case, we have two ways of using the same standard error: for a confidence interval estimate or for a significance test. The confidence interval is usually superior, because we not only demonstrate the existence of a difference but also have some idea of its size. This is of particular value when the difference is not significant. For example, in the same study only

27 children were reported to have phlegm during the day or at night. These had mean PEFR of 298.0 litre/min and standard deviation 53.9 litre/min, hence a standard error for the mean of 10.4 litre/min. This is greater than the standard error for the mean for those with cough, because the sample size is smaller. The 1 708 children not reported to have this symptom had mean 312.6 litre/min and standard deviation 55.4 litre/min, giving standard error 1.3 litre/min. Hence the difference between the means was –14.6, with standard error given by $\sqrt{10.4^2 + 1.3^2} = 10.5$. The test statistic is

$$\frac{-14.6}{10.5} = -1.4$$

This has a probability of about 0.16, and so the data are consistent with the null hypothesis. However, the 95% confidence interval for the difference is $-14.6 - 1.96 \times 10.5$ to $-14.6 + 1.96 \times 10.5$ giving –35 to 6 litre/min. We see that the difference could be just as great as for cough. Because the size of the smaller sample is not so great, the test is less likely to detect a difference for the phlegm comparison than for the cough comparison. The advantages of confidence intervals over tests of significance are discussed by Gardner and Altman (1986).

9.8 Comparison of two proportions

Suppose we wish to compare two proportions p_1 and p_2, estimated from large independent samples size n_1 and n_2. The null hypothesis is that the proportion in the populations from which the samples are drawn are the same, p say. Since under the null hypothesis the proportions for the two groups are the same, we can get one common estimate of the proportion and use it to estimate the standard errors. We estimate the common proportion from the data by

$$p = \frac{r_1 + r_2}{n_1 + n_2}$$

where $p_1 = r_1/n_1$, $p_2 = r_2/n_2$. We want to make inferences from the difference between sample proportions, $p_1 - p_2$, so we require the standard error of this difference.

$$\text{SE}(p_1) = \sqrt{\frac{p(1-p)}{n_1}} \qquad \text{SE}(p_2) = \sqrt{\frac{p(1-p)}{n_2}}$$

$$\text{SE}(p_1 - p_2) = \sqrt{\text{SE}(p_1)^2 + \text{SE}(p_2)^2}$$

since the samples are independent. Hence

$$\text{SE}(p_1 - p_2) = \sqrt{\frac{p(1-p)}{n_1} + \frac{p(1-p)}{n_2}} = \sqrt{p(1-p)\left(\frac{1}{n_1} + \frac{1}{n_2}\right)}$$

As p is based on more subjects than either p_1 or p_2, if the null hypothesis were true then standard errors would be more reliable than those estimated in §8.6 using p_1 and p_2 separately. We then find the test statistic

$$z = \frac{p_1 - p_2}{\text{SE}(p_1 - p_2)} = \frac{p_1 - p_2}{\sqrt{p(1-p)\left(\frac{1}{n_1} + \frac{1}{n_2}\right)}}$$

Under the null hypothesis this has mean zero. Because the sample is large, we assume that p is sufficiently well estimated for $\sqrt{p(1-p)(1/n_1 + 1/n_2)}$ to be a good estimate of the standard deviation of the distribution from which $p_1 - p_2$ comes, i.e. the standard error, and $p_1 - p_2$ can be assumed to come from a Normal distribution. Hence, if the null hypothesis were true, the test statistic would be from a Standard Normal distribution.

In §8.6, we looked at the proportions of children with bronchitis in infancy and with no such history who were reported to have respiratory symptoms in later life. We had 273 children with a history of bronchitis before age 5 years, 26 of whom were reported to have day or night cough at age 14. We had 1046 children with no bronchitis before age 5 years, 44 of whom were reported to have day or night cough at age 14. We shall test the null hypothesis that the prevalence of the symptom is the same in both populations, against the alternative that it is not.

bronchitis	no bronchitis
$n_1 = 273$	$n_2 = 1\,046$
$p_1 = 26/273 = 0.095\,24$	$p_2 = 44/1\,046 = 0.042\,07$

$$p = \frac{26 + 44}{273 + 1\,046} = 0.053\,07$$

$$p_1 - p_2 = 0.095\,24 - 0.042\,07 = 0.053\,17$$

$$\text{SE}(p_1 - p_2) = \sqrt{p(1-p)\left(\frac{1}{n_1} + \frac{1}{n_2}\right)}$$

$$= \sqrt{0.053\,07 \times (1 - 0.053\,07) \times \left(\frac{1}{273} + \frac{1}{1\,046}\right)} = 0.015\,24$$

$$\frac{p_1 - p_2}{\text{SE}(p_1 - p_2)} = \frac{0.053\,17}{0.015\,24} = 3.49$$

Referring this to Table 7.2 of the Normal distribution, we find the probability of such an extreme value is less than 0.01, so we conclude that the data are not consistent with the null hypothesis. There is good evidence that children with a history of bronchitis are more likely to be reported to have day or night cough at age 14.

Note that the standard error used here is not the same as that found in §8.6. It is only correct if the null hypothesis is true. The formula of §8.6 should be used for finding the confidence interval. Thus the standard error used for testing is not identical to that used for estimation, as was the case for the comparison of two means. It is possible for the test to be significant and the confidence interval include zero.

This is a large sample method, and is equivalent to the chi-squared test for a 2 by 2 table (§13.1–2). How small the sample can be and methods for small samples are discussed in §13.3–6.

Note that we do not need a different test for the ratio of two proportions, as the null hypothesis that the ratio in the population is one is the same as the null hypothesis that the difference in the population is zero.

9.9 The power of a test

The test for comparing means in §9.7 is more likely to detect a large difference between two populations than a small one. The probability that a test will produce a significant difference at a given significance level is called the **power** of the test. For a given test, this will depend on the true difference between the populations compared, the sample size and the significance level chosen. We have already noted in §9.4 that we are more likely to obtain a significant difference with a significance level of 0.05 than with one of 0.01. We have greater power if the P value chosen to be considered as significant is larger.

For example, we can calculate the power of the Normal comparison of two means quite easily. The sample difference $\bar{x}_1 - \bar{x}_2$ is an observation from a Normal distribution with mean $\mu_1 - \mu_2$ and standard deviation $\sqrt{{\sigma_1}^2/n_1 + {\sigma_2}^2/n_2}$, the standard error of the difference, which we shall denote by se_{diff}. The test statistic to test the null hypothesis that $\mu_1 = \mu_2$ is $(\bar{x}_1 - \bar{x}_2)/se_{diff}$. The test will be significant at the 0.05 level if the test statistic is further from zero than 1.96. If $\mu_1 > \mu_2$, it is very unlikely that we will find $\bar{x}_1$ significantly less than $\bar{x}_2$, so for a significant difference we must have $(\bar{x}_1 - \bar{x}_2)/se_{diff} > 1.96$. Subtracting $(\mu_1 - \mu_2)/se_{diff}$ from each side:

$$\begin{aligned}\frac{\bar{x}_1 - \bar{x}_2}{se_{diff}} - \frac{\mu_1 - \mu_2}{se_{diff}} &> 1.96 - \frac{\mu_1 - \mu_2}{se_{diff}} \\ \frac{\bar{x}_1 - \bar{x}_2 - (\mu_1 - \mu_2)}{se_{diff}} &> 1.96 - \frac{\mu_1 - \mu_2}{se_{diff}}\end{aligned}$$

$(\bar{x}_1 - \bar{x}_2 - (\mu_1 - \mu_2))/se_{diff}$ is an observation from a Standard Normal distribution, because we subtract from $\bar{x}_1 - \bar{x}_2$ its expected value, $\mu_1 - \mu_2$, and divide by its standard deviation, se_{diff}. We can find the probability that this exceeds any particular value z from $1 - \Phi(z)$ in Table 7.1. So the power of the test, the probability of getting a significant result, is $1 - \Phi(z)$ where $z = 1.96 - (\mu_1 - \mu_2)/se_{diff}$.

For the comparison of PEFR in children with and without phlegm (§9.7), for example, suppose that the population means were in fact $\mu_1 = 310$ and $\mu_2 = 295$ litre/min, and each population had standard deviation 55 litre/min. The sample sizes were $n_1 = 1\,708$ and $n_2 = 27$, so the standard error of the difference would be

$$se_{diff} = \sqrt{\frac{55^2}{1\,708} + \frac{55^2}{27}} = 10.67 \text{ litre/min}$$

The population difference we want to be able to detect is $\mu_1 - \mu_2 = 310 - 295 = 15$, and so

$$1.96 - \frac{\mu_1 - \mu_2}{se_{diff}} = 1.96 - \frac{15}{10.67} = 1.96 - 1.41 = 0.55$$

From Table 7.1, $\Phi(0.55)$ is between 0.691 and 0.726, about 0.71. The power of the test would be $1 - 0.71 = 0.29$. If these were the population means and standard deviation, our test would have had a poor chance of detecting the difference in means, even though it existed. The test would have low power. Figure 9.3 shows how the power of this test changes with the difference between population means. As the difference gets larger, the power increases, getting closer and closer to 1. The power is not zero even when the population difference is zero, because there is always the possibility of a significant difference, even when the null hypothesis is true.

9.10 Multiple significance tests

If we test a null hypothesis which is in fact true, using 0.05 as the critical significance level, we have a probability of 0.95 of coming to a 'not significant' (i.e. correct) conclusion. If we test two independent true null hypotheses, the probability that neither test will be significant is $0.95 \times 0.95 = 0.90$ (§6.2). If we test twenty such hypotheses the probability that none will be significant is $0.95^{20} = 0.36$. This gives a probability of $1 - 0.36 = 0.64$ of getting at least one significant result; we are more likely to get one than not. The expected number of spurious significant results is $20 \times 0.05 = 1$.

Many medical research studies are published with large numbers of significance tests. These are not usually independent, being carried out on the same set of subjects, so the above calculations do not apply exactly. However, it is clear that if we go on testing long enough we will find something which is 'significant'. We must beware of attaching too much importance

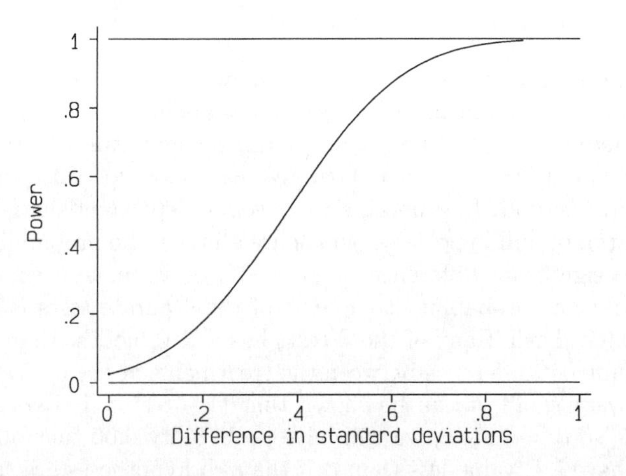

Fig. 9.3. Power curve for a comparison of two means from samples of size 1 708 and 27

to a lone significant result among a mass of non-significant ones. It may be the one in twenty which we should get by chance alone.

This is particularly important when we find that a clinical trial or epidemiological study gives no significant difference overall, but does so in a particular subset of subjects, such as women aged over 60. For example, Lee *et al.* (1980) simulated a clinical trial of the treatment of coronary artery disease by allocating 1073 patient records from past cases into two 'treatment' groups at random. They then analysed the outcome as if it were a genuine trial of two treatments. The analysis was quite detailed and thorough. As we would expect, it failed to show any significant difference in survival between those patients allocated to the two 'treatments'. Patients were then subdivided by two variables which affect prognosis, the number of diseased coronary vessels and whether the left ventricular contraction pattern was normal or abnormal. A significant difference in survival between the two 'treatment' groups was found in those patients with three diseased vessels (the maximum) and abnormal ventricular contraction. As this would be the subset of patients with the worst prognosis, the finding would be easy to account for by saying that the superior 'treatment' had its greatest advantage in the most severely ill patients! The moral of this story is that if there is no difference between the treatments overall, significant differences in subsets are to be treated with the utmost suspicion. This method of looking for a difference in treatment effect between subgroups of subjects is incorrect. A correct approach would be to use a multifactorial analysis, as described in Chapter 17, with treatment and group as two factors, and test for an interaction between groups and treatments. The power for detecting such interactions is quite low, and we need a larger sample than would be needed simply to show a difference overall.

This spurious significant difference comes about because, when there is no real difference, the probability of getting no significant differences in six subgroups is $0.95^6 = 0.74$, not 0.95. We can allow for this effect by the **Bonferroni** method. In general, if we have k independent significant tests, at the α level, of null hypotheses which are all true, the probability that we will get no significant differences is $(1-\alpha)^k$. If we make α small enough, we can make the probability that none of the separate tests is significant equal to 0.95. Then if any of the k tests has a P value less than α, we will have a significant difference between the treatments at the 0.05 level. Since α will be very small, it can be shown that $(1-\alpha)^k \approx 1 - k\alpha$. If we put $k\alpha = 0.05$, so $\alpha = 0.05/k$, we will have probability 0.05 that one of the k tests will have a P value less than α if the null hypotheses are true. Thus, if in a clinical trial we compare two treatments within 5 subsets of patients, the treatments will be significantly different at the 0.05 level if there is a P value less than 0.01 within any of the subsets. This is the Bonferroni method. Note that they are not significant at the 0.01 level, but at only the 0.05 level.

We can do the same thing by multiplying the observed P value from the significance tests by the number of tests, k, any kP which exceeds one being ignored. Then if any kP is less than 0.05, the two treatments are significant at the 0.05 level.

For example, Williams *et al.* (1992) randomly allocated elderly patients discharged from hospital to two groups. The intervention group received timetabled visits by health visitor assistants, the control patients group were not visited unless there was perceived need. Soon after discharge and after one year, patients were assessed for physical, disability and mental state using questionnaire scales. There were no significant differences overall between the intervention and control groups, but among women aged 75–79 living alone the control group showed significantly greater deterioration in physical score than did the intervention group (P=0.04), and among men over 80 years the control group showed significantly greater deterioration in disability score than did the intervention group (P=0.03). The authors stated that 'Two small sub-groups of patients were possibly shown to have benefited from the intervention. ... These benefits, however, have to be treated with caution, and may be due to chance factors.' Subjects were cross-classified by age groups, whether living alone, and sex, so there were at least eight subgroups, if not more. Thus even if we consider the three scales separately, only a P value less than $0.05/8 = 0.006$ would provide evidence of a treatment effect. Alternatively, the true P values are $8 \times 0.04 = 0.32$ and $8 \times 0.03 = 0.24$.

A similar problem arises if we have multiple outcome measurements. For example, Newnham *et al.* (1993) randomized pregnant women to receive a series of Doppler ultrasound blood flow measurements or to control. They found a significantly higher proportion of birthweights below the 10th and 3rd centiles (P=0.006 and P=0.02). These were only two of many comparisons, however, and one would suspect that there may be some spurious significant differences among so many. At least 35 were reported in the paper, though only these two were reported in the abstract (birthweight was not the intended outcome variable for the trial). These tests are not independent, because they are all on the same subjects, using variables which may not be independent. The proportions of birthweights below the 10th and 3rd centiles are clearly not independent, for example. The probability that two correlated variables both give non-significant differences when the null hypothesis is true is greater than $(1-\alpha)^2$, because if the first test is not significant, the second now has a probability greater than $1-\alpha$ of being not significant also. (Similarly, the probability that both are significant exceeds α^2, and the probability that only one is significant is reduced.) For k tests the probability of no significant differences is greater than $(1-\alpha)^k$ and so greater than $1-k\alpha$. Thus if we carry out each test at the $\alpha = 0.05/k$ level, we will still have a probability of no significant differences which is greater than 0.95. A P value less than α for any variable, or $k\text{P} < 0.05$, would mean that the treatments were significantly different. For the example, we have $\alpha = 0.05/35 = 0.0014$ and so by the Bonferroni criterion the treatment groups are not significantly different. Alternatively, the P values could be adjusted by $35 \times 0.006 = 0.21$ and $35 \times 0.02 = 0.70$.

Because the probability of obtaining no significant differences if the null hypotheses are all true is greater than the 0.95 which we want it to be, the overall P value is actually smaller than the nominal 0.05, by an unknown amount which depends on the lack of independence between the tests. The power of the test, its ability to detect true differences in the population, is correspondingly diminished. In statistical terms, the test is conservative.

Other multiple testing problems arise when we have more than two groups of subjects and wish to compare each pair of groups (§10.9), when we have a series of observations over time, such as blood pressure every 15 minutes after administration of a drug, where there may be a temptation to test each time point separately (§10.7), and when we have relationships between many variables to examine, as in a survey. For all these problems, the multiple tests are highly correlated and the Bonferroni method is inappropriate, as it will be highly conservative and may miss real differences.

9M Multiple choice questions 44 to 49

(Each branch is either true or false)

44. In a case control study, patients with a given disease drank coffee more frequently than did controls, and the difference was highly significant. We can conclude that:

(a) drinking coffee causes the disease;

(b) there is evidence of a real relationship between the disease and coffee drinking in the sampled population;

(c) the disease is not related to coffee drinking;

(d) eliminating coffee would prevent the disease;

(e) coffee and the disease always go together.

45. When comparing the means of two large samples using the Normal test:

(a) the null hypothesis is that the sample means are equal;

(b) the null hypothesis is that the means are not significantly different;

(c) standard error of the difference is the sum of the standard errors of the means;

(d) the standard errors of the means must be equal;

(e) the test statistic is the ratio of the difference to its standard error.

46. In a comparison of two methods of measuring PEFR, 6 of 17 subjects had higher readings on the Wright peak flow meter, 10 had higher readings on the mini peak flow meter and one had the same on both. If the difference between the instruments is tested using a sign test:

(a) the test statistic may be the number with the higher reading on the Wright meter;

(b) the null hypothesis is that there is no tendency for one instrument to read higher than the other;

(c) a one tailed test of significance should be used;

(d) the test statistic should follow the Binomial distribution ($n = 16$ and $p = 0.5$) if the null hypothesis were true;

(e) the instruments should have been presented in random order.

47. In a small randomized double blind trial of a new treatment in acute myocardial infarction, the mortality in the treated group was half that in the control group, but the difference was not significant. We can conclude that:

(a) the treatment is useless;

(b) there is no point in continuing to develop the treatment;

(c) the reduction in mortality is so great that we should introduce the treatment immediately;

(d) we should keep adding cases to the trial until the Normal test for comparison of two proportions is significant;

(e) we should carry out a new trial of much greater size.

48. In a large sample comparison between two groups, increasing the sample size will:

(a) improve the approximation of the test statistic to the Normal distribution;

(b) decrease the chance of an error of the first kind;

(c) decrease the chance of an error of the second kind;

(d) increase the power against a given alternative;

(e) make the null hypothesis less likely to be true.

49. In a study of breast feeding and intelligence (Lucas *et al.* 1992), 300 children who were very small at birth were given their mother's breast milk or infant formula, at the choice of the mother. At the age of 8 years the IQ of these children was measured. The mean IQ in the formula group was 92.8, compared to a mean of 103.0 in the breast milk group. The difference was significant, $P < 0.001$:

(a) there is good evidence that formula feeding of very small babies reduces IQ at age eight;

(b) there is good evidence that choosing to express breast milk is related to higher IQ in the child at age eight;

(c) type of milk has no effect on subsequent IQ;

(d) the probability that type of milk affects subsequent IQ is less than 0.1%;

(e) if type of milk were unrelated to subsequent IQ, the probability of getting a difference in mean IQ as big as that observed is less than 0.001.

9E Exercise: Crohn's disease and cornflakes

The suggestion that cornflakes may cause Crohn's disease arose in the study of James (1977). Crohn's disease is an inflammatory disease, usually of the last part of the small intestine. It can cause a variety of symptoms, including vague pain, diarrhoea, acute pain and obstruction. Treatment may be by drugs or surgery, but many patients have had the disease for many years. James' initial hypothesis was that foods taken at breakfast may be asso-

Table 9.2. Numbers of Crohn's disease patients and controls who ate various cereals regularly (at least once per week) (James 1977)

		Patients	Controls	Significance test
Cornflakes	Regularly	23	17	$P < 0.0001$
	Rarely or never	11	51	
Wheat	Regularly	16	12	$P < 0.01$
	Rarely or never	18	56	
Porridge	Regularly	11	15	$0.5 > P > 0.1$
	Rarely or never	23	53	
Rice	Regularly	8	10	$0.5 > P > 0.1$
	Rarely or never	26	56	
Bran	Regularly	6	2	$P = 0.02$
	Rarely or never	28	66	
Muesli	Regularly	4	3	$P = 0.17$
	Rarely or never	30	65	

ciated with Crohn's disease. James studied 16 men and 18 women with Crohn's disease, aged 19–64 years, mean time since diagnosis 4.2 years. These were compared to controls, drawn from hospital patients without major gastro-intestinal symptoms. Two controls were chosen per patient, matched for age and sex. James interviewed all cases and controls himself. Cases were asked whether they ate various foods for breakfast before the onset of symptoms, and controls were asked whether they ate various foods before a corresponding time (Table 9.2). There was a significant excess of eating of cornflakes, wheat and bran among the Crohn's patients. The consumption of different cereals was inter-related, people reporting one cereal being likely to report others. In James' opinion the principal association of Crohn's disease was with cornflakes, based on the apparent strength of the association. Only one case had never eaten cornflakes.

Several papers soon appeared in which this study was repeated, with variations. None was identical in design to James' study and none appeared to support his findings. Mayberry *et al.* (1978) interviewed 100 patients with Crohn's disease, mean duration nine years. They obtained 100 controls, matched for age and sex, from patients and their relatives attending a fracture clinic. Cases and controls were interviewed about their *current* breakfast habits (Table 9.3). The only significant difference was an excess of fruit juice drinking in controls. Cornflakes were eaten by 29 cases compared to 22 controls, which was not significant. In this study there was no particular tendency for cases to report more foods than controls. The authors also asked cases whether they knew of an association between food (unspecified) and Crohn's disease. The association with cornflakes was reported by 29, and 12 of these had stopped eating them, having previously eaten them regularly. In their 29 matched controls, 3 were past cornflakes eaters. Of the 71 Crohn's patients who were unaware of the association,

Table 9.3. Number of patients and controls regularly consuming certain foods at least twice weekly (Mayberry *et al.* 1978)

Foods at breakfast	Crohn's patients ($n = 100$)	Controls ($n = 100$)	Significance test
Bread	91	86	
Toast	59	64	
Egg	31	37	
Fruit or fruit juice	14	30	$P < 0.02$
Porridge	20	18	
Weetabix, Shreddies or Shredded Wheat	21	19	
Cornflakes	29	22	
Special K	4	7	
Rice Krispies	6	6	
Sugar Puffs	3	1	
Bran or All Bran	13	12	
Muesli	3	10	
Any Cereal	55	55	

21 had discontinued eating cornflakes compared to 10 of their 71 controls. The authors remarked 'seemingly patients with Crohn's disease had significantly reduced their consumption of cornflakes compared with controls, irrespective of whether they were aware of the possible association'.

1. Are the cases and controls comparable in either of these studies?
2. What other sources of bias could there be in these designs?
3. What is the main point of difference in design between the study of James and that of Mayberry *et al.* ?
4. In the study of Mayberry *et al.* how many Crohn's cases and how many controls had ever been regular eaters of cornflakes? How does this compare with James' findings?
5. Why did James think that eating cornflakes was particularly important?
6. For the data of Table 9.2, calculate the percentage of cases and controls who said that they ate the various cereals. Now divide the proportion of cases who said that they had eaten the cereal by the proportion of controls who reported eating it. This tells us, roughly, how much more likely cases were to report the cereal than were controls. Do you think eating cornflakes is particularly important?
7. If we have an excess of all cereals when we ask what was ever eaten, and none when we ask what is eaten now, what possible factors could account for this?

10

Comparing the means of small samples

10.1 The t distribution

We have seen in Chapters 8 and 9 how the Normal distribution can be used to calculate confidence intervals and to carry out tests of significance for the means of large samples. In this chapter we shall see how similar methods may be used when we have small samples, using the t distribution, and go on to compare several means.

So far, the probability distributions we have used have arisen because of the way data were collected, either from the way samples were drawn (Binomial distribution), or from the mathematical properties of large samples (Normal distribution). The distribution did not depend on any property of the data themselves. To use the t distribution we must make an assumption about the distribution from which the observations themselves are taken, the distribution of the variable in the population, which we must assume to be a Normal distribution. As we saw in Chapter 7, many naturally occurring variables have been found to follow a Normal distribution closely. I shall discuss the effects of any deviations from the Normal later.

I have already mentioned the t distribution (§7A), as one of those derived from the Normal. I shall now look at it in more detail. Suppose we have a random sample of observations from a Normal distribution with mean μ and variance σ^2. We estimate μ and σ^2 from the data by the sample mean and variance, $\bar{x}$ and s^2. The distribution of all possible sample means, i.e. of all possible $\bar{x}$s, has a standard deviation (§8.1), the standard error of the sample mean, estimated by $\sqrt{s^2/n}$ (§8.2). If we had a large sample, we would then say that the mean $\bar{x}$ comes from a Normal distribution and that $\sqrt{s^2/n}$ is a good estimate of its standard deviation. The ratio $(\bar{x} - \mu)/\sqrt{s^2/n}$ would follow a Normal distribution with mean 0 and standard deviation 1, the Standard Normal distribution. This is not true for a small sample. The estimated standard deviation, s, may vary from sample to sample. Samples with small standard deviations will give very large ratios and the distribution will have much longer tails than the Normal.

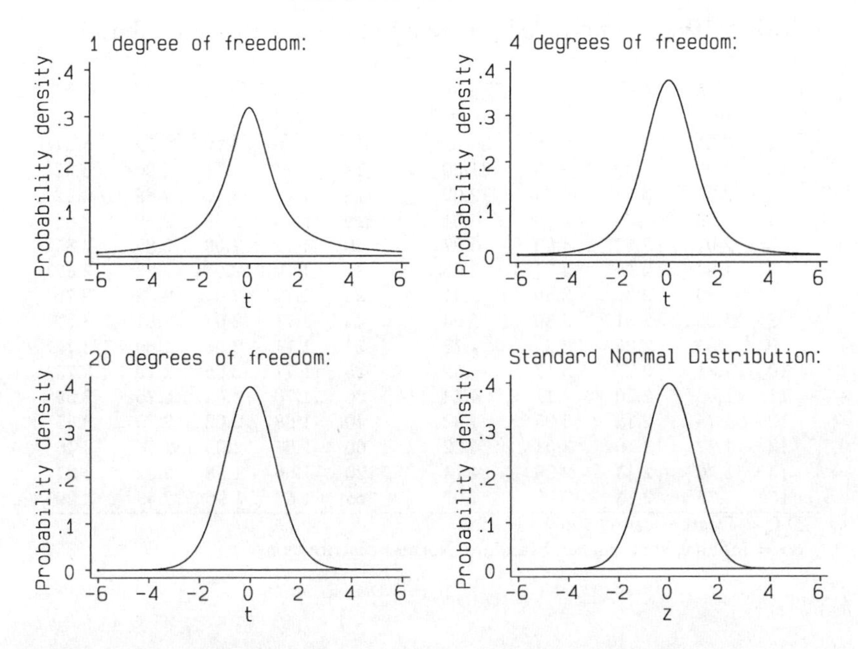

Fig. 10.1. Student's t distribution with 1, 4 and 20 degrees of freedom, showing convergence to the Standard Normal distribution

The distribution of the mean over standard error calculated from a small sample depends on the distribution from which the original observations come. As so many variables follow a Normal distribution, it is worth looking at what happens when the observations are Normal. Provided our observations are from a Normal distribution, $\bar{x}$ is too. But we cannot assume that $\sqrt{s^2/n}$ is a good estimate of its standard deviation. We must allow for the variation of s^2 from sample to sample. It can be shown that, provided the observations come from a Normal distribution, the sampling distribution of $(\bar{x}-\mu)/\sqrt{s^2/n}$ is Student's t distribution with $n-1$ degrees of freedom (§10A). We can therefore replace the Normal distribution by the t distribution in confidence intervals and significance tests for small samples. In fact, when we divide anything which is Normally distributed with mean zero, such as $\bar{x}-\mu$, by its standard error which is based on a single sum of squares of Normally distributed data, we get a t distribution.

Figure 10.1 shows the t distribution with 1, 4 and 20 degrees of freedom. It is symmetrical, with longer tails than the Normal distribution. For example, with 4 d.f. the probability of t being greater than 2.78 is 2.5%, whereas for the Standard Normal distribution the probability of being greater than 2.78 is only 0.3%. This is what we should expect, as in the expression $(\bar{x}-\mu)/\sqrt{s^2/n}$ the variation in s^2 from sample to sample will

Table 10.1. Two tailed probability points of the t distribution

D.f.	Probability				D.f.	Probability			
	0.10	0.05	0.01	0.001		0.10	0.05	0.01	0.001
	10%	5%	1%	0.1%		10%	5%	1%	0.1%
1	6.31	12.70	63.66	636.62	16	1.75	2.12	2.92	4.01
2	2.92	4.30	9.93	31.60	17	1.74	2.11	2.90	3.97
3	2.35	3.18	5.84	12.92	18	1.73	2.10	2.88	3.92
4	2.13	2.78	4.60	8.61	19	1.73	2.09	2.86	3.88
5	2.02	2.57	4.03	6.87	20	1.72	2.09	2.85	3.85
6	1.94	2.45	3.71	5.96	21	1.72	2.08	2.83	3.82
7	1.89	2.36	3.50	5.41	22	1.72	2.07	2.82	3.79
8	1.86	2.31	3.36	5.04	23	1.71	2.07	2.81	3.77
9	1.83	2.26	3.25	4.78	24	1.71	2.06	2.80	3.75
10	1.81	2.23	3.17	4.59	25	1.71	2.06	2.79	3.73
11	1.80	2.20	3.11	4.44	30	1.70	2.04	2.75	3.65
12	1.78	2.18	3.05	4.32	40	1.68	2.02	2.70	3.55
13	1.77	2.16	3.01	4.22	60	1.67	2.00	2.66	3.46
14	1.76	2.14	2.98	4.14	120	1.66	1.98	2.62	3.37
15	1.75	2.13	2.95	4.07	∞	1.64	1.96	2.58	3.29

D.f. = Degrees of freedom
∞ = infinity, same as the Standard Normal distribution

produce some samples with low values of s^2 and so large values of t. As the degrees of freedom, and hence the sample size, increase, s^2 will tend to be closer to its expected value of σ^2. The variation in s^2 will be less, and hence the variation in t will be less. This means that extreme values of t will be less likely, and so the tails of the distribution, which contain the probability associated with extreme values of t, will be smaller. We have already seen that for large samples $(\bar{x}-\mu)/\sqrt{s^2/n}$ follows a Standard Normal distribution. The t distribution gets more and more like the Standard Normal distribution as the degrees of freedom increase.

Like the Normal distribution, the t distribution function cannot be integrated algebraically and its numerical values have been tabulated. Because the t distribution depends on the degrees of freedom, it is not usually tabulated in full like the Normal distribution in Table 7.1. Instead, probability points are given for different degrees of freedom. Table 10.1 shows two sided probability points for selected degrees of freedom. Thus, with 4 degrees of freedom, we can see that, with probability 0.05, t will be 2.78 or more from its mean, zero.

Because only certain probabilities are quoted, we cannot usually find the exact probability associated with a particular value of t. For example, suppose we want to know the probability of t on 9 degrees of freedom being further from zero than 3.7. From Table 10.1 we see that the 0.01 point is 3.25 and the 0.001 point is 4.78. We therefore know that the required probability lies between 0.01 and 0.001. We could write this as $0.001 < \text{P} < 0.01$. Often the lower bound, 0.001, is omitted and we write

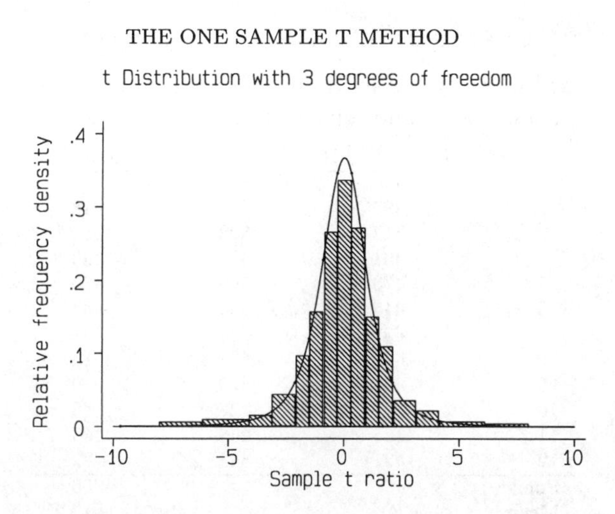

Fig. 10.2. Sample t ratios derived from 750 samples of 4 human heights, after Student (1908)

P < 0.01. With a computer it is possible to calculate the exact probability every time, so this common practice is due to disappear.

The name 'Student's t distribution' often puzzles newcomers to the subject. It is not, as may be thought, an easy method suitable for students to use. The origin of the name is part of the folk lore of statistics. The distribution was discovered by W. S. Gossett, an employee of the Guinness brewery in Dublin. At that time, the company would not allow its employees to publish the results of their work, lest it should lose some commercial advantage. Gossett therefore submitted his paper under the pseudonym 'Student' (Student 1908). In this paper he not only presented the mathematical derivation of the distribution, but also gave the results of a sampling experiment like those described in §4.7 and §8.2. He took the heights of 3000 criminals, wrote each onto a piece of card, then drew 750 samples of size 4 to give 750 $(\bar{x} - \mu)/\sqrt{s^2/n}$ statistics. Figure 10.2 shows the very good agreement which he obtained.

10.2 The one sample t method

We can use the t distribution to find confidence intervals for means estimated from a small sample from a Normal distribution. We do not usually have small samples in sample surveys, but we often do find them in clinical studies. For example, we can use the t distribution to find confidence intervals for the size of difference between two treatment groups, or between measurements obtained from subjects under two conditions. I shall deal with the latter, single sample problem first.

The population mean, μ, is unknown and we wish to estimate it using a

Table 10.2. PEFR (litre/min) measured by Wright meter and mini meter, female subjects

Subject	Wright PEFR	Mini PEFR	Difference
1	490	525	−35
2	397	415	−18
3	512	508	4
4	401	444	−43
5	470	500	−30
6	415	460	−45
7	431	390	41
8	429	432	−3
9	420	420	0
10	275	227	48
11	165	268	−103
12	421	443	−22
Sum			−206
Mean			−17.2
Sum of squares about the mean			17889.7
Variance			1626.3
Standard error of mean			11.6

95% confidence interval. We can see that, for 95% of samples, the difference between $\bar{x}$ and μ is at most t standard errors, where t is the value of the t distribution such that 95% of observations will be closer to zero than t. For a large sample this will be 1.96 as for the Normal distribution. For small samples we must use Table 10.1. In this table, the probability that the t distribution is further from zero than t is given, so we must first find one minus our desired probability, 0.95. We have $1 - 0.95 = 0.05$, so we use the 0.05 column of the table to get the value of t. We then have the 95% confidence interval: $\bar{x} - t$ standard errors to $\bar{x} - t$ standard errors.

Consider the data of Table 10.2. These are results from a comparison of two instruments for measuring PEFR, a Wright Peak Flow Meter and a Mini Peak Flow Meter. The subjects were family and colleagues, and so not a random sample. Each gave two readings on each instrument in random order. Table 10.2 shows the second reading on each. We shall measure the amount of bias between the instruments, the amount by which one tends to read above the other. The first step is to find the differences (Wright − mini). We then find the mean difference and its standard error, as described in §8.2.

To find the 95% confidence interval for the mean difference we must suppose that the differences follow a Normal distribution. To calculate the interval, we first require the relevant point of the t distribution from Table 10.1. There are 12 differences and hence $n - 1 = 11$ degrees of freedom associated with s^2. We want a probability of 0.95 of being closer to zero than t, so we go to Table 10.1 with probability $= 1 - 0.95 = 0.05$. Using the 11 d.f. row, we get $t = 2.20$. Hence the difference between a sample mean and the population mean is less than 2.20 standard errors for 95% of samples, and the 95% confidence interval is $-17.2 - 2.20 \times 11.6$ to $-17.2 + 2.20 \times 11.6 = -42.7$ to 8.3 litre/min. In the large sample case, we would use the Normal distribution instead of the t distribution, putting 1.96 instead of 2.20. We would not then need the differences themselves to follow a Normal distribution.

On the basis of these data, the mini meter could tend to over-read by as much as 43 litre/min, or to under-read by as much as 8 litre/min. An error of 43 litre/min is quite substantial, and we may have a problem. We would need a much larger sample to obtain a more precise estimate if we thought this were required.

We can also use the t distribution to test the null hypothesis that the mean difference is zero. If the null hypothesis were true, and the differences follow a Normal distribution, the test statistic mean/standard error would be from a t distribution with $n - 1$ degrees of freedom. This is because the null hypothesis is that the mean difference $\mu = 0$, hence the numerator $\bar{x} - \mu = \bar{x}$. We have the usual 'estimate over standard error' formula. For the example, we have

$$\frac{\bar{x}}{\sqrt{\frac{s^2}{n}}} = \frac{-17.2}{11.6} = -1.48$$

If we go to the 11 d.f. row of Table 10.1, we find that the probability of such an extreme value arising is greater than 0.10, the 0.10 point of the distribution being 1.80. Using a computer we would find P = 0.17. The data are consistent with the null hypothesis and we have failed to demonstrate the existence of a bias. Note that the confidence interval is more informative than the significance test.

We could also use the sign test to test the null hypothesis of no bias. This gives us 3 positives out of 11 differences (one difference, being zero, gives no useful information) which gives a two sided probability of 0.23, similar to the result of the t test. Provided the assumption of a Normal distribution is true, the t test is preferred because it is the most powerful test, and so most likely to detect differences should they exist.

The validity of the methods described above depends on the assumption that the differences are from a Normal distribution. We can check the assumption of a Normal distribution by a Normal plot (§7.5). Figure 10.3 shows a Normal plot for the differences and also for the Wright meter reading. The Normal plot for the differences deviates only slightly from a straight line, one point in particular, subject 11, appearing rather out. The Wright meter readings show a clear kink in the line and are unlikely to be from a Normal distribution. At first sight this is surprising, as we have remarked before that PEFR tends to be Normally distributed. However,

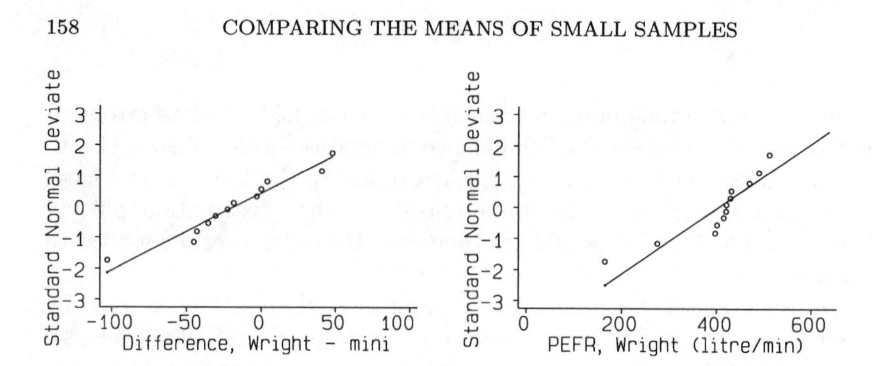

Fig. 10.3. Normal plots for the data of Table 10.2

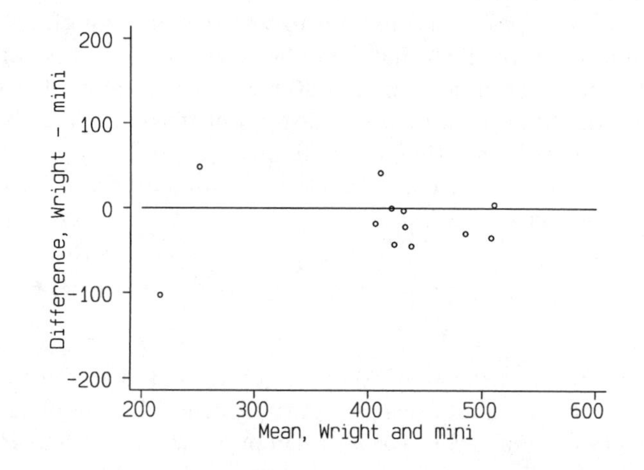

Fig. 10.4. Plot of difference against mean for the data of Table 10.4

this is not a sample from a population of similar age, or from the adult population in general. Most of these subjects were aged between 20 and 30 years, but subjects 10 and 11 (my mother and mother-in-law!) were in an older age group and so produced much lower PEFRs.

Another plot which is a useful check here is the difference against the subject mean (Figure 10.4). If the difference depends on magnitude, then we should be careful of drawing any conclusion about the mean difference. We may want to investigate this further, perhaps by transforming the data (§10.4). In this case the difference between the two readings does not appear to be related to the level of PEFR and we need not be concerned about this. Incidentally, we should be wary of drawing any conclusions about the Mini Wright Peak Flow Meter from this sample of one instrument. It had been used in a field survey of wheezy children, being kept in their homes for periods of a fortnight (Johnston *et al.* 1984). It may have been battered.

Despite the clear non-Normal distribution of the PEFR, the differences look like a fairly good fit to the Normal. There are two reasons for this: the subtraction removes variability between subjects (e.g. due to age and height), leaving the measurement error which is more likely to be Normal, and the two measurement errors are then added, producing the tendency of sums to the Normal seen in the Central Limit Theorem (§7.3). The assumption of a Normal distribution for the one sample case is quite likely to be met. I discuss this further in §10.5. When the one sample t test is used with differences, as in the PEFR meter example, it is also known as the **paired t test**.

10.3 The means of two independent samples

Suppose we have two samples from Normally distributed populations, with which we want to estimate the difference between the population means. If the samples were large, the 95% confidence interval for the difference would be the observed difference – 1.96 standard errors to observed difference + 1.96 standard errors. Unfortunately, we cannot simply replace 1.96 by a number from Table 10.1. This is because the standard error does not have the simple form described in §10.2. It is not the square root of a constant times a sum of squares, but rather is the square root of the sum of two constants multiplied by two sums of squares. Hence, it does not follow the square root of the Chi-squared distribution as required for the denominator of a t distributed random variable (§7A). In order to use the t distribution we must make a further assumption about the data. Not only must the samples be from Normal distributions, they must be from Normal distributions with the same variance. This is not as unreasonable an assumption as it may sound. A difference in mean but not in variability is a common phenomenon. The PEFR data for children with and without symptoms analysed in §8.5 and §9.6 show the characteristic very clearly.

We now estimate the common variance, s^2. First we find the sum of squares about the sample mean for each sample, which we can label SS_1 and SS_2. We form a combined sum of squares by $SS_1 + SS_2$. The sum of squares for the first group, SS_1, has $n_1 - 1$ degrees of freedom and, the second, SS_2, has $n_2 - 1$ degrees of freedom. The total degrees of freedom is therefore $n_1 - 1 + n_2 - 1 = n_1 + n_2 - 2$. We have lost 2 degrees of freedom because we have a sum of squares about two means. The combined estimate of variance is

$$s^2 = \frac{SS_1 + SS_2}{n_1 + n_2 - 2}$$

The standard error of $\bar{x}_1 - \bar{x}_2$ is

$$\sqrt{\frac{s^2}{n_1} + \frac{s^2}{n_2}} = \sqrt{s^2\left(\frac{1}{n_1} + \frac{1}{n_2}\right)}$$

Now we have a standard error related to the square root of the Chi-squared distribution and we can get a t distributed variable by

$$\frac{\bar{x}_1 - \bar{x}_2 - (\mu_1 - \mu_2)}{\sqrt{s^2\left(\frac{1}{n_1} + \frac{1}{n_2}\right)}}$$

having $n_1 + n_2 - 2$ degrees of freedom. The 95% confidence interval for the difference between means is

$$\bar{x}_1 - \bar{x}_2 - t\sqrt{s^2(1/n_1 + 1/n_2)} \text{ to } \bar{x}_1 - \bar{x}_2 + t\sqrt{s^2(1/n_1 + 1/n_2)}$$

where t is the 0.05 point with n_1+n_2-2 degrees of freedom from Table 10.1. Alternatively, we can test the null hypothesis that in the population the difference is zero, i.e. that $\mu_1 = \mu_2$, using the test statistic

$$\frac{\bar{x}_1 - \bar{x}_2}{\sqrt{s^2\left(\frac{1}{n_1} + \frac{1}{n_2}\right)}}$$

which would follow the t distribution with $n_1 + n_2 - 2$ d.f. if the null hypothesis were true.

For a practical example, Table 10.3 shows the data for male subjects corresponding to that in Table 10.2. We shall estimate the difference between the biases (i.e. mean difference between Wright and mini meter) for females and males. We have already noted the approximate Normal distribution of the differences. We must now consider the similarity of their variances. It is clear that the sample variance for the males is much smaller than that for females, and the assumption that in the populations the variances are the same is in question. However, the disparity is not too great to be due to chance, as can be shown using the F test (§10.8). We will accept it for the moment and consider its effect later.

First we find the common variance estimate, s^2. The sums of squares about the two sample means are 17 889.7 and 1 351.2. This gives the combined sum of squares about the sample means to be $17\,889.7 + 1\,351.2 = 19\,240.9$. The combined degrees of freedom are $n_1+n_2-2 = 12+5-2 = 15$. Hence $s^2 = 19\,240.9/15 = 1\,282.73$. The standard error of the difference between means is

$$\sqrt{s^2\left(\frac{1}{n_1} + \frac{1}{n_2}\right)} = \sqrt{1\,282.73\left(\frac{1}{12} + \frac{1}{5}\right)} = 19.06$$

The value of the t distribution for the 95% confidence interval is found from the 0.05 column and 15 d.f. row of Table 10.1 to be 2.13. The difference

Table 10.3. PEFR (litre/min) measured by Wright meter and mini meter, male subjects

Subject	Wright PEFR	Mini PEFR	Difference
13	611	625	−14
14	638	642	−4
15	633	605	28
16	492	467	25
17	372	370	2
Sum			37
Mean			7.4
Sum of squares about the mean			1351.2
Variance			337.8
Standard error of mean			8.2

between means is $-17.2 - 7.4 = -24.6$. Hence the 95% confidence interval is $-24.6-2.13\times19.06$ to $-24.6+2.13\times19.06$, giving –65.2 to 16.0 litre/min. Hence there could be quite a large difference between the response of males and females, from this very small sample, or there may be none at all.

To test the null hypothesis that in the population the male–female difference is zero, the test statistic is difference over standard error, $-24.6/19.06 = -1.29$. If the null hypothesis were true, this would be an observation from the t distribution with 15 degrees of freedom. From Table 10.1, the probability of such an extreme value is greater than 0.10. Hence the data are consistent with the null hypothesis and we cannot conclude that the bias is different for males and females. Again, we can see the advantage which estimation by confidence interval has over significance tests.

What happens if we do not make the assumption of uniform variance? There is an approximate solution based on the t distribution (e.g. see Davies and Goldsmith 1972; Snedecor and Cochran 1980) using the standard error formula of §8.5, $\sqrt{s_1^2/n_1 + s_2^2/n_2}$. For our data this standard error is 14.3. The difference between the variances leads to a rather complicated reduction in the degrees of freedom, in this case to 14. For this example we obtain a confidence interval of –55.1 to 6.0 litres/min, or a t test statistic of -1.7 with 14 degrees of freedom, $P = 0.11$. This is similar to what we obtained by the standard method. There are several other approaches based on the t test (see Armitage and Berry 1987). Another approach is to abandon the use of variance altogether and use the Mann Whitney U test (§12.2).

10.4 The use of transformations

We have already seen (§7.4) that some variables which are not Normally distributed can be made so by a suitable transformation. There are several transformations which can be used for this purpose. The most commonly used is the logarithm, which is suitable for data which are quite highly

Table 10.4. Biceps skinfold thickness (mm) in two groups of patients

Crohn's disease				Coeliac disease	
1.8	2.8	4.2	6.2	1.8	3.8
2.2	3.2	4.4	6.6	2.0	4.2
2.4	3.6	4.8	7.0	2.0	5.4
2.5	3.8	5.6	10.0	2.0	7.6
2.8	4.0	6.0	10.4	3.0	

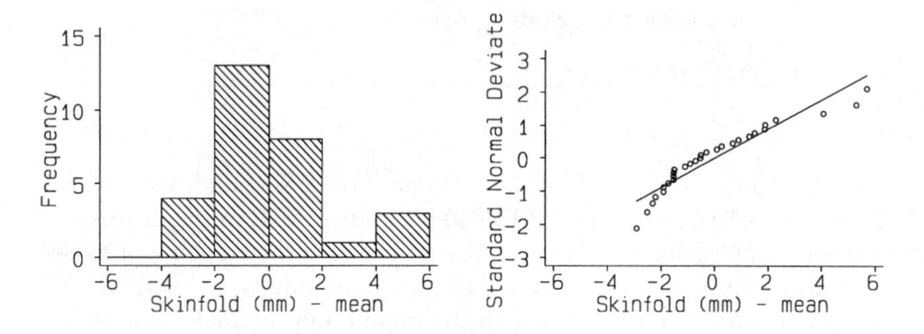

Fig. 10.5. Histogram and Normal plot for the biceps skinfold data

skewed, or when the standard deviations of different samples from the population are proportional to their means. A typical variable for this treatment is the serum triglyceride of Figure 7.11 and this is often true of such serum measurements. The square root transformation is useful when data are not so highly skewed and when the variance of a sample is proportional to its mean. Poisson variables have this property, for example. The reciprocal can be used when the standard deviation is proportional to the square of the mean, and data are very highly skewed indeed. Survival times tend to behave like this.

In large data sets, there are fairly good methods of determining the appropriate transformation (see Healy 1968). For small samples it is a matter of experience, trial and error. Table 10.4 shows some data from a study of anthropometry and diagnosis in patients with intestinal disease (Maugdal *et al.* 1985). We were interested in differences in anthropometrical measurements between patients with different diagnoses, and here we have the biceps skinfold measurements for 20 patients with Crohn's disease and 9 patients with coeliac disease. The data have been put into order of magnitude and it is fairly obvious that the distribution is skewed to the right. Figure 10.5 shows this clearly. I have subtracted the group mean from each observation, giving what is called the **within-group residuals**, and then found both the frequency distribution and Normal plot. The distribution is clearly skew, and this is reflected in the Normal plot, which shows a

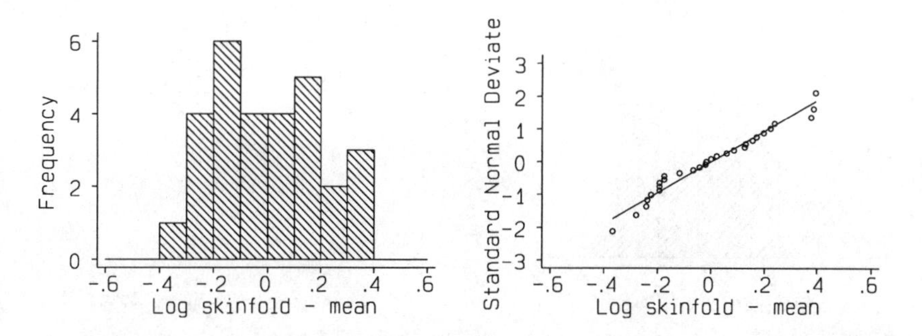

Fig. 10.6. Histogram and Normal plot for the biceps skinfold data, log transformed

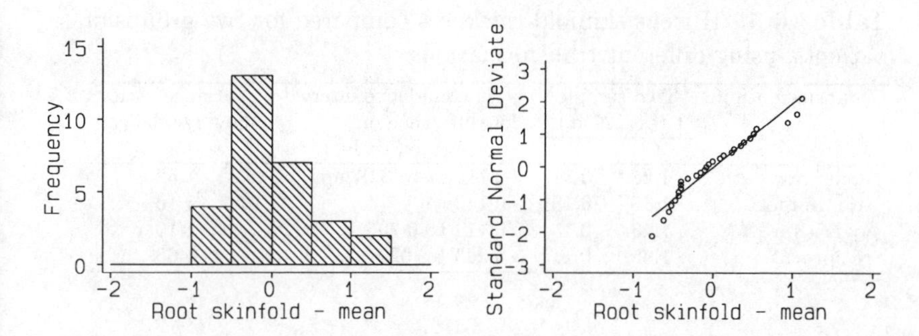

Fig. 10.7. Histogram and Normal plot for the biceps skinfold data, square root transformed

pronounced curvature.

We need a Normalizing transformation, if one can be found. The first guess is the log transform, and Figure 10.6 shows the histogram and Normal plot for the residuals after transformation. (These are natural logarithms rather than to base 10. It makes no difference to the final result and the calculations are the same to the computer.) The fit to the Normal distribution is not perfect, but much better than in Figure 10.5. We could use the two sample t method on these data quite happily. Compare Figure 10.7, which shows the result of a square root transformation. The skewness is still apparent, though less than in the untransformed data. Figure 10.8 shows the results of the reciprocal transformation. The results are if anything marginally worse than those for the log transformation, though it is not easy to choose between them. In practice I would select the log transformation, as the resulting statistics are easier to interpret. The reciprocal transformation changes the sign of the difference, for example.

Table 10.5 shows the results of the two sample t method used with the raw, untransformed data and with each transformation. The t test statistic

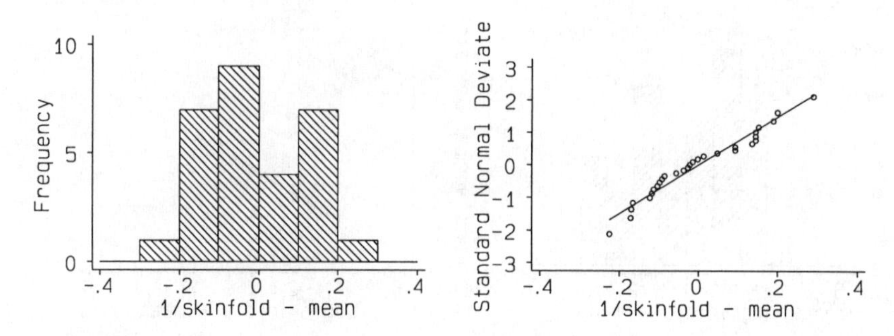

Fig. 10.8. Histogram and Normal plot for the biceps skinfold data, reciprocal transformed

Table 10.5. Biceps skinfold thickness compared for two groups of patients, using different transformations

Transformation	Two sample t test, 27 d.f.		95% confidence interval for difference on transformed scale	Variance ratio, larger/smaller
	t	P		
none, raw data	1.28	0.21	–0.71mm to 3.07mm	1.52
square root	1.38	0.18	–0.140 to 0.714	1.16
logarithm	1.48	0.15	–0.114 to 0.706	1.10
reciprocal	–1.65	0.11	–0.203 to .022	1.63

increases and its associated probability decreases as we move closer to a Normal distribution, reflecting the increasing power of the t test as its assumptions are more closely met. Table 10.5 also shows the ratio of the variances in the two samples. We can see that, as the transformed data gets closer to a Normal distribution, the variances tend to become more equal also.

The transformed data clearly gives a better test of significance than the raw data. The confidence intervals for the transformed data are more difficult to interpret, however, so the gain here is not so apparent. The confidence limits for the difference cannot be transformed back to the original scale. If we try it, the square root and reciprocal limits give ludicrous results. The log gives interpretable results (0.89 to 2.03) but these are not limits for the difference in millimetres. How could they be, for they do not contain zero yet the difference is not significant? They are in fact the 95% confidence limits for the ratio of the Crohn's disease mean to the coeliac disease mean. If there were no difference, of course, the expected value of this ratio would be one, not zero, and so lies within the limits. The reason is that when we take the difference between the logarithms of two numbers, we get the logarithm of their ratio, not of their difference (§5A). However, when we take the mean of the logarithms of several numbers we do get the logarithm of a mean of sorts, the geometric mean. The **geometric mean** of n numbers is the nth root of their product.

10.5 Deviations from the assumptions of t methods

The methods described in this chapter depend on some strong assumptions about the distributions from which the data come. This often worries users of statistical methods, who feel that these assumptions must limit greatly the use of t distribution methods and find the attitude of many statisticians, who often use methods based on Normal assumptions almost as a matter of course, rather sanguine. We shall look at some consequences of deviations from the assumptions.

First we shall consider a non-Normal distribution. As we have seen, some variables conform very closely to the Normal distribution, others do not. Deviations occur in two main ways: grouping and skewness. **Grouping** occurs when a continuous variable, such as human height, is measured in units which are fairly large relative to the range. This happens, for example, if we measure human height to the nearest inch. The heights in Figure 10.2 were to the nearest inch, and the fit to the t distribution is very good. This was a very coarse grouping, as the standard deviation of heights was 2.5 inches and so 95% of the 3000 observations had values over a range of 10 inches, only 10 or 11 possible values in all. We can see from this that if the underlying distribution is Normal, rounding the measurement is not going to affect the application of the t distribution by much.

Skewness, on the other hand, can invalidate methods based on the t distribution. For small samples of highly skew data, the t distribution does not fit the distribution of $(\bar{x}-\mu)/(\sqrt{s^2/n})$ at all well. When we have paired data this is not so important, because we have the Normalizing effect of the subtraction (§10.2). Skewness affects the two-sample t statistic of §10.3, but not so much as for one sample. In general for two equal sized samples the t method is very resistant to deviations from the Normal distribution, though as the samples become less equal in size the approximation becomes less good. The most likely effect of skewness is that we lose power and may fail to detect differences which exist or have confidence intervals which are too wide. We are unlikely to get spurious significant differences. This means that we need not worry about small departures from the Normal. If there is an obvious departure from the Normal, we should try to transform the data to the Normal before we apply the t distribution.

The other assumption of the two sample t method is that the variances in the two populations are the same. If this is not correct, the t distribution will not necessarily apply. The effect is usually small if the two populations are from a Normal distribution. This is unusual because, for samples from the same population, mean and variance are independent if the distribution is Normal (§7A). There is an approximate t method, as we noted in §10.3.

However, unequal variance is more often associated with skewness in the data, in which case a transformation designed to correct one fault often tends to correct the other as well.

Both the paired and two sample t methods are robust to most deviations from the assumptions. Only large deviations are going to have much effect on these methods. The main problem is with skewed data in the one sample method, but for reasons given in §10.2, the paired test will usually provide differences with a reasonable distribution. If the data do appear to be non-Normal, then a Normalizing transformation will improve matters. If this does not work, then we must turn to methods which do not require these assumptions (§9.2, §12.2, §12.3).

10.6 What is a large sample?

In this chapter we have looked at small sample versions of the large sample methods of §8.5 and §9.7. There we ignored both the distribution of the variable and the variability of s^2, on the grounds that they did not matter provided the samples were large. How small can a large sample be? This question is critical to the validity of these methods, but seldom seems to be discussed in text books.

Provided the assumptions of the t test apply, the question is easy enough to answer. Inspection of Table 10.1 will show that for 30 degrees of freedom the 5% point is 2.04, which is so close to the Normal value of 1.96 that it makes little difference which is used. So for Normal data with uniform variance we can forget the t distribution when we have more than 30 observations.

When the data are not in this happy state, things are not so simple. If the t method is not valid, we cannot assume that a large sample method which approximates to it will be valid. I recommend the following rough guide. First, if in doubt, treat the sample as small. Second, transform to a Normal distribution if possible. In the paired case you should transform *before* subtraction. Third, the more non-Normal the data, the larger the sample needs to be before we can ignore errors in the Normal approximation.

There is no simple answer to the question: 'how large is a large sample?'. We should be reasonably safe with inferences about means if the sample is greater than 100 for a single sample, or if both samples are greater than 50 for two samples. The application of statistical methods is a matter of judgement as well as knowledge.

10.7 Serial data

Table 10.6 shows levels of zidovudine (AZT) in the blood of AIDS patients at several times after administration of the drug, for patients with normal

Table 10.6. Blood zidovudine levels at times after administration of the drug by presence of fat malabsorption

Malabsorption patients:

Time since administration of zidovudine											
0	15	30	45	60	90	120	150	180	240	300	360
0.08	13.15	5.70	3.22	2.69	1.91	1.72	1.22	1.15	0.71	0.43	0.32
0.08	0.08	0.14	2.10	6.37	4.89	2.11	1.40	1.42	0.72	0.39	0.28
0.08	0.08	3.29	3.47	1.42	1.61	1.41	1.09	0.49	0.20	0.17	0.11
0.08	0.08	1.33	1.71	3.30	1.81	1.16	0.69	0.63	0.36	0.22	0.12
0.08	6.69	8.27	5.02	3.98	1.90	1.24	1.01	0.78	0.52	0.41	0.42
0.08	4.28	4.92	1.22	1.17	0.88	0.34	0.24	0.37	0.09	0.08	0.08
0.08	0.13	9.29	6.03	3.65	2.32	1.25	1.02	0.70	0.43	0.21	0.18
0.08	0.64	1.19	1.65	2.37	2.07	2.54	1.34	0.93	0.64	0.30	0.20
0.08	2.39	3.53	6.28	2.61	2.29	2.23	1.97	0.73	0.41	0.15	0.08

Normal absorption patients:

Time since administration of zidovudine											
0	15	30	45	60	90	120	150	180	240	300	360
0.08	3.72	16.02	8.17	5.21	4.84	2.12	1.50	1.18	0.72	0.41	0.29
0.08	6.72	5.48	4.84	2.30	1.95	1.46	1.49	1.34	0.77	0.50	0.28
0.08	9.98	7.28	3.46	2.42	1.69	0.70	0.76	0.47	0.18	0.08	0.08
0.08	1.12	7.27	3.77	2.97	1.78	1.27	0.99	0.83	0.57	0.38	0.25
0.08	13.37	17.61	3.90	5.53	7.17	5.16	3.84	2.51	1.31	0.70	0.37

fat absorption or fat malabsorption. A line graph of these data was shown in Figure 5.6. One common approach to such data is to carry out a two sample t test at each time separately, and researchers often ask at what time the difference becomes significant. This is a misleading question, as significance is a property of the sample rather than the population. The difference at 15 minutes may not be significant because the sample is small and the difference to be detected is small, not because there is no difference in the population. Further, if we do this for each time point we are carrying out multiple significance tests (§9.10) and each test only uses a small part of the data so we are losing power (§9.9). It is better to ask whether there is any evidence of a difference between the response of normal and malabsorption subjects over the whole period of observation.

The simplest approach is to reduce the data for a subject to one number. We can use the highest value attained by the subject, the time at which this peak value was reached, or the area under the curve. The first two are self-explanatory. The area under the curve is found by joining all the points for the subject, drawing vertical lines to the horizontal axis at the start and end times, and then calculating the area of the resulting polygon. Figure 10.9 shows this for the first subject in Table 10.6. The 'curve' is actually formed by a series of straight lines and the area can be calculated by taking each straight line segment and calculating the area under this. This is the base multiplied by the average of the two vertical heights. We

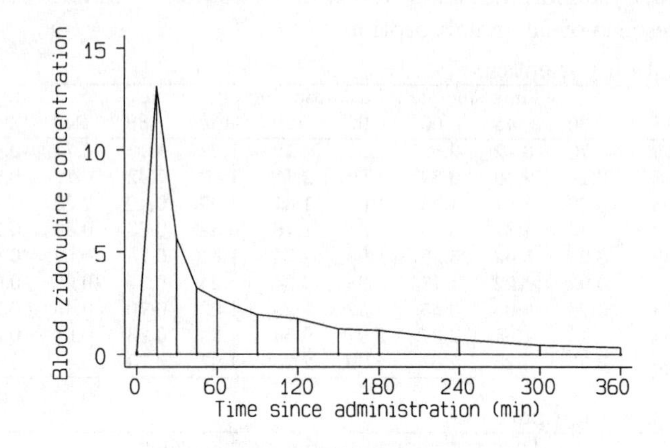

Fig. 10.9. Calculation of the area under the curve for one subject

Table 10.7. Area under the curve for data of Table 10.6

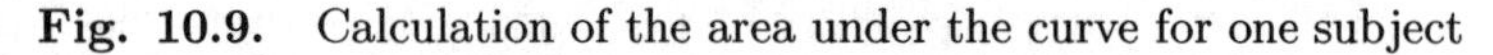

Malabsorption patients		Normal patients
667.425	256.275	919.875
569.625	527.475	599.850
306.000	388.800	499.500
298.200	505.875	472.875
617.850		1377.975

calculate this for each line segment, i.e. between each pair of adjacent time points, and add. Thus for the first subject we get $(15-0)\times(0.08+13.15)/2+(30-15)\times(13.15+5.70)/2+\cdots+(360-300)\times(0.43+0.32)/2 = 667.425$. This can be done fairly easily by most statistical computer packages. The area for each subject is shown in Table 10.7.

We can now compare the mean area by the two sample t method. As Figure 10.10 shows, the log area gives a better fit to the Normal distribution than does the area itself. Using the log area we get $n_1 = 9$, $\bar{x}_1 = 2.639\,541$, $s_1 = 0.153\,376$ for malabsorption subjects and $n_2 = 5$, $\bar{x}_2 = 2.850\,859$, $s_2 = 0.197\,120$ for the normal subjects. The common variance is $s^2 = 0.028\,635$, standard error of the difference between the means is $\sqrt{0.028\,635\times(1/9+1/5)} = 0.094\,385$, and the t statistic is $t = (2.639\,541 - 2.850\,859)/0.094\,385 = -2.24$ which has 12 degrees of freedom, P = 0.04. The 95% confidence interval for the difference is $2.639\,541 - 2.850\,859 \pm 2.18 \times 0.094\,385$, giving $-0.417\,078$ to $-0.005\,558$, and if we antilog this we get 0.38 to 0.99. Thus the area under the curve for malabsorption subjects is between 0.38 and 0.99 of that for normal AIDS patients, and we conclude that malabsorption inhibits uptake of the drug

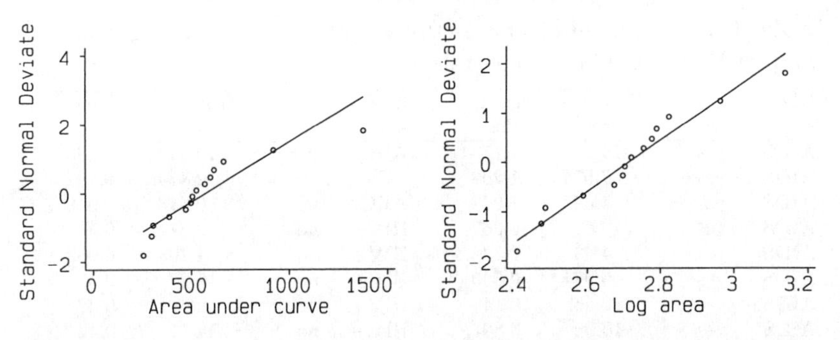

Fig. 10.10. Normal plots for area under the curve and log area for the data of Table 10.6

by this route. A fuller discussion of the analysis of serial data is given by Matthews *et al.* (1990).

10.8 Comparing two variances by the F test

We can test the null hypothesis that two population variances are equal using the F distribution. Provided the data are from a Normal distribution, the ratio of two independent estimates of the same variance will follow a F distribution (§7A), the degrees of freedom being the degrees of freedom of the two estimates. The F distribution is defined as that of the ratio of two independent Chi-squared variables divided by their degrees of freedom:

$$F_{m,n} = \frac{\chi^2_m/m}{\chi^2_n/n}$$

where m and n are the degrees of freedom (§7A). For Normal data the distribution of a sample variance s^2 from n observations is that of $\sigma^2\chi^2_{n-1}/(n-1)$ and when we divide one estimate of varaince by another to give the F ratio, the σ^2 cancels out. Like other distributions derived from the Normal, the F distribution cannot be integrated and so we must use a table. Because it has two degrees of freedom, the table is cumbersome, covering several pages, and I shall omit it. Most F methods are done using computer programs which calculate the probability directly. The table is usually only given as the upper percentage points.

To test the null hypothesis, we divide the larger variance by the smaller. For the PEFR data of §10.3, the variances are 1 626.3 with 11 d.f. for the females and 337.8 with 4 d.f. for the males, giving $F = 1\,626.3/338.7 = 4.80$. The probability of this being exceeded by the F distribution with 11 and 4 degrees of freedom is 0.15, the 5% point being 5.93. Several variances can be compared by the Bartlett's test or the Levene test (see Armitage and Berry 1987; Snedecor and Cochran 1980).

Table 10.8. Mannitol and lactulose gut permeability tests in a group of HIV patients and controls

HIV status	Diarrhoea	%mannitol	%lactulose	HIV status	Diarrhoea	%mannitol	%lactulose
AIDS	yes	14.9	1.17	ARC	yes	10.212	0.323
AIDS	yes	7.074	1.203	ARC	no	2.474	0.292
AIDS	yes	5.693	1.008	ARC	no	0.813	0.018
AIDS	yes	16.82	0.367	HIV+	no	18.37	0.4
AIDS	yes	4.93	1.13	HIV+	no	4.703	0.082
AIDS	yes	9.974	0.545	HIV+	no	15.27	0.37
AIDS	yes	2.069	0.14	HIV+	no	8.5	0.37
AIDS	yes	10.9	0.86	HIV+	no	14.15	0.42
AIDS	yes	6.28	0.08	HIV+	no	3.18	0.12
AIDS	yes	11.23	0.398	HIV+	no	3.8	0.05
AIDS	no	13.95	0.6	HIV−	no	8.8	0.122
AIDS	no	12.455	0.4	HIV−	no	11.77	0.327
AIDS	no	10.45	0.18	HIV−	no	14.0	0.23
AIDS	no	8.36	0.189	HIV−	no	8.0	0.104
AIDS	no	7.423	0.175	HIV−	no	11.6	0.172
AIDS	no	2.657	0.039	HIV−	no	19.6	0.591
AIDS	no	19.95	1.43	HIV−	no	13.95	0.251
AIDS	no	15.17	0.2	HIV−	no	15.83	0.338
AIDS	no	12.59	0.25	HIV−	no	13.3	0.579
AIDS	no	21.8	1.15	HIV−	no	8.7	0.18
AIDS	no	11.5	0.36	HIV−	no	4.0	0.096
AIDS	no	10.5	0.33	HIV−	no	11.6	0.294
AIDS	no	15.22	0.29	HIV−	no	14.5	0.38
AIDS	no	17.71	0.47	HIV−	no	13.9	0.54
AIDS	yes	7.256	0.252	HIV−	no	6.6	0.159
AIDS	no	17.75	0.47	HIV−	no	16.5	0.31
ARC	yes	7.42	0.21	HIV−	no	9.989	0.398
ARC	yes	9.174	0.399	HIV−	no	11.184	0.186
ARC	yes	9.77	0.215	HIV−	no	2.72	0.045
ARC	no	22.03	0.651				

10.9 Comparing several means using analysis of variance

Consider the data of Table 10.8. These are measures of gut permeability obtained from four groups of subjects. We want to investigate the differences between the groups.

One approach would be to use the t test to compare each pair of groups. This has disadvantages. First, there are many comparisons, $m(m-1)/2$ where m is the number of groups. The more groups we have, the more likely it is that two of them will be far enough apart to produce a 'significant' difference when the null hypothesis is true and the population means are the same (§9.10). Second, when groups are small, there may not be many degrees of freedom for the estimate of variance. If we can use all the data to estimate variance we will have more degrees of freedom and hence a more powerful comparison.

Table 10.9. Some artificial data to illustrate how analysis of variance works

	Group 1	Group 2	Group 3	Group 4
	6	4	7	3
	7	5	9	5
	8	6	10	6
	8	6	11	6
	9	6	11	6
	11	8	13	8
Mean	8.167	5.833	10.167	5.667

To illustrate how the analysis of variance, or **anova**, works, I shall use some artificial data, as set out in Table 10.9. In practice, equal numbers in each group are unusual in medical applications. We start by estimating the common variance within the groups, just as we do in a two sample t test (§10.3). We find the sum of squares about the group mean for each group and add them. We call this the **within groups sum of squares.** For Table 10.9 this gives 57.833. For each group we estimate the mean from the data, so we have estimated 4 parameters and have $24 - 4 = 20$ degrees of freedom. In general, for m groups of size n each we have $nm - m$ degrees of freedom. This gives us an estimate of variance of

$$s^2 = \frac{57.833}{20} = 2.892$$

This is the **within groups variance** or **residual variance**. There is an assumption here. For a common variance, we assume that the variances are the same in the four populations represented by the four groups.

We can get two other estimates of variance from the data. We can find the variance of the whole data, ignoring the groups. The sum of squares, called the **total sum of squares**, is 139.958 and there are $24 - 1 = 23$ degrees of freedom. The estimated variance is $139.958/23 = 6.085$. This is much bigger than the within groups variance, because in this example there is a lot of variability between groups.

We can also find an estimate of variance from the group means. The variance of the four group means is 5.475. If there were no difference between the means in the population from which the sample comes, this variance would be the variance of the sampling distribution of the mean of n observations, which is s^2/n, the square of the standard error (§8.2). Thus n times this variance should be equal to the within groups variance. For the example, this is $5.475 \times 5 = 27.375$, which is much greater than the 2.892 actually observed. We express this by the ratio of one variance estimate to the other, between groups over within groups, which we call the variance ratio or F ratio. If the null hypothesis is true and if the observations are from a Normal distribution with uniform variance, this

Table 10.10. One-way analysis of variance for the data of Table 10.9

Source of variation	Degrees of freedom	Sum of squares	Mean square	Variance ratio (F)	Probability
Total	23	139.958			
Between groups	3	82.125	27.375	9.47	0.0004
Within groups	20	57.833	2.892		

Table 10.11. One-way analysis of variance for the mannitol data

Source of variation	Degrees of freedom	Sum of squares	Mean square	Variance ratio (F)	Probability
Total	58	1559.036			
Between groups	3	49.012	16.337	0.6	0.6
Residual	55	1510.024	27.455		

ratio follows a known distribution, the F distribution with $m-1$ and $n-1$ degrees of freedom (§10.8).

For the example we would have 3 and 20 degrees of freedom and

$$F_{3,20} = \frac{27.375}{2.892} = 9.47$$

If the null hypothesis is true, the expected value of this ratio to be 1.0. A large value gives us evidence of a difference between the means in the four populations. For the example we have a large value of 9.47 and the probability of getting a value as big as this if the null hypothesis were true would be 0.0004. Thus there is a significant difference between the four groups.

We can set these calculations out in an analysis of variance table, as shown in Table 10.10. The sum of squares in the 'between groups' row is the sum of squares of the group means times n. We call this the **between groups sum of squares**. Notice that in the 'degrees of freedom' and 'sum of squares' columns the 'within groups' and 'between groups' rows add up to the total. The within groups sum of squares is also called the **residual sum of squares**, because it is what is left when the group effect is removed, or the **error sum of squares**, because it measures the random variation or error remaining when all systematic effects have been removed.

Returning to the mannitol data, as so often happens the groups are of unequal size. The calculation of the between groups sum of squares becomes more complicated and we usually do it by subtracting the within groups sum of squares from the total sum of squares. Otherwise, the table is the same, as shown in Table 10.11. As these calculations are usually done by computer the extra complexity in calculation does not matter. Here there is no significant difference between the groups.

If we have only two groups, one-way analysis of variance is another way of doing a two sample t test. For example, from Table 10.12 we can

Table 10.12. One-way analysis of variance for the comparison of two groups from Table 10.8

Source of variation	Degrees of freedom	Sum of squares	Mean square	Variance ratio (F)	Probability
Total	32	987.860			
Between groups	1	34.176	34.176	1.11	0.3
Residual	31	953.684	30.764		

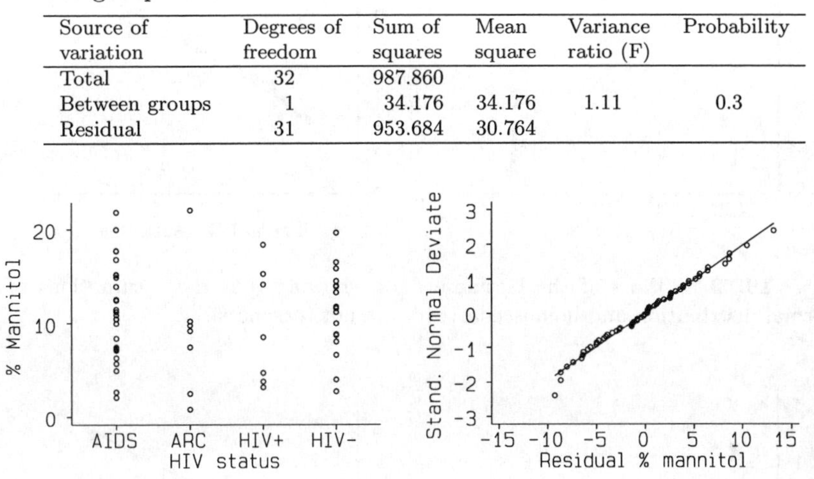

Fig. 10.11. Plots of the mannitol data, showing that the assumptions of Normal distribution and homoscedasticity are reasonable

compare mannitol excretion for the AIDS and ARC patients. Using the t test we get $t = 1.05$ with 31 degrees of freedom, $P = 0.3$. The analysis of variance table is shown in Table 10.12. The probability is the same and the F ratio, 1.11, is the square of the t statistic, 1.05. The residual mean square is the common variance of the t test.

10.10 Assumptions of the analysis of variance

There are two assumptions for analysis of variance: that data come from Normal distributions within the groups and that the variances of these distributions are the same. The technical term for uniformity of variance is **homoscedasticity**; lack of uniformity is **heteroscedasticity**. Heteroscedasticity can affect analyses of variance a lot and we try to guard against it.

We can examine these assumptions graphically. Figures 10.11 and 10.12 show graphical analyses for the mannitol and lactulose data. For the mannitol data, the scatter plot for the groups shows the spread of data in each group, suggesting that the assumption of uniform variance is met. This is not the case for the lactulose data, as Figure 10.12 illustrates. The group with the highest mean, AIDS, has the greatest spread.

The Normal plot of residuals shows how well the differences or deviations from the group means conform to a Normal distribution. For the mannitol data, this shows a fairly good fit to the straight line, indicating

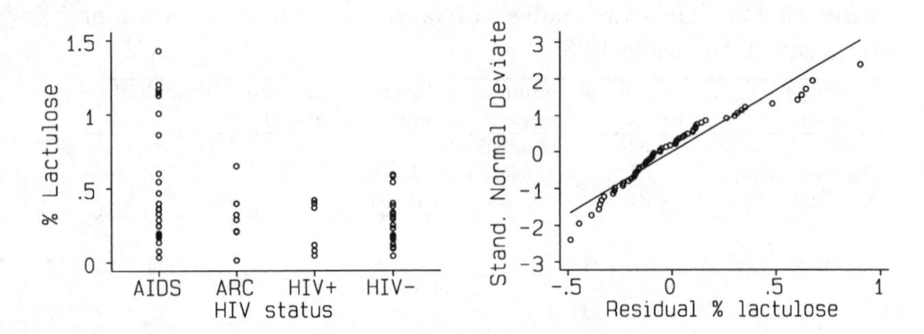

Fig. 10.12. Plots of the lactulose data, showing that the assumptions of Normal distribution and homoscedasticity are not reasonable

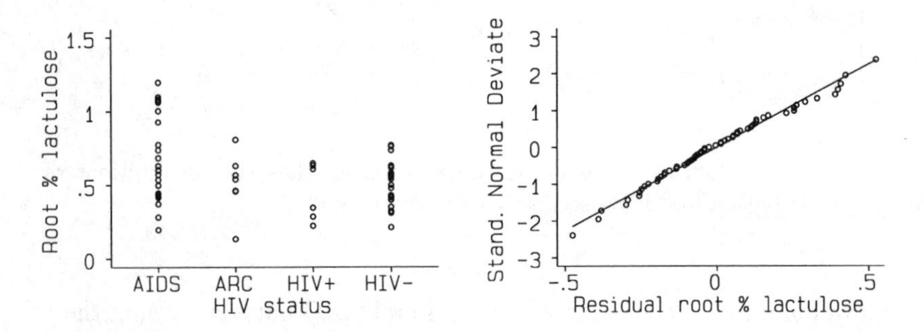

Fig. 10.13. Plots of the lactulose data after square root transformation

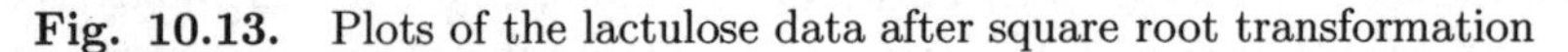

Table 10.13. One-way analysis of variance for the square root transformed lactulose data of Table 10.8

Source of variation	Degrees of freedom	Sum of squares	Mean square	Variance ratio (F)	Probability
Total	58	3.254 41			
HIV status	3	0.428 70	0.142 90	2.78	0.0495
Residual	55	2.825 71	0.051 38		

that they match the Normal distribution assumption reasonably well. The plot for lactulose shows pronounced curvature, indicating skewness. The square root transformation of the lactulose fits better (Figure 10.13). As Figure 10.14 shows, the log transform over-compensates for skewness, by producing skewness in the opposite direction, though the variances appear uniform. We could use either the square root or the logarithmic transformation. The square root has been used here. Table 10.13 shows the analysis of variance for square root transformed lactulose.

There are also significance tests which we can apply for Normal distribution and homoscedasticity. For example, Bartlett's test for homogeneity of variance gives, for the data of Table 10.9, $\chi^2 = 0.84$, d.f.=3, P = 0.8,

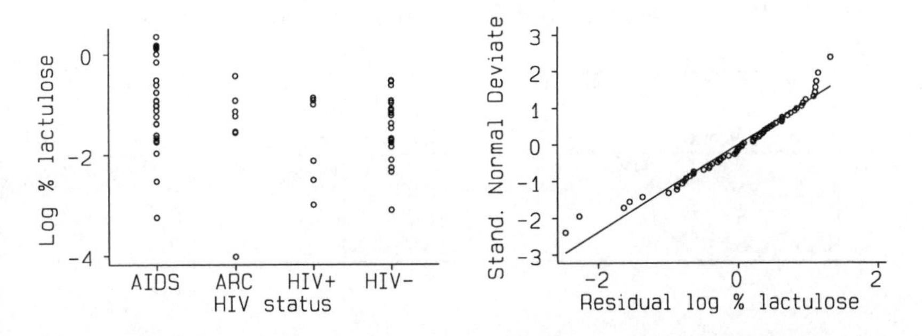

Fig. 10.14. Plots of the lactulose data after logarithmic transformation

Table 10.14. The Newman–Keuls test for the data of Table 10.9

S = significant at the 0.05 level, N = not significant.

Group	Group 1	2	3
2	S		
3	N	S	
4	S	N	S

S = significant at the 0.01 level, N = not significant.

Group	Group 1	2	3
2	N		
3	N	S	
4	N	N	S

indicating that the data are consistent with the assumptions. I shall omit the details.

10.11 Comparison of means after analysis of variance

Concluding from Tables 10.10 and 10.13 that the there is a significant difference between the means is rather unsatisfactory. We want to know which means differ from which. There are a number of ways of doing this, called **multiple comparisons procedures**. These are mostly designed to give only one type I error (§9.3) per 20 analyses when the null hypothesis is true, as opposed to doing t tests for each pair of groups, which gives one error per 20 comparisons when the null hypothesis is true. We shall not go into details, but look at a couple of examples. There are several tests which can be used when the numbers in each group are the same, Tukey's Honestly Significant Difference, the Newman–Keuls sequential procedure (both called Studentized range tests), Duncan's multiple range test, etc. The one you use will depend on which computer program you have. The results of the Newman–Keuls sequential procedure for the data of Table 10.9 are shown in Table 10.14. Group 1 is significantly different from groups 2 and 4, and group 3 from groups 2 and 4. At the 1% level, group 3 is significantly different from groups 2 and 4 only.

For unequal-sized groups, the choice of multiple comparison procedures

Table 10.15. Gabriel's test for the root transformed lactulose data

S = significant at the 0.05 level, N = not significant.

Group	Group AIDS	ARC	HIV+
ARC	N		
HIV+	S	N	
HIV−	S	N	N

S = significant at the 0.01 level, N = not significant.

Group	Group AIDS	ARC	HIV+
ARC	N		
HIV+	N	N	
HIV−	N	N	N

is more limited. Gabriel's test can be used with unequal sized groups. For the root transformed lactulose data, the results of Gabriel's test are shown in Table 10.15. This shows that the AIDS subjects are significantly different from the asymptomatic HIV+ patients and from the HIV− controls. For the mannitol data, most multiple comparison procedures will give no significant differences because they are designed to give only one type I error per analysis of variance, so when the F test is not significant, no group comparisons will be either.

10A Appendix: The ratio mean/standard error

We know that $\bar{x}$ has a Normal distribution with mean μ and variance σ^2/n. Hence $(\bar{x}-\mu)/\sqrt{\sigma^2/n}$ will be Normal with mean 0 and variance 1. The distribution of $(n-1)s^2/\sigma^2$ is Chi-squared with $n-1$ degrees of freedom (Appendix 7A). If we divide a Standard Normal variable by the square root of an independent Chi-squared variable over its degrees of freedom, we get the t distribution:

$$\frac{\frac{\bar{x}-\mu}{\sqrt{\sigma^2/n}}}{\sqrt{\frac{(n-1)s^2/\sigma^2}{n-1}}} = \frac{\frac{\bar{x}-\mu}{\sqrt{\sigma^2/n}}}{\sqrt{\frac{s^2}{\sigma^2}}} = \frac{\bar{x}-\mu}{\sqrt{\frac{\sigma^2}{n}}\times\sqrt{\frac{s^2}{\sigma^2}}} = \frac{\bar{x}-\mu}{\sqrt{\frac{\sigma^2}{n}\times\frac{s^2}{\sigma^2}}} = \frac{\bar{x}-\mu}{\sqrt{\frac{s^2}{n}}}$$

As if by magic, we have our sample mean over its standard error. I shall not bother to go into this detail for the other similar ratios which we shall encounter. Any Normally distributed quantity with mean zero (such as $\bar{x}-\mu$), divided by its standard error, will follow a t distribution provided the standard error is based on one sum of squares and hence is related to the Chi-squared distribution.

10M Multiple choice questions 50 to 56

(Each branch is either true or false)

50. The paired t test is:

(a) impractical for large samples;

(b) useful for the analysis of qualitative data;

(c) suitable for very small samples;

(d) used for independent samples;

(e) based on the Normal distribution.

51. Which of the following conditions must be met for a valid t test between the means of two samples:

(a) the numbers of observations must be the same in the two groups;

(b) the standard deviations must be approximately the same in the two groups;

(c) the means must be approximately equal in the two groups;

(d) the observations must be from approximately Normal distributions;

(e) the samples must be small.

52. In a two sample clinical trial, one of the outcome measures was highly skewed. To test the difference between the levels of this measure in the two groups of patients, possible approaches include:

(a) a standard t test using the observations;

(b) a Normal approximation if the sample is large;

(c) transforming the data to a Normal distribution and using a t test;

(d) a sign test;

(e) the standard error of the difference between two proportions.

53. In the two sample t test, deviation from the Normal distribution by the data may seriously affect the validity of the test if:

(a) the sample sizes are equal;

(b) the distribution followed by the data is highly skewed;

(c) one sample is much larger than the other;

(d) both samples are large;

(e) the data deviate from a Normal distribution because the measurement unit is large and only a few values are possible.

Table 10.16. Semen analyses for successful and unsuccessful sperm donors (Paraskevaides *et al.* 1991)

	Successful donors			Unsuccessful donors		
	n	mean	(sd)	n	mean	(sd)
Volume (ml)	17	3.14	(1.28)	19	2.91	(0.91)
Semen count (10^6/ml)	18	146.4	(95.7)	19	124.8	(81.8)
% motility	17	60.7	(9.7)	19	58.5	(12.8)
% abnormal morphology	13	22.8	(8.4)	16	20.3	(8.5)

All differences not significant, t test

54. Table 10.16 shows a comparison of successful (i.e. fertile) and unsuccessful artificial insemination donors. The authors concluded that 'Conventional semen analysis may be too insensitive an indicator of high fertility [in AID]'

(a) The table would be more informative if P values were given;

(b) The t test is important to the conclusion given;

(c) It is likely that semen count follows a Normal distribution;

(d) If the null hypothesis were true, the sampling distribution of the t test statistic for semen count would approximate to a t distribution;

(e) If the null hypothesis were false, the power of the t test for semen count could be increased by a log transformation.

55. If we take samples of size n from a Normal distribution and calculate the sample mean $\bar{x}$ and variance s^2:

(a) samples with large values of $\bar{x}$ will tend to have large s^2;

(b) the sampling distribution of $\bar{x}$ will be Normal;

(c) the sampling distribution of s^2 will be related to the Chi-squared distribution with $n-1$ degrees of freedom;

(d) the ratio $\bar{x}/\sqrt{s^2/n}$ will be from a t distribution with $n-1$ degrees of freedom;

(e) the sampling distribution of s will be approximately Normal if $n > 20$.

56. In the one-way analysis of variance table for the comparison of three groups:

(a) the group mean square + the error mean square = the total mean square;

(b) there are two degrees of freedom for groups;

(c) the group sum of squares + the error sum of squares = the total sum of squares;

(d) the numbers in each group must be equal;

(e) the group degrees of freedom + the error degrees of freedom = the total degrees of freedom.

Table 10.17. $p_a(O_2)$ and compliance for two inspiratory flow waveforms

Patient	$p_a(O_2)$ (kpa) Waveform		Compliance (ml/cm H_2O) Waveform	
	Constant	Decelerating	Constant	Decelerating
1	9.1	10.8	65.4	72.9
2	5.6	5.9	73.7	94.4
3	6.7	7.2	37.4	43.3
4	8.1	7.9	26.3	29.0
5	16.2	17.0	65.0	66.4
6	11.5	11.6	35.2	36.4
7	7.9	8.4	24.7	27.7
8	7.2	10.0	23.0	27.5
9	17.7	22.3	133.2	178.2
10	10.5	11.1	38.4	39.3
11	9.5	11.1	29.2	31.8
12	13.7	11.7	28.3	26.9
13	9.7	9.0	46.6	45.0
14	10.5	9.9	61.5	58.2
15	6.9	6.3	25.7	25.7
16	18.1	13.9	48.7	42.3

10E Exercise: The paired t method

Table 10.17 shows the total static compliance of the respiratory system and the arterial oxygen tension ($p_a(O_2)$) in 16 patients in intensive care (Al-Saady, personal communication). The patients' breathing was assisted by a respirator and the question was whether their respiration could be improved by varying the characteristics of the air flow. Table 10.17 compares a constant inspiratory flow waveform with a decelerating inspiratory flow waveform. We shall examine the effect of waveform on compliance.

1. Calculate the changes in compliance. Find a stem and leaf plot. (Hint: you will need both a zero and a minus zero row).
2. As a check on the validity of the t method, plot the difference against the subject's mean compliance. Do they appear to be related?
3. Calculate the mean, variance, standard deviation and standard error of the mean for the compliance differences.
4. Even though the compliance differences are far from a Normal distribution, calculate the 95% confidence interval using the t distribution. We will compare this with that for transformed data.
5. Find the logarithms of the compliance and repeat steps 1 to 3. Do the assumptions of the t distribution method apply more closely?
6. Calculate the 95% confidence interval for the log difference and transform back to the original scale. What does this mean and how does it compare to that based on the untransformed data?
7. What can be concluded about the effect of inspiratory waveform on static compliance in intensive care patients?

11

Regression and correlation

11.1 Scatter diagrams

In this chapter I shall look at methods of analysing the relationship between two quantitative variables. Consider Table 11.1, which shows data collected by a group of medical students in a physiology class. Inspection of the data suggests that there may be some relationship between FEV1 and height. Before trying to quantify this relationship, we can plot the data and get an idea of its nature. The usual first plot is a scatter diagram, §5.6. Which variable we choose for which axis depends on our ideas as to the underlying relationship between them, as discussed below. Figure 11.1 shows the scatter diagram for FEV1 and height.

Inspection of Figure 11.1 suggests that FEV1 increases with height. The next step is to try and draw a line which best represents the relationship. The simplest line is a straight one; I shall consider more complicated relationships in Chapter 17.

The equation of a straight line relationship between variables x and y is $y = a + bx$, where a and b are constants. The first, a, is called the **intercept**. It is the value of y when x is 0. The second, b, is called the **slope** or **gradient** of the line. It is the increase in y corresponding to an increase of one unit in x. Their geometrical meaning is shown in Figure 11.2. We can find the values of a and b which best fit the data by regression analysis.

Table 11.1. FEV1 and height for 20 male medical students

Height (cm)	FEV1 (litres)	Height (cm)	FEV1 (litres)	Height (cm)	FEV1 (litres)
164.0	3.54	172.0	3.78	178.0	2.98
167.0	3.54	174.0	4.32	180.7	4.80
170.4	3.19	176.0	3.75	181.0	3.96
171.2	2.85	177.0	3.09	183.1	4.78
171.2	3.42	177.0	4.05	183.6	4.56
171.3	3.20	177.0	5.43	183.7	4.68
172.0	3.60	177.4	3.60		

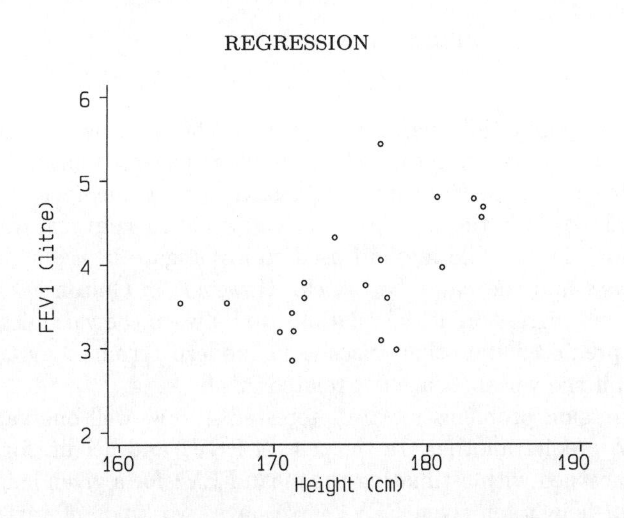

Fig. 11.1. Scatter diagram showing the relationship between FEV1 and height for a group of male medical students

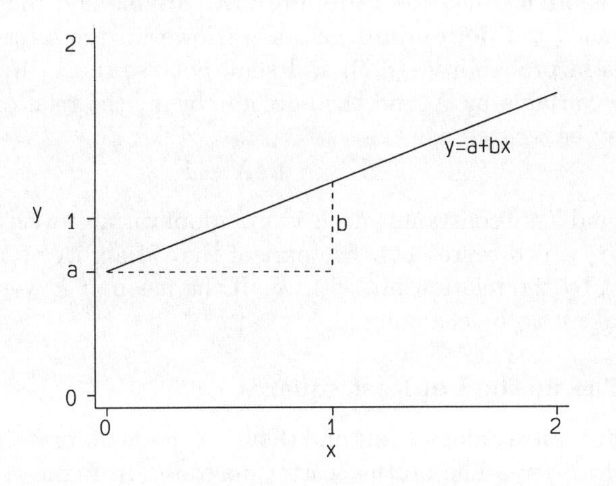

Fig. 11.2. Coefficients of a straight line

11.2 Regression

Regression is a method of estimating the numerical relationship between variables. For example, we would like to know what is the mean or expected FEV1 for students of a given height, and what increase in FEV1 is associated with a unit increase in height.

The name 'regression' is due to Galton (1886), who developed the technique to investigate the relationship between the heights of children and of their parents. He observed that if we choose a group of parents of a given height, the mean height of their children will be closer to the mean height of the population than is the given height. In other words, tall parents tend to be taller than their children, short parents tend to be shorter. Galton termed this phenomenon 'regression towards mediocrity', meaning 'going back towards the average'. It is now called **regression towards the mean** (§11.4). The method used to investigate it was called regression analysis and the name has stuck. However, in Galton's terminology there was 'no regression' if the relationship between the variables was such that one predicted the other exactly; in modern terminology there is no regression if the variables are not related at all.

In regression problems we are interested in how well one variable can be used to predict another. In the case of FEV1 and height, for example, we are concerned with estimating the mean FEV1 for a given height rather than mean height for given FEV1. We have two kinds of variables: the **outcome** variable which we are trying to predict, in this case FEV1, and the **predictor** or **explanatory** variable, in this case height. The predictor variable is often called the **independent** variable and the outcome variable is called the **dependent** variable. However, these terms have other meanings in probability (§6.2), so I shall not use them. If we denote the predictor variable by X and the outcome by Y, the relationship between them may be written as

$$Y = a + bX + E$$

where a and b are constants and E is a random variable with mean 0, called the **error**, which represents that part of the variability of Y which is not explained by the relationship with X. If the mean of E were not zero, we could make it so by changing a.

11.3 The method of least squares

If the points all lay along a line and there was no random variation, it would be easy to draw a line on the scatter diagram. In Figure 11.1 this is not the case. There are many possible values of a and b which could represent the data and we need a criterion for choosing the best line. Figure 11.3 shows the deviation of a point from the line, the distance from the point to the line in the Y direction. The line will fit the data well if the deviations from it are small, and will fit badly if they are large. These deviations represent the error E, that part of the variable Y not explained by X. One solution to the problem of finding the best line is to choose that which leaves the minimum amount of the variability of Y unexplained, by making the variance of E a minimum. This will be achieved by making the sum of squares of the deviations about the line a minimum. This is called the **method of least squares** and the line found is the **least squares line**.

The method of least squares is the best method if the deviations from the line are Normally distributed with uniform variance along the line.

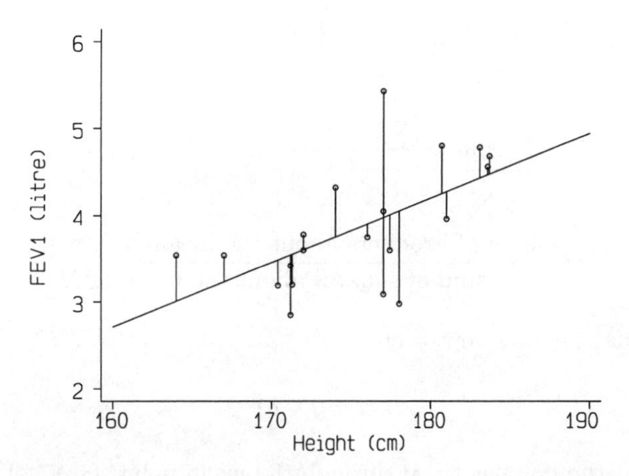

Fig. 11.3. Deviations from the line in the y direction

This is likely to be the case, as the regression tends to remove from Y the variability between subjects and leave the measurement error, which is likely to be Normal. I shall deal with deviations from this assumption in §11.8.

Many users of statistics are puzzled by the minimization of variation in one direction only. Usually both variables are measured with some error and yet we seem to ignore the error in X. Why not minimize the perpendicular distances to the line rather than the vertical? There are two reasons for this. First, we are finding the best prediction of Y from the observed values of X, not from the 'true' values of X. The measurement error in both variables is one of the causes of deviations from the line, and is included in these deviations measured in the Y direction. Second, the line found in this way depends on the units in which the variables are measured. For the data of Table 11.1 the line found by this method is

$$\text{FEV1 (litre)} = -9.33 + 0.075 \times \text{height (cm)}$$

If we measure height in metres instead of centimetres, we get

$$\text{FEV1 (litre)} = -34.70 + 22.0 \times \text{height (m)}$$

Thus by this method the predicted FEV1 for a student of height 170 cm is 3.42 litres, but for a student of height 1.70 m it is 2.70 litres. This is clearly unsatisfactory and we will not consider this approach further.

Returning to Figure 11.3, the equation of the line which minimizes the sum of squared deviations from the line in the outcome variable is found quite easily (§11A). The solution is:

$$\begin{aligned} b &= \frac{\sum(x_i - \bar{x})(y_i - \bar{y})}{\sum(x_i - \bar{x})^2} \\ &= \frac{\sum x_i y_i - \frac{(\sum x_i)(\sum y_i)}{n}}{\sum x_i^2 - \frac{(\sum x_i)^2}{n}} \\ &= \frac{\text{sum of products about the mean of } X \text{ and } Y}{\text{sum of squares about the mean of } X} \end{aligned}$$

We then find the intercept a by

$$a = \bar{y} - b\bar{x}$$

Notice that the line has to go through the mean point, $(\bar{x}, \bar{y})$. The numerator, the **sum of products about the mean** is similar to the sum of squares about the mean as used in the calculation of variance. The second form, which is easier for calculator work, is found as in §4B. We shall say more about the properties of the **sum of products**, as it is usually termed, when we discuss correlation. Fitting a straight line by this method is called **simple linear regression.**

The equation $Y = a + bX$ is called the **regression equation of Y on** X, Y being the outcome variable and X the predictor. The gradient, b, is also called the **regression coefficient**. We shall calculate it for the data of Table 11.1. We have

$$\sum x_i = 3\,507.6 \qquad \sum x_i^2 = 615\,739.24 \qquad n = 20$$
$$\sum y_i = 77.12 \qquad \sum y_i^2 = 306.813\,4 \qquad \sum x_i y_i = 13\,568.18$$
$$\bar{x} = 3\,507.6/20 = 175.38 \qquad \bar{y} = 77.12/20 = 3.856$$

$$\text{sum of squares } X = \sum x_i^2 - \frac{\left(\sum x_i\right)^2}{n} = 615\,739.24 - \frac{3\,507.6^2}{20} = 576.352$$

$$\text{sum of squares } Y = \sum y_i^2 - \frac{\left(\sum y_i\right)^2}{n} = 306.813\,4 - \frac{77.12^2}{20} = 9.438\,68$$

$$\begin{aligned} \text{sum of products about mean} &= \sum x_i y_i - \frac{\left(\sum x_i\right)\left(\sum y_i\right)}{n} \\ &= 13\,568.18 - \frac{3\,507.6 \times 77.12}{20} = 42.874\,4 \end{aligned}$$

We do not need the sum of squares for Y yet, but we shall later.

$$b = \frac{42.874\,4}{576.352} = 0.074\,389 \text{ litre/cm}$$

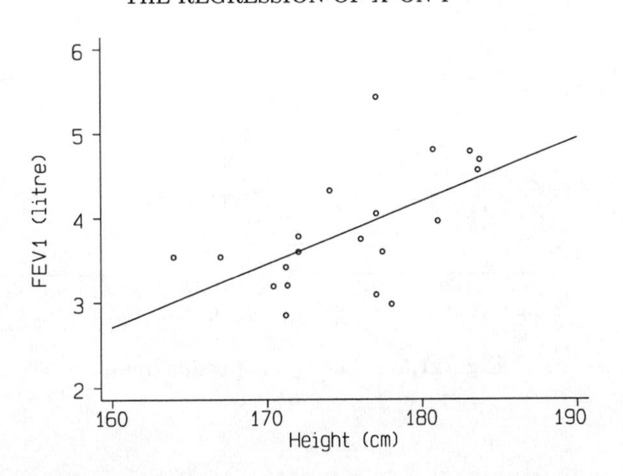

Fig. 11.4. The regression of FEV1 on height

$$a = \bar{y} - b\bar{x} = 3.856 - 0.074\,389 \times 175.38 = -9.19 \text{ litre}$$

Hence the regression equation of FEV1 on height is

$$\text{FEV} = -9.19 + 0.0744 \times \text{height}$$

Figure 11.4 shows the line drawn on the scatter diagram.

The coefficients a and b have dimensions, depending on those of X and Y. If we change the units in which X and Y are measured we also change a and b, but we do not change the line. For example, if height is measured in metres we divide the x_i by 100 and we find that b is multiplied by 100 to give $b = 7.4389$ litres/m. The line is

$$\text{FEV1 (litres)} = -9.19 + 7.44 \times \text{height (m)}$$

This is exactly the same line on the scatter diagram.

11.4 The regression of X on Y

What happens if we change our choice of outcome and predictor variables? The regression equation of height on FEV1 is

$$\text{height} = 158 + 4.54 \times \text{FEV1}$$

This is not the same line as the regression of FEV1 on height. For if we rearrange this equation by dividing each side by 4.54 we get

$$\text{FEV1} = -34.8 + 0.220 \times \text{height}$$

The slope of the regression of height on FEV1 is greater than that of FEV1 on height (Figure 11.5). In general, the slope of the regression of X on Y

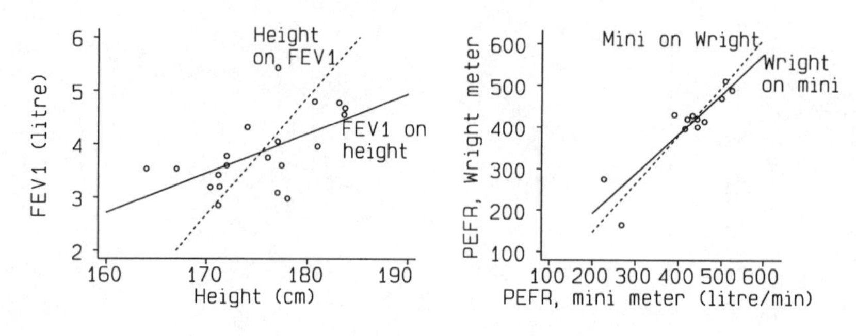

Fig. 11.5. The two regression lines

is greater than that of Y on X, when X is the horizontal axis. Only if all the points lie exactly on a straight line are the two equations the same.

Figure 11.5 also shows the two PEFR measurements of Table 10.2, with the two regression lines. The regression equations are Wright $=$ $1.54 + 0.96 \times$ mini and mini $= 73.54 + 0.86 \times$ Wright. Each regression coefficient is less than one. This means that for subjects with any given mini meter measurement, the predicted Wright meter measurement will be closer to the mean than the mini meter measurement, and for any given Wright meter measurement, the predicted mini meter measurement will be closer to the mean than the Wright meter measurement. This is regression towards the mean (§11.2). Regression towards the mean is a purely statistical phenomenon, produced by the selection of the given value of the predictor and the imperfect relationship between the variables. Regression towards the mean may manifest itself in many ways. For example, if we measure the blood pressure of an unselected group of people, then select subjects with high blood pressure, e.g. diastolic>95 mm Hg, then measured the selected group again, the mean diastolic pressure for the selected group will be less on the second occasion than on the first, without any intervention or treatment. The apparent fall is caused by the initial selection.

11.5 The standard error of the regression coefficient

In any estimation procedure, we want to know how reliable our estimates are. We do this by finding their standard errors and hence confidence intervals. We can also test hypotheses about the coefficients, for example, the null hypothesis that in the population the slope is zero and there is no linear relationship. The details are given in §11C. We first find the sum of squares of the deviations from the line, that is, the difference between the observed y_i and the values predicted by the regression line. This is

$$\sum(y_i - \bar{y})^2 - b^2 \sum(x_i - \bar{x})^2$$

$\sum(y_i - \bar{y})^2$ is of course the total sum of squares about the mean of y_i. The term $b^2 \sum(x_i - \bar{x})^2$ is called the **sum of squares due to the regression on X**. The difference between them is the **residual sum of squares** or **sum of squares about the regression.** The sum of squares due to the regression divided by the total sum of squares is called the **proportion of variability explained by the regression.**

In order to estimate the variance we need the degrees of freedom with which to divide the sum of squares. We have estimated not one parameter from the data, as for the sum of squares about the mean (§4.6), but two, a and b. We lose two degrees of freedom, leaving us with $n-2$. Hence the variance of Y about the line, called the **residual variance**, is

$$s^2 = \frac{1}{n-2}\left(\sum(y_i - \bar{y})^2 - b^2 \sum(x_i - \bar{x})^2\right)$$

For the FEV1 data the sum of squares due to the regression is $0.074\,389^2 \times 576.352 = 3.189\,37$ and the sum of squares about the regression is $9.438\,68 - 3.189\,37 = 6.249\,31$. There are $20 - 2 = 18$ degrees of freedom, so the variance about the regression is $s^2 = 6.249\,3/18 = 0.347\,18$. The standard error of b is given by

$$\text{SE}(b) = \sqrt{\frac{s^2}{\sum(x_i - \bar{x})^2}} = \sqrt{\frac{0.347\,18}{576.352}} = 0.024\,54 \text{ litre/cm}$$

We have already assumed that the error E is Normally distributed, so b must be, too. The standard error is based on a single sum of squares, so $b/\text{SE}(b)$ is an observation from the t distribution with $n-2$ degrees of freedom (§10.1). We can find a 95% confidence interval for b by taking t standard errors on either side of the estimate. For the example, we have 18 degrees of freedom. From Table 10.1, the 5% point of the t distribution is 2.10, so the 95% confidence interval for b is $0.074\,389 - 2.10 \times 0.024\,54$ to $0.074\,389 + 2.10 \times 0.024\,54$ or 0.02 to 0.13 litres/cm. We can see that FEV1 and height are related, though the slope is not very well estimated.

We can also test the null hypothesis that, in the population, the slope $= 0$ against the alternative that the slope is not equal to 0, a relationship in either direction. The test statistic is $b/\text{SE}(b)$ and if the null hypothesis is true this will be from a t distribution with $n-2$ degrees of freedom. For the example,

$$t = \frac{b}{\text{SE}(b)} = \frac{0.074\,389}{0.024\,54} = 3.03$$

From Table 10.1 this has two tailed probability of less than 0.01. The computer tells us that the probability is about 0.007. Hence the data are inconsistent with the null hypothesis and the data provide fairly good evidence that a relationship exists. If the sample were much larger, we could dispense with the t distribution and use the Standard Normal distribution in its place.

11.6 Using the regression line for prediction

We can use the regression equation to predict the mean or expected Y for any given value of X. This is called the **regression estimate** of Y. We can use this to say whether any individual has an observed Y greater or less than would be expected given X. For example, the predicted FEV1 for students with height 177 cm is $-9.19 + 0.0744 \times 177 = 3.98$ litres. Three subjects had height 177 cm. The first had observed FEV1 of 5.43 litres, 1.45 litres above that expected. The second had a rather low FEV1 of 3.09 litres, 0.89 litres below expectation, while the third with an FEV1 of 4.05 litres was very close to that predicted. We can use this clinically to adjust a measured lung function for height and thus get a better idea of the patient's status. We would, of course, use a much larger sample to establish a precise estimate of the regression equation. We can also use a variant of the method (§17.1) to adjust FEV1 for height in comparing different groups, where we can both remove variation in FEV1 due to variation in height and allow for differences in mean height between the groups. We may wish to do this to compare patients with respiratory disease on different therapies, or to compare subjects exposed to different environmental factors, such as air pollution, cigarette smoking, etc.

As with all sample estimates, the regression estimate is subject to sampling variation. We estimate its precision by standard error and confidence interval in the usual way. The standard error of the expected Y for an observed value x is

$$\text{SE(mean } Y \text{ given } X) = \sqrt{s^2\left(\frac{1}{n} + \frac{(x-\bar{x})^2}{\sum(x_i-\bar{x})^2}\right)}$$

We need not go into the algebraic details of this. It is very similar to that in §11C. For $x = 177$ we have

$$\begin{aligned}\text{SE(mean } Y \text{ given } X = 177) &= \sqrt{0.347\,18^2\left(\frac{1}{20} + \frac{(177-175.38)^2}{576.352}\right)} \\ &= 0.138\end{aligned}$$

This gives a 95% confidence interval of $3.98 - 2.10 \times 0.138$ to $3.98 + 2.10 \times 0.138$ giving 3.69 litres to 4.27 litres. Here 3.98 is the estimate and 2.10 is the 5% point of the t distribution with $n - 2 = 18$ degrees of freedom.

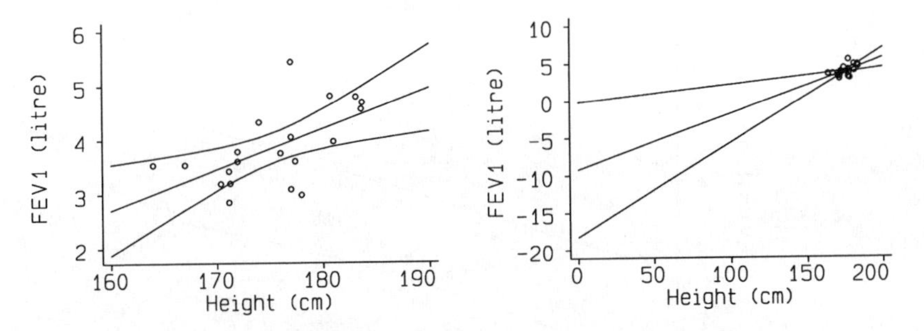

Fig. 11.6. Confidence intervals for the regression estimate

The standard error is a minimum at $x = \bar{x}$, and increases as we move away from $\bar{x}$ in either direction. It can be useful to plot the standard error and 95% confidence interval about the line on the scatter diagram. Figure 11.6 shows this for the FEV1 data. Notice that the lines diverge considerably as we reach the extremes of the data. It is very dangerous to extrapolate beyond the data. Not only do the standard errors become very wide, but we often have no reason to suppose that the straight line relationship would persist if we could.

The intercept a, the predicted value of Y when $X = 0$, is a special case of this. Clearly, we cannot actually have a medical student of height zero and with FEV1 of -9.19 litres. Figure 11.6 also shows the confidence interval for the regression estimate with a much smaller scale, to show the intercept. The confidence interval is very wide at height = 0, and this does not take account of any breakdown in linearity.

We can also use the regression equation of Y on X to predict X from Y. However this is much less accurate than predicting Y from X. For example, if we use the regression of height on FEV1 (Figure 11.5) to predict the FEV1 of subjects with height 177 cm, we get a prediction of 4.21 litres, with standard error 0.255. This is almost twice the standard error obtained from the regression of FEV1 on height. Thus we can see that if in doubt about the choice of outcome and predictor variables, the outcome variable should be the one we wish to predict. Only if there is no possibility of deviations in X fulfilling the assumptions of Normal distribution and uniform variance, and so no way of fitting $X = a + bY$, should we consider predicting X from the regression of Y on X. This might happen if X is fixed in advance, e.g. the dose of a drug.

Rather than predict the expected Y for a given value of X, we may wish to predict the value of Y which we would observe for a given X. In other words we may wish to use the value of X for a subject to estimate that subject's value of Y, a callibration problem. The estimate is the same as the regression estimate, but the standard error is much greater:

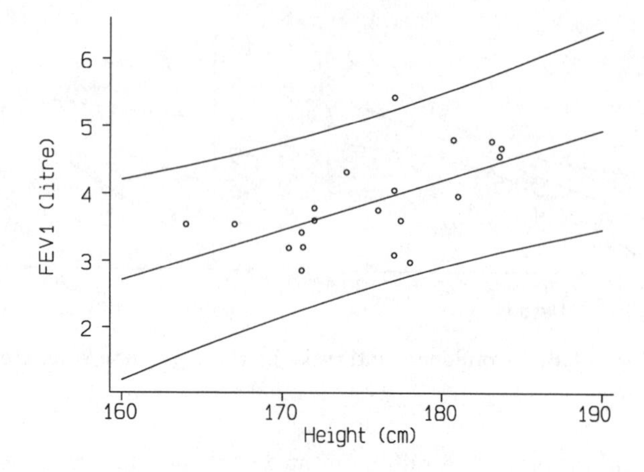

Fig. 11.7. Confidence interval for a further observation

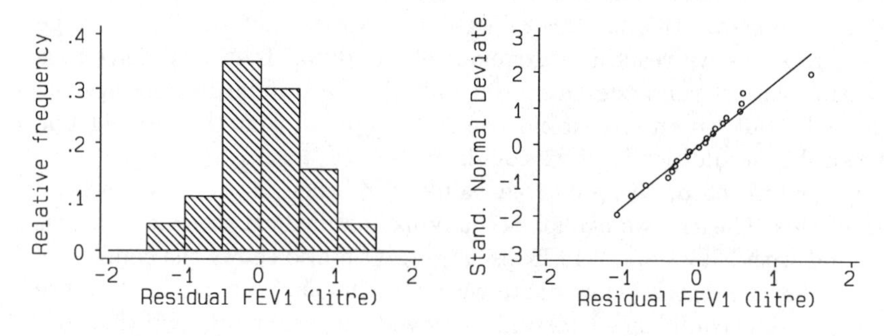

Fig. 11.8. Distribution of residuals for the FEV1 data

$$\text{SE}(Y \text{ given } X) = \sqrt{s^2\left(1 + \frac{1}{n} + \frac{(x-\bar{x})^2}{\sum(x_i-\bar{x})^2}\right)}$$

For a student with a height of 177 cm, the predicted FEV1 is 3.98 litres, with standard error 0.605. Figure 11.7 shows the precision of the prediction of a further observation. As we might expect, the 95% confidence intervals include all but one of the 20 observations. This is only going to be a useful prediction when the residual variance s^2 is small.

11.7 Analysis of residuals

It is often very useful to examine the residuals, the differences between the observed and predicted Y. This is best done graphically. We can assess the assumption of a Normal distribution by looking at the histogram or Normal plot (§7.5). Figure 11.8 shows these for the FEV1 data. The fit is quite good.

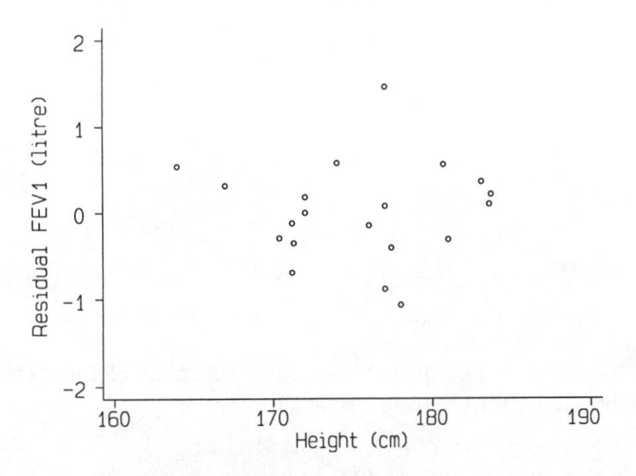

Fig. 11.9. Residuals against height for the FEV1 data

Figure 11.9 shows a plot of residuals against the predictor variable. This plot enables us to examine deviations from linearity. For example, if the true relationship were quadratic, so that Y increases more and more rapidly as X increases, we should see that the residuals are related to X. Large and small X would tend to have positive residuals whereas central values would have negative residuals. Figure 11.9 shows no relationship between the residuals and height, and the linear model seems to be an adequate fit to the data.

Figure 11.9 shows something else, however. One point stands out as having a rather larger residual than the others. This may be an **outlier**, a point which may well come from a different population. It is often difficult to know what to do with such data. At least we have been warned to double check this point for transcription errors. It is all too easy to transpose adjoining digits when transferring data from one medium to another. This may have been the case here, as an FEV1 of 4.53, rather than the 5.43 recorded, would have been more in line with the rest of the data. If this happened at the point of recording, there is not much we can do about it. We could try to measure the subject again, or exclude him and see whether this makes any difference. I think that, on the whole, we should work with all the data unless there are very good reasons for not doing so. I have retained this case here.

11.8 Deviations from assumptions in regression

Both the appropriateness of the method of least squares and the use of the t distribution for confidence intervals and tests of significance depend on the assumption that the residuals are Normally distributed. This assumption

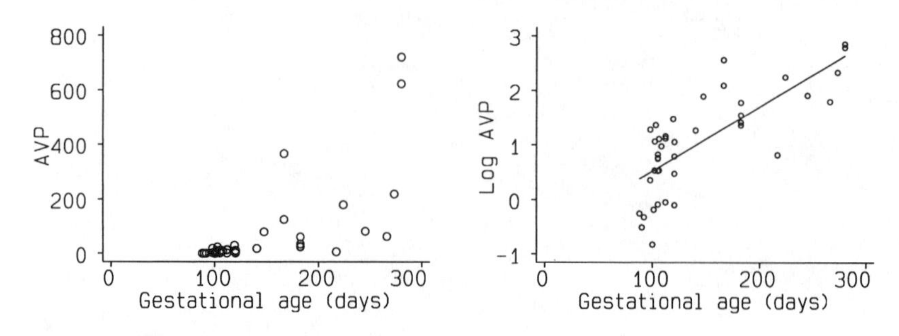

Fig. 11.10. Data which do not meet the conditions of the method of least squares, before and after log transformation

is easily met, for the same reasons that it is in the paired t test (§10.2). The removal of the variation due to X tends to remove some of the variation between individuals, leaving the measurement error. Problems can arise, however, and it is always a good idea to plot the original scatter diagram and the residuals to check that there are no gross departures from the assumptions of the method. Not only does this help preserve the validity of the statistical method used, but it may also help us learn more about the structure of the data.

Figure 11.10 shows the relationship between gestational age and cord blood levels of AVP, the antidiuretic hormone, in a sample of male fetuses. The variability of the outcome variable AVP depends on the actual value of the variable, being larger for large values of AVP. The assumptions of the method of least squares do not apply. However, we can use a transformation as we did for the comparison of means in §10.4. Figure 11.10 also shows the data after AVP has been log transformed, together with the least squares line.

11.9 Correlation

The regression method tells us something about the nature of the relationship between two variables, how one changes with the other, but it does not tell us how close that relationship is. To do this we need a different coefficient, the correlation coefficient. The correlation coefficient is based on the sum of products about the mean of the two variables, so we shall start by considering the properties of the sum of products and why it is a good indicator of the closeness of the relationship.

Take the scatter diagram of Figure 11.1 and draw two new axes through the mean point (Figure 11.11). The distances of the points from these axes represent the deviations from the mean. In the top right section of Figure 11.11, the deviations from the mean of both variables, FEV1 and

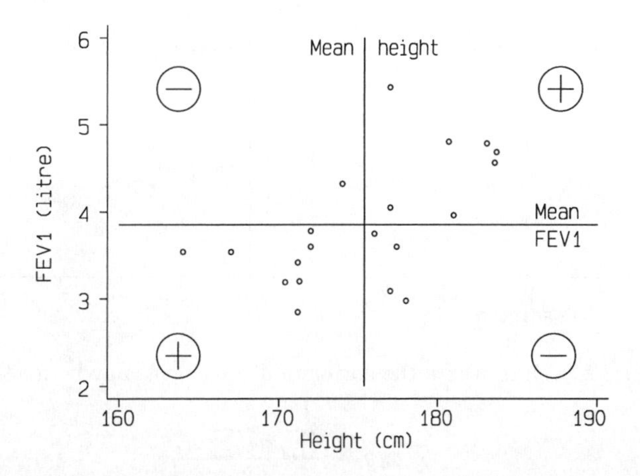

Fig. 11.11. Scatter diagram with axes through the mean point

height, are positive. Hence, their products will be positive. In the bottom left section, the deviations from the mean of the two variables will both be negative. Again, their product will be positive. In the top left section of Figure 11.11, the deviations of FEV1 will be positive, and the deviation of height from its mean will be negative. The product of these will be negative. In the bottom right section, the product will again be negative. So in Figure 11.11 nearly all these products will be positive, and their sum will be positive. We say that there is a **positive correlation** between the two variables; as one increases so does the other. If one variable decreased as the other increased, we would have a scatter diagram where most of the points lay in the top left and bottom right sections. In this case the sum of the products is negative and there is a **negative correlation** between the variables. When the two variables are not related, we have a scatter diagram with roughly the same number of points in each of the sections. In this case, there are as many positive as negative products, and the sum is zero. There is **zero correlation** or **no correlation**. The variables are said to be **uncorrelated**.

The value of the sum of products depends on the limits in which the two variables are measured. We can find a dimensionless coefficient if we divide the sum of products by the square roots of the sums of squares of X and Y. This gives us the **product moment correlation coefficient**, or the **correlation coefficient** for short, usually denoted by r.

If the n pairs of observations are denoted by (x_i, y_i), then r is given by

$$r = \frac{\sum(x_i - \bar{x})(y_i - \bar{y})}{\sqrt{\left(\sum(x_i - \bar{x})^2\right)\left(\sum(y_i - \bar{y})^2\right)}}$$

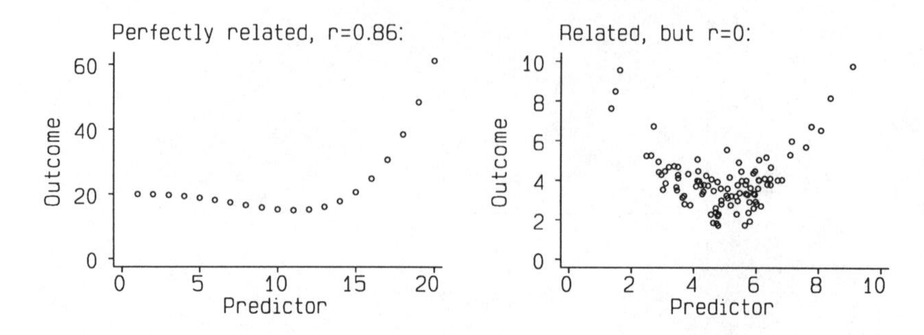

Fig. 11.12. Data where the correlation coefficient may be misleading

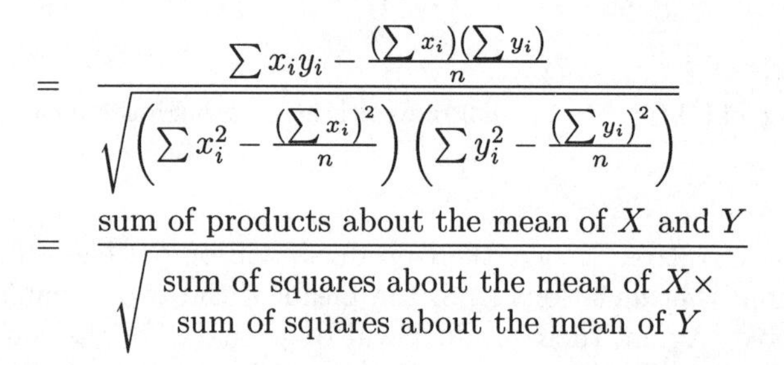

$$= \frac{\sum x_i y_i - \frac{(\sum x_i)(\sum y_i)}{n}}{\sqrt{\left(\sum x_i^2 - \frac{(\sum x_i)^2}{n}\right)\left(\sum y_i^2 - \frac{(\sum y_i)^2}{n}\right)}}$$

$$= \frac{\text{sum of products about the mean of } X \text{ and } Y}{\sqrt{\begin{array}{c}\text{sum of squares about the mean of } X \times \\ \text{sum of squares about the mean of } Y\end{array}}}$$

For the FEV1 and height we have

$$r = \frac{42.8744}{\sqrt{576.352 \times 9.43868}} = 0.58$$

The effect of dividing the sum of products by the root sum of squares of deviations of each variable is to make the correlation coefficient lie between -1.0 and $+1.0$. When all the points lie exactly on a straight line such that Y increases as X increases, $r = 1$. This can be shown by putting $a + bx_i$ in place of y_i in the equation for r. When all the points lie exactly on a straight line with negative slope, $r = -1$. When there is no relationship at all, $r = 0$, because the sum of products is zero. The correlation coefficient describes the closeness of the linear relationship between two variables. It does not matter which variable we take to be Y and which to be X. There is no choice of predictor and outcome variable, as there is in regression.

The correlation coefficient measures how close the points are to a straight line. Even if there is a perfect mathematical relationship between X and Y, the correlation coefficient will not be exactly 1 unless this is of the form $Y = a + bX$. For example, Figure 11.12 shows two variables which are perfectly related but have $r = 0.86$. Figure 11.12 also shows two variables which are clearly related but have zero correlation. This shows again the importance of plotting the data and not relying on summary statistics such as the correlation coefficient only. In practice, relationships like those of

Figures 11.12 are rare in medical data, although the possibility is always there. More often, there is so much random variation that it is not easy to discern any relationship at all.

The correlation coefficient r is related to the regression coefficient b in a simple way. If $Y = a + bX$ is the regression of Y on X, and $X = a' + b'Y$ is the regression of X on Y, then $r^2 = bb'$. This arises from the formulae for r and b. For the FEV1 data, $b = 0.074\,389$ and $b' = 4.542\,4$, so $bb' = 0.074\,389 \times 4.542\,4 = 0.337\,90$, the square root of which is 0.581 29, the correlation coefficient. We also have

$$\begin{aligned} r^2 &= \frac{(\text{sum of products about mean})^2}{\text{sum of squares of } X \times \text{sum of squares of } Y} \\ &= \frac{(\text{sum of products about mean})^2}{(\text{sum of squares of X})^2} \times \frac{\text{sum of squares of } X}{\text{sum of squares of } Y} \\ &= \frac{b^2 \times \text{sum of squares of } X}{\text{sum of squares of } Y} \end{aligned}$$

This is the proportion of variability explained, described in §11.5.

11.10 Significance test for the correlation coefficient

The correlation coefficient is unusual among sample statistics in having a most awkward sampling distribution. Even when X and Y are both Normally distributed, r does not itself approach a Normal distribution until the sample size is in the thousands. Furthermore, its distribution is rather sensitive to deviations from the Normal in X and Y. However, Fisher's z transformation gives a Normally distributed variable whose mean and variance are known in terms of the population correlation coefficient which we wish to estimate. From this a confidence interval can be found. **Fisher's z transformation** is

$$z = \frac{1}{2}\log_e\left(\frac{1+r}{1-r}\right)$$

which follows a Normal distribution with mean

$$z_\rho = \frac{1}{2}\log_e\left(\frac{1+\rho}{1-\rho}\right) + \frac{\rho}{2(n-1)}$$

and variance $1/(n-3)$ approximately, where ρ is the population correlation coefficient and n is the sample size. The 95% confidence interval for z will be approximately $z \pm 1.96\sqrt{1/(n-3)}$. For the FEV1 data, $r = 0.58$ and $n = 20$.

$$z = \frac{1}{2}\log_e\left(\frac{1+0.58}{1-0.58}\right) = 0.6625$$

The 95% confidence interval will be $0.6625 \pm 1.96\sqrt{1/17}$, giving 0.1871 to 1.1379. The transformation back from the z scale to the correlation coefficient scale is

$$r = \frac{\exp(2z)-1}{\exp(2z)+1}$$

so for the lower limit we have

$$\frac{\exp(2 \times 0.1871)) - 1}{\exp(2 \times 0.1871) + 1} = 0.18$$

and for the upper limit

$$\frac{\exp(2 \times 1.1379)) - 1}{\exp(2 \times 1.1379) + 1} = 0.81$$

and the 95% confidence interval is 0.18 to 0.81. This is very wide, reflecting the sampling variation which the correlation coefficient has for small samples. Correlation coefficients must be treated with some caution when derived from small samples.

When it comes to testing the null hypothesis that $r = 0$, or that there is no linear relationship, things are much simpler. The test is numerically equivalent to testing the null hypothesis that $b = 0$, and the test is valid provided at least one of the variables is from a Normal distribution. This condition is the same as that for testing b, where the residuals in the Y direction must be Normal. If $b = 0$, the residuals in the Y direction are simply the deviations from the mean, and these will only be Normally distributed if Y is. If the condition is not met, we can use a transformation (§11.8), or one of the rank correlation methods (§12.4–5).

Because the correlation coefficient does not depend on the means or variances of the observations, the distribution of the sample correlation coefficient when the population coefficient is zero is easy to tabulate. Table 11.2 shows the correlation coefficient at the 5% and 1% level of significance. For the example we have $r = 0.58$ from 20 observations. The 1% point for 20 observations is 0.56, so we have $P < 0.01$, and the correlation is unlikely to have arisen if there were no linear relationship in the population. Note that the values of r which can arise by chance with small samples are quite high. With 10 points r would have to be greater than 0.63 to be significant. On the other hand with 1000 points very small values of r, as low as 0.06, will be significant.

Table 11.2. Two-sided 5% and 1% points of the distribution of the correlation coefficient, r, under the null hypothesis

n	5%	1%	n	5%	1%	n	5%	1%
3	1.00	1.00	16	0.50	0.62	29	0.37	0.47
4	0.95	0.99	17	0.48	0.61	30	0.36	0.46
5	0.88	0.96	18	0.47	0.59	40	0.31	0.40
6	0.81	0.92	19	0.46	0.58	50	0.28	0.36
7	0.75	0.87	20	0.44	0.56	60	0.25	0.33
8	0.71	0.83	21	0.43	0.55	70	0.24	0.31
9	0.67	0.80	22	0.42	0.54	80	0.22	0.29
10	0.63	0.77	23	0.41	0.53	90	0.21	0.27
11	0.60	0.74	24	0.40	0.52	100	0.20	0.25
12	0.58	0.71	25	0.40	0.51	200	0.14	0.18
13	0.55	0.68	26	0.39	0.50	500	0.09	0.12
14	0.53	0.66	27	0.38	0.49	1000	0.06	0.08
15	0.51	0.64	28	0.37	0.48			

n = number of observations.

The ease of the significance test compared to the relative complexity of the confidence interval calculation has meant that in the past a significance test was usually given for the correlation coefficient. The increasing availability of computers with well-written statistical packages should lead to correlation coefficients appearing with confidence intervals in the future.

11.11 Uses of the correlation coefficient

The correlation coefficient has several uses. Using Table 11.2, it provides a simple test of the null hypothesis that the variables are not linearly related, with less calculation than the regression method. It is also useful as a summary statistic for the strength of relationship between two variables. This is of great value when we are considering the inter-relationships between a large number of variables. We can set up a square array of the correlations of each pair of variables, called the **correlation matrix**. Examination of the correlation matrix can be very instructive, but we must bear in mind the possibility of non-linear relationships. There is no substitute for plotting the data. The correlation matrix also provides the starting point for a number of methods for dealing with a large number of variables simultaneously.

Of course, for the reasons discussed in Chapter 3, the fact that two variables are correlated does not mean that one causes the other.

11.12 Using repeated observations

In clinical research we are often able to take several measurements on the same patient. We may want to investigate the relationship between two variables, and take pairs of readings with several pairs from each of several

Table 11.3. Simulated data showing 10 pairs of measurements of two independent variables for four subjects

	Subject 1		Subject 2		Subject 3		Subject 4	
	x	y	x	y	x	y	x	y
	47	51	49	52	51	46	63	64
	46	53	50	56	46	48	70	62
	50	57	42	46	46	47	63	66
	52	54	48	52	45	55	58	64
	46	55	60	53	52	49	59	62
	36	53	47	49	54	61	61	62
	47	54	51	52	48	53	67	58
	46	57	57	50	47	48	64	62
	36	61	49	50	47	50	59	67
	44	57	49	49	54	44	61	59
Means	45.0	55.2	50.2	50.9	49.0	50.1	62.5	62.6
	$r = -0.33$		$r = 0.49$		$r = 0.06$		$r = -0.39$	
	P = 0.35		P = 0.15		P = 0.86		P = 0.27	

patients. The analysis of such data is quite complex. This is because the variability of measurements made on different subjects is usually much greater than the variability between measurements on the same subject, and we must take these two kinds of variability into account. What we must *not* do is to put all the data together, as if they were one sample.

Consider the simulated data of Table 11.3. The data were generated from random numbers, and there is no relationship between X and Y at all. First values of X and Y were generated for each 'subject', then a further random number was added to make the individual 'observation'. For each subject separately, there was no significant correlation between X and Y. For the subject means, the correlation coefficient was $r = 0.77$, P = 0.23. However, if we put all 40 observations together we get $r = 0.53$, P = 0.0004. Even though the coefficient is smaller than that between subject means, because it is based on 40 pairs of observations rather than 4 it becomes significant. The data are plotted if Figure 11.13, with three other simulations. As the null hypothesis is always true in these simulated data, the correlations for each 'subject' and for the means are not significant. Because the numbers of observations are small, they vary greatly. As Table 11.2 shows, large correlation coefficients can arise by chance in small samples. However, the overall correlation is 'significant' in three of the four simulations, though in different directions.

We only have four subjects and only four points. By using the repeated data, we are not increasing the number of subjects, but the statistical calculation is done as if we have, and so the number of degrees of freedom for the significance test is incorrectly increased and a spurious significant correlation produced.

There are two simple ways to approach this type of data, and which is

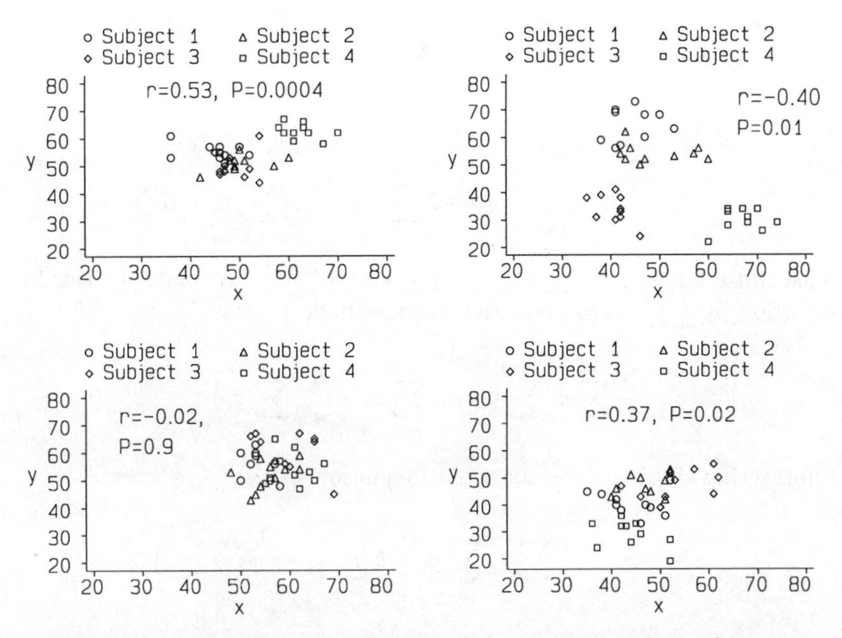

Fig. 11.13. Simulations of 10 pairs of observations on four subjects

chosen depends on the question being asked. If we want to know whether subjects with a high value of X tend to have a high value of Y also, we use the subject means and find the correlation between them. If we have different numbers of observations for each subject, we can use a weighted analysis, weighted by the number of observations for the subject. If we want to know whether changes in one variable in the same subject are parallelled by changes in the other, we need to use multiple regression, taking subjects out as a factor (§17.1, §17.6). In either case, we should not mix observations from different subjects indiscriminately.

11A Appendix: The least squares estimates

This section requires knowledge of calculus. We want to find a and b so that the sum of squares about the line $y = a + bx$ is a minimum. We therefore want to minimize $\sum(y_i - a - bx_i)^2$. This will have a minimum when the partial differentials with respect to a and b are both zero.

$$\begin{aligned}
\frac{\partial \sum(y_i - a - bx_i)^2}{\partial a} &= \sum 2(y_i - a - bx_i)(-1) \\
&= -2\sum y_i + 2a\sum 1 + 2b\sum x_i \\
&= -2\sum y_i + 2an + 2b\sum x_i
\end{aligned}$$

This must equal 0 so $\sum y_i = na + b\sum x_i$.

$$\begin{aligned}\frac{\partial \sum(y_i - a - bx_i)^2}{\partial b} &= \sum 2(y_i - a - bx_i)(-x_i) \\ &= -2\sum x_i y_i + 2a\sum x_i + 2b\sum x_i^2\end{aligned}$$

This must equal 0 so $\sum x_i y_i = a\sum x_i + b\sum x_i^2$ We multiply the first equation by $\frac{1}{n}\sum x_i$, to make the coefficients of a equal.

$$\frac{1}{n}\sum x_i \sum y_i = a\sum x_i + \frac{b}{n}\left(\sum x_i\right)^2$$

Subtracting this from the the second equation we get

$$\sum x_i y_i - \frac{1}{n}\sum x_i \sum y_i = b\sum x_i^2 - \frac{b}{n}\left(\sum x_i\right)^2$$

$$\sum x_i y_i - \frac{1}{n}\sum x_i \sum y_i = b\left(\sum x_i^2 - \frac{1}{n}\left(\sum x_i\right)^2\right)$$

This gives us

$$b = \frac{\sum x_i y_i - \frac{\sum x_i \sum y_i}{n}}{\sum x_i^2 - \frac{(\sum x_i)^2}{n}}$$

If we divide the first equation by n we get the formula for a:

$$\frac{1}{n}\sum y_i = a + \frac{b}{n}\sum x_i$$

$$a = \bar{y} - b\bar{x}$$

11B Appendix: Variance about the regression line

We find the formula for the variance about the regression line, s^2, as follows. The regression model is $Y = a + bX + E$, and a and b are constants. We are predicting Y for given X, so there is no random variation in X; all the random variation is in E. Hence $s^2 = \text{VAR}(Y \text{ given } X) = \text{VAR}(E)$. We have seen in §11.2 that the error E is the random variable which stands for the deviations from the line in the Y direction. These deviations are $y_i - (a + bx_i)$, since $a + bx_i$ is the Y value for the line at $X = x_i$. The sum of squares of these deviations is found by a mathematical trick, replacing a by $\bar{y} - b\bar{x}$.

$$\begin{aligned}
\sum (y_i - (a + bx_i))^2 &= \sum (y_i - (\bar{y} - b\bar{x} + bx_i))^2 \\
&= \sum (y_i - \bar{y} - (bx_i - b\bar{x}))^2 \\
&= \sum (y_i - \bar{y} - b(x_i - \bar{x}))^2 \\
&= \sum ((y_i - \bar{y})^2 - 2b(y_i - \bar{y})(x_i - \bar{x}) + b^2(x_i - \bar{x})^2) \\
&= \sum (y_i - \bar{y})^2 - 2b\sum (y_i - \bar{y})(x_i - \bar{x}) + b^2\sum (x_i - \bar{x})^2 \\
&= \sum (y_i - \bar{y})^2 - 2b \times b\sum (x_i - \bar{x})^2 + b^2\sum (x_i - \bar{x})^2 \\
&= \sum (y_i - \bar{y})^2 - b^2\sum (x_i - \bar{x})^2
\end{aligned}$$

This is because $b = \left(\sum (y_i - \bar{y})(x_i - \bar{x})\right) / \left(\sum (x_i - \bar{x})^2\right)$, so $b\sum (y_i - \bar{y})(x_i - \bar{x}) = \sum (x_i - \bar{x})^2$.

11C Appendix: The standard error of b

To find the standard error of b, we must bear in mind that in our regression model all the random variation is in Y. We first rewrite the sum of products:

$$\begin{aligned}
\sum (x_i - \bar{x})(y_i - \bar{y}) &= \sum ((x_i - \bar{x})y_i - (x_i - \bar{x})\bar{y}) \\
&= \sum (x_i - \bar{x})y_i - \sum (x_i - \bar{x})\bar{y} \\
&= \sum (x_i - \bar{x})y_i - \bar{y}\sum (x_i - \bar{x}) \\
&= \sum (x_i - \bar{x})y_i
\end{aligned}$$

This is because $\bar{y}$ is the same for all i and so comes out of the summation, and $\sum (x_i - \bar{x}) = 0$. We now find the variance of the sampling distribution of b by

$$\begin{aligned}
\text{VAR}(b) &= \text{VAR}\left(\frac{\sum (x_i - \bar{x})(y_i - \bar{y})}{\sum (x_i - \bar{x})^2}\right) \\
&= \text{VAR}\left(\frac{\sum (x_i - \bar{x})y_i}{\sum (x_i - \bar{x})^2}\right) \\
&= \frac{1}{\left(\sum (x_i - \bar{x})^2\right)^2}\text{VAR}\sum (x_i - \bar{x})y_i
\end{aligned}$$

The variance of a constant times a random variable is the square of the constant times the variance of the random variable (§6.6). The x_i are constants, not random variables, so

$$\text{VAR}(b) = \frac{1}{\left(\sum(x_i - \bar{x})^2\right)^2} \sum(x_i - \bar{x})^2 \text{VAR}(y_i)$$

$\text{VAR}(y_i)$ is the same for all y_i, say $\text{VAR}(y_i) = s^2$. Hence

$$\text{VAR}(b) = \frac{s^2}{\sum(x_i - \bar{x})^2}$$

The standard error of b is the square root of this.

11M Multiple choice questions 57 to 61

(Each branch is either true or false)

57. In Figure 11.14(a):

(a) predictor and outcome are independent;

(b) predictor and outcome are uncorrelated;

(c) the correlation between predictor and outcome is less than 1;

(d) predictor and outcome are perfectly related;

(e) the relationship is best estimated by simple linear regression.

58. In Figure 11.14(b):

(a) predictor and outcome are independent random variables;

(b) the correlation between predictor and outcome is close to zero;

(c) outcome increases as predictor increases;

(d) predictor and outcome are linearly related;

(e) the relationship could be made linear by a logarithmic transformation of the outcome.

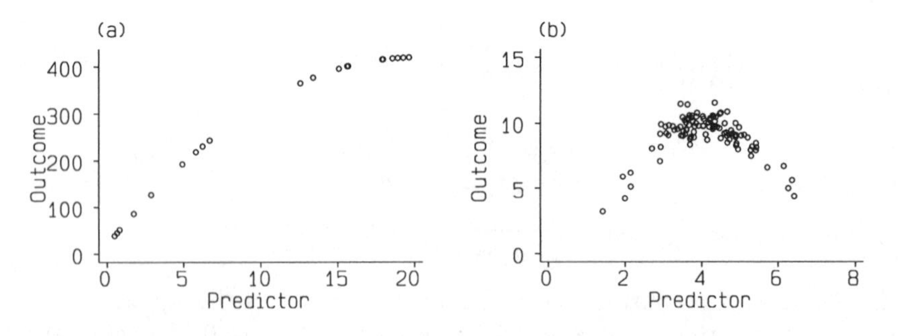

Fig. 11.14. Scatter diagrams

59. A simple linear regression equation:

(a) describes a line which goes through the origin;

(b) describes a line with zero slope;

(c) is not affected by changes of scale;

(d) describes a line which goes through the mean point;

(e) is affected by the choice of dependent variable.

60. If a t test is used to test the significance of the slope of a regression line:

(a) deviations from the line in the independent variable must follow a Normal distribution;

(b) deviations from the line in the dependent variable must follow a Normal distribution;

(c) the variance about the line is assumed to be the same throughout the range of the predictor variable;

(d) the y variable must be log transformed;

(e) all the points must lie on the line.

61. The product moment correlation coefficient, r:

(a) must lie between -1 and +1;

(b) can only have a valid significance test carried out when at least one of the variables is from a Normal distribution;

(c) is 0.5 when there is no relationship;

(d) depends on the choice of dependent variable;

(e) measures the magnitude of the change in one variable associated with a change in the other.

11E Exercise: Comparing two regression lines

Table 11.4 and Figure 11.15 show the PEFR and heights of a sample of male and female medical students. Table 11.5 shows the sums of squares and products for these data.

1. Estimate the slopes of the regression lines for females and males.
2. Estimate the standard errors of the slopes.
3. Find the standard error for the difference between the slopes, which are independent. Calculate a 95% confidence interval for the difference.
4. Use the standard error to test the null hypothesis that the slopes are the same in the population from which these data come.

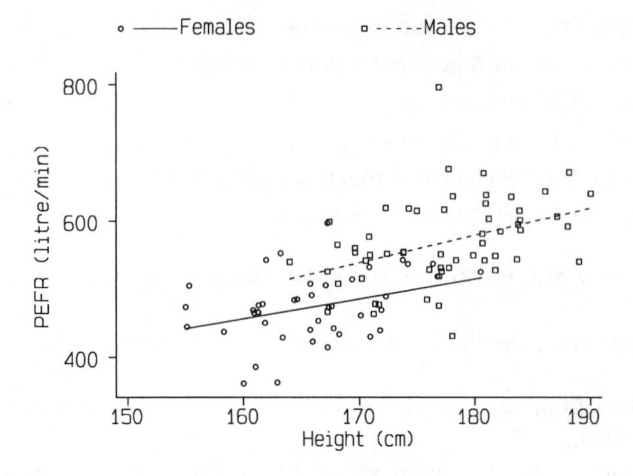

Fig. 11.15. PEFR and height for female and male medical students

Table 11.4. Height and PEFR in a sample of medical students

Females						Males					
Ht	PEFR	Ht	PEFR	Ht	PEFR	Ht	PEFR	Ht	PEFR	Ht	PEFR
148	418	162	439	170	505	162	578	177	650	182	550
152	400	162	495	170	415	164	572	177	640	182	592
152	470	163	460	171	455	168	555	177	528	183	660
156	405	163	480	171	482	169	600	178	655	183	550
158	405	163	416	175	470	169	650	178	560	183	560
158	453	163	492	175	535	170	600	178	495	183	560
158	355	163	512	175	545	173	580	178	560	185	571
159	495	164	490	175	500	174	516	178	657	185	525
159	360	165	460	176	479	174	595	179	615	185	598
160	435	165	535	177	425	174	450	179	620	186	570
160	513	165	480	177	473	175	493	179	595	187	665
160	494	166	500	180	530	175	565	180	483	187	700
161	438	166	450	180	635	175	548	180	648	187	690
161	410	167	455	181	585	176	540	180	645	188	610
161	455	167	425	182	620	176	540	181	503	190	610
161	370	168	430	183	590	176	580	181	590	194	530
161	457	168	490	183	540	176	570	181	515	197	640
161	540	169	580	187	700	177	550	182	523	199	570
161	465	169	430	190	665						
161	435	169	572	192	640						
162	510	170	480								

Table 11.5. Summary statistics for height and PEFR in a sample of medical students

	Females	Males
Number	43	58
Sum of squares, height	1 444.6	2 267.5
Sum of squares, PEFR	101 107.6	226 873.5
Sum of products about mean	4 206.9	9 045.4

12

Methods based on rank order

12.1 Non-parametric methods

In Chapters 10 and 11 I described a number of methods of analysis which relied on the assumption that the data came from a Normal distribution. To be more precise, we could say the data come from one of the Normal family of distributions, the particular Normal distribution involved being defined by its mean and standard deviation, the parameters of the distribution. These methods are called **parametric** because we estimate the parameters of the underlying Normal distribution. Methods which do not assume a particular family of distributions for the data are said to be **non-parametric**. In this and the next chapter I shall consider some non-parametric tests of significance. There are many others, but these will illustrate the general principal. We have already met one non-parametric test, the sign test (§9.2). The large sample Normal test could also be regarded as non-parametric.

It is useful to distinguish between three types of measurements scales. On an **interval scale**, the size of the difference between two values on the scale has a consistent meaning. For example, the difference in temperature between 1°C and 2°C is the same as the difference between 31°C and 32°C. On an **ordinal scale**, observations are ordered, but differences may not have a meaning. For example, anxiety is often measured using sets of questions, the number of positive answers giving the anxiety scale. A set of 36 questions would give a scale from 0 to 36. The difference in anxiety between scores of 1 and 2 is not necessarily the same as the difference between scores 31 and 32. On a **nominal scale**, we have a qualitative or categorical variable, where individuals are grouped but not necessarily ordered. Eye colour is a good example. When categories are ordered, we can treat the scale as either ordered or nominal, as appropriate.

All the methods of Chapters 10 and 11 apply to interval data, being based on differences of observations from the mean. Most of the methods in this chapter apply to ordinal data. Any interval scale which does not meet the requirements of Chapters 10 and 11 may be treated as ordinal, since it, is, of course, ordered. This is the more common application in medical work.

General texts such as Armitage and Berry (1987), Snedecor and Cochran (1980) and Colton (1974) tend not to go into a lot of detail about rank and related methods, and more specialized books are needed (Siegel 1956; Conover 1980).

12.2 The Mann–Whitney U test

This is the non-parametric analogue of the two sample t test (§10.3). It works like this. Consider the following artifical data showing observations of a variable in two independent groups, A and B:

A	7	4	9	17
B	11	6	21	14

We want to know whether there is any evidence that A and B are drawn from populations with different levels of the variable. The null hypothesis is that there is no tendency for members of one population to exceed members of the other. The alternative is that there is such a tendency, in one direction or the other.

First we arrange the observations in ascending order, i.e. we rank them:

4	6	7	9	11	14	17	21
A	B	A	A	B	B	A	B

We now choose one group, say A. For each A, we count how many Bs precede it. For the first A, 4, no Bs precede. For the second, 7, one B precedes, for the third A, 9, one B, for the fourth, 17, three Bs. We add these numbers of preceding Bs together to give $U = 0 + 1 + 1 + 3 = 5$. Now, if U is very small, nearly all the As are less than nearly all the Bs. If U is large, nearly all As are greater than nearly all Bs. Moderate values of U mean that As and Bs are mixed. The minimum U is 0, when all Bs exceed all As, and maximum U is $n_1 \times n_2$ when all As exceed all Bs. The magnitude of U has a meaning, because U/n_1n_2 is an estimate of the probability that an observation drawn at random from population A would exceed an observation drawn at random from population B.

There is another possible U, which we will call U', obtained by counting the number of As before each B, rather than the number of Bs before each A. This would be $1 + 3 + 3 + 4 = 11$. The two possible values of U and U' are related by $U + U' = n_1n_2$. So we subtract U' from n_1n_2 to give $4 \times 4 - 11 = 5$.

If we know the distribution of U under the null hypothesis that the samples come from the same population, we can say with what probability these data could have arisen if there were no difference. We can carry out

Table 12.1. Distribution of the Mann–Whitney U statistic, for two samples of size 4

U	probability	U	probability	U	probability
0	0.014	6	0.100	12	0.071
1	0.014	7	0.100	13	0.043
2	0.029	8	0.114	14	0.029
3	0.043	9	0.100	15	0.014
4	0.071	10	0.100	16	0.014
5	0.071	11	0.071		

the test of significance. The distribution of U under the null hypothesis can be found easily. The two sets of four observations can be arranged in 70 different ways, from AAAABBBB to BBBBAAAA (8!/4!4!, §6A). Under the null hypothesis these arrangements are all equally likely and, hence, have probability 1/70. Each has its value of U, from 0 to 16, and by counting the number of arrangements which give each value of U we can find the probability of that value. For example, $U = 0$ only arises from the order AAAABBBB and so has probability $1/70 = 0.014$. $U = 1$ only arises from AAABABBB and so has probability $1/70 = 0.014$ also. $U = 2$ can arise in two ways: AAABBABB and AABAABBB. It has probability 2/70 = 0.029. The full set of probabilities is shown in Table 12.1.

We apply this to the example. For groups A and B, $U = 5$ and the probability of this is 0.071. As we did for the sign test (§9.2) we consider the probability of more extreme values of U, $U = 5$ or less, which is 0.071 + 0.071 + 0.043 + 0.029 + 0.014 + 0.014 = 0.242. This gives a one sided test. For a two sided test, we must consider the probabilities of a difference as extreme in the opposite direction. We can see from Table 12.1 that the distribution of U is symmetrical, so the probability of an equally extreme value in the opposite direction is also 0.242, hence the two sided probability is 0.242 + 0.242 = 0.484. This is clearly likely to have happened by chance and so the two samples could have come from the same population.

In practice, there is no need to carry out the summation of probabilities described above, as these are already tabulated. Table 12.2 shows the 5% points of U for each combination of sample sizes n_1 and n_2 up to 20. For our groups A and B, $U = 5$, we find the $n_2 = 4$ column and the $n_1 = 4$ row. From this we see that the 5% point for U is 0, and so $U = 5$ is not significant. If we had calculated the larger of the two values of U, 11, we can use Table 12.2 by finding the lower value, $n_1 n_2 - U = 16 - 11 = 5$.

We can now turn to the practical analysis of some real data. Consider the biceps skinfold thickness data of Table 10.4, reproduced as Table 12.3. We will analyse these using the Mann–Whitney U test. Denote the Crohn's disease group by A and the coeliac group by B. The joint order is as follows:

Table 12.2. Two sided 5% points for the distribution of the smaller value of U in the Mann–Whitney U test

	n_2																		
n_1	2	3	4	5	6	7	8	9	10	11	12	13	14	15	16	17	18	19	20
2	–	–	–	–	–	–	0	0	0	0	1	1	1	1	1	2	2	2	2
3	–	–	–	0	1	1	2	2	3	3	4	4	5	5	6	6	7	7	8
4	–	–	0	1	2	3	4	4	5	6	7	8	9	10	11	11	12	13	13
5	–	0	1	2	3	5	6	7	8	9	11	12	13	14	15	17	18	19	20
6	–	1	2	3	5	6	8	10	11	13	14	16	17	19	21	22	24	25	27
7	–	1	3	5	6	8	10	12	14	16	18	20	22	24	26	28	30	32	34
8	0	2	4	6	8	10	13	15	17	19	22	24	26	29	31	34	36	38	41
9	0	2	4	7	10	12	15	17	20	23	26	28	31	34	37	39	42	45	48
10	0	3	5	8	11	14	17	20	23	26	29	33	36	39	42	45	48	52	55
11	0	3	6	9	13	16	19	23	26	30	33	37	40	44	47	51	55	58	62
12	1	4	7	11	14	18	22	26	29	33	37	41	45	49	53	57	61	65	69
13	1	4	8	12	16	20	24	28	33	37	41	45	50	54	59	63	67	72	76
14	1	5	9	13	17	22	26	31	36	40	45	50	55	59	64	67	74	78	83
15	1	5	10	14	19	24	29	34	39	44	49	54	59	64	70	75	80	85	90
16	1	6	11	15	21	26	31	37	42	47	53	59	64	70	75	81	86	92	98
17	2	6	11	17	22	28	34	39	45	51	57	63	67	75	81	87	93	99	105
18	2	7	12	18	24	30	36	42	48	55	61	67	74	80	86	93	99	106	112
19	2	7	13	19	25	32	38	45	52	58	65	72	78	85	92	99	106	113	119
20	2	8	13	20	27	34	41	48	55	62	69	76	83	90	98	105	112	119	127

If U is less than or equal to the tabulated value the difference is significant.

Table 12.3. Biceps skinfold thickness (mm) in two groups of patients

Crohn's Disease				Coeliac Disease	
1.8	2.8	4.2	6.2	1.8	3.8
2.2	3.2	4.4	6.6	2.0	4.2
2.4	3.6	4.8	7.0	2.0	5.4
2.5	3.8	5.6	10.0	2.0	7.6
2.8	4.0	6.0	10.4	3.0	

1.8	1.8	2.0	2.0	2.0	2.2	2.4	2.5	2.8	2.8
A	B	B	B	B	A	A	A	A	A

3.0	3.2	3.6	3.8	3.8	4.0	4.2	4.2	4.4	4.8
B	A	A	A	B	A	B	A	A	A

5.4	5.6	6.0	6.2	6.6	7.0	7.6	10.0	10.4
B	A	A	A	A	A	B	A	A

Let us count the As before each B. Immediately we have a problem. The first A and the first B have the same value. Does the first A come before the first B or after it? We resolve this dilemma by counting one half for the tied A. The ties between the second, third and fourth Bs do not matter, as we can count the number of As before each without difficulty. We have for the U statistic:

$$U = 0.5 + 1 + 1 + 1 + 6 + 8.5 + 10.5 + 13 + 18 = 59.5$$

This is the lower value, since $n_1 n_2 = 9 \times 20 = 180$ and so the middle value is 90. We can therefore refer it to Table 12.2. The critical value at the 5% level for groups size 9 and 20 is 48, which our value exceeds. Hence the difference is not significant at the 5% level and the data are consistent with the null hypothesis that there is no tendency for members of one population to exceed members of the other. This is the same as the result of the t test of §10.4.

For larger values of n_1 and n_2 calculation of U can be rather tedious. A simple formula for U can be found using the ranks. The rank of the lowest observation is 1, of the next is 2, and so on. If a number of observations are tied, each having the same value and hence the same rank, we give each the average of the ranks they would have were they ordered. For example, in the skinfold data the first two observations are each 1.8. They each receive rank $(1+2)/2 = 1.5$. The third, fourth and fifth are tied at 2.0, giving each of them rank $(3+4+5)/3 = 4$. The sixth, 2.2, is not tied and so has rank 6. The ranks for the skinfold data are as follows:

skinfold	1.8	1.8	2.0	2.0	2.0	2.2	2.4	2.5	2.8	2.8
group	A	B	B	B	B	A	A	A	A	A
rank	1.5	1.5	4	4	4	6	7	8	9.5	9.5
		r_1	r_2	r_3	r_4					

skinfold	3.0	3.2	3.6	3.8	3.8	4.0	4.2	4.2	4.4	4.8
group	B	A	A	A	B	A	A	B	A	A
rank	11	12	13	14.5	14.5	16	17.5	17.5	19	20
	r_5				r_6			r_7		

skinfold	5.4	5.6	6.0	6.2	6.6	7.0	7.6	10.0	10.4
group	B	A	A	A	A	A	B	A	A
rank	21	22	23	24	2	2	27	28	29
	r_8						r_9		

We denote the ranks of the B group by $r_1, r_2, \ldots, r_{n_1}$. The number of As preceding the first B must be $r_1 - 1$, since there are no Bs before it and it is the r_1th observation. The number of As preceding the second B is $r_2 - 2$, since it is the r_2th observation, and one preceding observation is a B. Similarly, the number preceding the third B is $r_3 - 3$, and the number preceding the ith B is $r_i - i$. Hence we have:

$$\begin{aligned} U &= \sum_{i=1}^{n_1}(r_i - i) \\ &= \sum_{i=1}^{n_1} r_i - \sum_{i=1}^{n_1} i \\ &= \sum_{i=1}^{n_1} r_i - \frac{n_1(n_1+1)}{2} \end{aligned}$$

That is, we add together the ranks of all the n_1 observations, subtract $n_1(n_1+1)/2$ and we have U. For the example, we have

$$\begin{aligned} U &= 1.5 + 4 + 4 + 4 + 11 + 14.5 + 17.5 + 21 + 27 - \frac{9 \times (9+1)}{2} \\ &= 104.5 - 45 \\ &= 59.5 \end{aligned}$$

as before. This formula is sometimes written

$$U' = n_1 n_2 + \frac{n_1(n_1+1)}{2} - \sum_{i=1}^{n_1} r_i$$

But this is simply based on the other group, since $U + U' = n_1 n_2$. For testing we use the smaller value, as before.

As n_1 and n_2 increase, the calculation of the exact probability distribution becomes more difficult. When we cannot use Table 12.2, we use a large sample approximation instead. Because U is found by adding together a number of independent, identically distributed random variables, the central limit theorem (§7.2) applies. If the null hypothesis is true, the distribution of U approximates to a Normal distribution with mean $n_1 n_2/2$ and standard deviation is $\sqrt{n_1 n_2(n_1+n_2+1)/12}$. Hence

$$\frac{U - \frac{n_1 n_2}{2}}{\sqrt{\frac{n_1 n_2(n_1+n_2+1)}{12}}}$$

is an observation from a Standard Normal distribution. For the example, $n_1 = 9$ and $n_2 = 20$, we have

$$\begin{aligned} \frac{U - \frac{n_1 n_2}{2}}{\sqrt{\frac{n_1 n_2(n_1+n_2+1)}{12}}} &= \frac{59.5 - \frac{9\times 20}{2}}{\sqrt{\frac{9\times 20\times(9+20+1)}{12}}} \\ &= -1.44 \end{aligned}$$

From Table 7.1 this gives two sided probability = 0.15, similar to that found by the two sample t test (§10.3).

Neither Table 12.2 nor the above formula for the standard deviation of U take ties into account; both assume the data can be fully ranked. Their use for data with ties is an approximation. For small samples we must accept this. For the Normal approximation, ties can be allowed for using the following formula for the standard deviation of U when the null hypothesis is true:

$$\sqrt{\frac{n_1 n_2}{(n_1+n_2)(n_1+n_2-1)} \sum_{i=1}^{n_1+n_2} r_i^2 - \frac{n_1 n_2 (n_1+n_2+1)^2}{4(n_1+n_2-1)}}$$

where $\sum_{i=1}^{n_1+n_2} r_i^2$ is the sum of the squared ranks for all observations, i.e. for both groups (see Conover 1980). The Mann–Whitney U test is not free of assumptions which may be violated. We assume that the data can be fully ordered, which in the case of ties is not so.

The Mann–Whitney U test is a non-parametric analogue of the two sample t test. The advantage over the t test is that the only assumption about the distribution of the data is that the observations can be ranked, whereas for the t test we must assume the data are from Normal distributions with uniform variance. There are disadvantages. For data which are Normally distributed, the U test is less powerful than the t test, i.e. the t test, when valid, can detect smaller differences for given sample size. The U test is almost as powerful for moderate and large sample sizes, and this difference is important only for small samples. For very small samples, e.g. two groups of three observations, the test is useless as all possible values of U have probabilities above 0.05 (Table 12.2). The U test is primarily a test of significance. The t method also enables us to estimate the size of the difference and gives a confidence interval. Although as noted above U/n_1n_2 has an interpretation, we cannot, so far as I know, find a confidence interval for it. A confidence interval for the difference between means or between medians based on the U test can be calculated for interval data (Campbell and Gardner 1989) but we must assume that the groups come from distributions which are identical in shape, the only difference being in the location, i.e. in the mean. The distributions therefore have equal variances, which is very unlikely if the data do not follow a Normal distribution (§7A), so the t method could be used anyway.

There are other non-parametric tests which test the same or similar null hypotheses. Two of these, the Wilcoxon two sample test and the Kendall Tau test, are different versions of the Mann–Whitney U test which were developed around the same time and later shown to be identical. These names are sometimes used interchangeably. The test statistics and tables are not the same, and the user must be very careful that the calculation of the test statistic being used corresponds to the table to which it is

Table 12.4. Results of a trial of pronethalol for the prevention of angina pectoris (Pritchard *et al.* 1963), in rank order of differences

Number of attacks while on		Difference placebo −	Rank of difference		
placebo	pronethalol	pronethalol	All	positive	negative
2	0	2	1.5	1.5	
17	15	2	1.5	1.5	
3	0	3	3	3	
7	2	5	4	4	
8	1	7	6	6	
14	7	7	6	6	
23	16	7	6	6	
34	25	9	8	8	
79	65	14	9	9	
60	41	19	10	10	
323	348	−25	11		11
71	29	42	12	12	
Sum of ranks				67	11

referred. Another difficulty with tables is that some are drawn so that for a significant difference U must be less than or equal to the tabulated value (as in Table 12.2), for others U must be strictly less than the tabulated value. For more than two groups, the rank analogue of one way analysis of variance (§10.9) is the Kruskal–Wallis test, see Conover (1980) and Siegel (1956).

12.3 The Wilcoxon matched pairs test

This test is an analog of the paired t test. We have a sample measured under two conditions and the null hypothesis is that there is no tendency for the outcome on one condition to be higher or lower than the other. The alternative hypothesis is that the outcome on one condition tends to be higher or lower than the other. As the test is based on the magnitude of the differences, the data must be interval.

Consider the data of Table 12.4, previously discussed in §2.6 and §9.2, where we used the sign test for the analysis. In the sign test, we have ignored the magnitude of differences, and only considered their signs. If we can use information about the magnitude, we would hope to have a more powerful test. Clearly, we must have interval data to do this. To avoid making assumptions about the distribution of the differences, we use their rank order in a similar manner to the Mann–Whitney U test.

First, we rank the differences by their absolute values, i.e. ignoring the sign. As in §12.2, tied observations are given the average of their ranks. We now sum the ranks of the positive differences, 67, and the ranks of the negative differences, 11 (Table 12.4). If the null hypothesis were true and there was no difference, we would expect the rank sums for positive and

Table 12.5. Two sided 5% and 1% points of the distribution of T (lower value) in the Wilcoxon one sample test

Sample size	Probability that $T \leq$ the tabulated value		Sample size	Probability that $T \leq$ the tabulated value	
n	5%	1%	n	5%	1%
5	–	–	16	30	19
6	1	–	17	35	23
7	2	–	18	40	28
8	4	0	19	46	32
9	6	2	20	52	37
10	8	3	21	59	43
11	11	5	22	66	49
12	14	7	23	73	55
13	17	10	24	81	61
14	21	13	25	90	68
15	25	16			

negative differences to be about the same, equal to 39 (their average). The test statistic is the lesser of these sums, T. The smaller T is, the lower the probability of the data arising by chance.

The distribution of T when the null hypothesis is true can be found by enumerating all the possibilities, as described for the Mann–Whitney U statistic. Table 12.5 gives the 5% and 1% points for this distribution, for sample size n up to 25. For the example, $n = 12$ and so the difference would be significant at the 5% level if T were less than or equal to 14. We have $T = 11$, so the data are not consistent with the null hypothesis. The data support the view that there is a real tendency for patients to have fewer attacks while on the active treatment.

From Table 12.5, we can see that the probability that $T \leq 11$ lies between 0.05 and 0.01. This is greater than the probability given by the sign test, which was 0.006 (§9.2). Usually we would expect greater power, and hence lower probabilities when the null hypothesis is false, when we use more of the information. In this case, the greater probability reflects the fact that the one negative difference, -25, is large. Examination of the original data shows that this individual had very large numbers of attacks on both treatments, and it seems possible that he may belong to a different population from the other eleven.

Like Table 12.2, Table 12.5 is based on the assumption that the differences can be fully ranked and there are no ties. Ties may occur in two ways in this test. Firstly, ties may occur in the ranking sense. In the example we had two differences of $+2$ and three of $+7$. These were ranked equally: 1.5 and 1.5, and 6, 6 and 6. When ties are present between negative and positive differences, Table 12.5 only approximates to the distribution of T.

Ties may also occur between the paired observations, where the observed difference is zero. In the same way as for the sign test, we omit zero differences (§9.2). The test is done using the number of non-zero differences only to enter Table 12.5. This seems odd, in that a lot of zero differences would appear to support the null hypothesis. For example, if in Table 12.4 we had another dozen patients with zero differences, the calculation and conclusion would be the same. However, the mean difference would be smaller and the Wilcoxon test tells us nothing about the size of the difference, only its existence. This illustrates the danger of allowing significance tests to outweigh all other ways of looking at the data.

As n increases, the distribution of T under the null hypothesis tends towards a Normal distribution, as does that of Mann–Whitney U statistic. The sum of all the ranks, irrespective of sign, is $n(n+1)/2$, so the expected value of T under the null hypothesis is $n(n+1)/4$, since the two sums should be equal. If the null hypothesis is true, the standard deviation of T is $\sqrt{\frac{1}{4}\sum r_i^2}$, where r_i is the rank of the ith difference, which is $\sqrt{n(n+1)(2n+1)/24}$ when there are no ties. Hence

$$\frac{T - \frac{n(n+1)}{4}}{\sqrt{\frac{n(n+1)(2n+1)}{24}}}$$

is from a Standard Normal distribution if the null hypothesis is true. For the example of Table 12.4, we have:

$$\frac{T - \frac{n(n+1)}{4}}{\sqrt{\frac{n(n+1)(2n+1)}{24}}} = \frac{11 - \frac{12\times 13}{4}}{\sqrt{\frac{12\times 13\times 25}{24}}} = -2.197$$

From Table 7.1 this gives a two tailed probability of 0.028, similar to that obtained from Table 12.5.

We have three possible tests for paired data, the Wilcoxon, sign and paired t methods. If the differences are Normally distributed, the t test is the most powerful test. The Wilcoxon test is almost as powerful, however, and in practice the difference is not great except for small samples. Like the Mann–Whitney U test, the Wilcoxon is useless for very small samples. The sign test is similar in power to the Wilcoxon for very small samples, but as the sample size increases the Wilcoxon test becomes much more powerful. This might be expected since the Wilcoxon test uses more of the information. The Wilcoxon test uses the magnitude of the differences, and hence requires interval data. This means that, as for t methods, we will get different results if we transform the data. For truly ordinal data we should use the sign test. The paired t method also gives a confidence interval for the difference. The Wilcoxon test is purely a test of significance, but a confidence interval for the median difference can be found using the

Binomial method described in §15.5, or in more detail by Conover (1980) or Campbell and Gardner (1989).

12.4 Spearman's rank correlation coefficient, ρ

We noted in Chapter 11 the sensitivity to assumptions of Normality of the product moment correlation coefficient, r. This led to the development of non-parametric approaches based on ranks. Spearman's approach was direct. First we rank the observations, then calculate the product moment correlation of the ranks, rather than of the observations themselves. The resulting statistic has a distribution which does not depend on the distribution of the original variables. It is usually denoted by the Greek letter ρ, pronounced 'rho', or by r_S.

Table 12.6 shows data from a study of the geographical distribution of a tumour, Kaposi's sarcoma, in mainland Tanzania. The incidence rates were calculated from cancer registry data and there was considerable doubt that all cases were notified. The degree of reporting of cases may have been related to population density or availability of health services. In addition, incidence was closely related to age and sex (where recorded) and so could be related to the age and sex distribution in the region. To check that none of these were producing artefacts in the geographical distribution, I calculated the rank correlation of disease incidence with each of the possible explanatory variables. Table 12.6 shows the relationship of incidence to the percentage of the population living within 10 km of a health centre. Figure 12.1 shows the scatter diagram of these data. The percentage within 10 km of a health centre is very highly skewed, whereas the disease incidence appears somewhat bimodal. The assumption of the product moment correlation do not appear to be met, so rank correlation was preferred.

The calculation of Spearman's ρ proceeds as follows. The ranks for the two variables are found (Table 12.6). We apply the formula for the product moment correlation (§11.9) to these ranks. We define:

$$\rho = \frac{\text{sum of products about mean of ranks}}{\sqrt{\begin{array}{l}\text{sum of squares of ranks for first variable}\\ \times\ \text{sum of squares of ranks for second variable}\end{array}}}$$

The calculation is as described in §11.9, giving $\rho = 0.38$. We can now test the null hypothesis that the variables are independent, the alternative being that either one variable increases as the other increases, or that one decreases as the other increases. As usual with ranking statistics, the distribution of ρ for small samples can be found by listing all the possible permutations and their values of ρ. For a sample size of n there are, of course, $n!$ possibilities. Table 12.7 shows the critical value of ρ for

Table 12.6. Incidence of Kaposi's sarcoma and access of population to health centres for each region of mainland Tanzania (Bland *et al.* 1977)

Region	Incidence per million per year	Percent population within 10 km of health centre	Rank order	
			Incidence	Pop %
Coast	1.28	4.0	1	3
Shinyanga	1.66	9.0	2	7
Mbeya	2.06	6.7	3	6
Tabora	2.37	1.8	4	1
Arusha	2.46	13.7	5	13
Dodoma	2.60	11.1	6	10
Kigoma	4.22	9.2	7	8
Mara	4.29	4.4	8	4
Tanga	4.54	23.0	9	16
Singida	6.17	10.8	10	9
Morogoro	6.33	11.7	11	11
Mtwara	6.40	14.8	12	14
Westlake	6.60	12.5	13	12
Kilimanjaro	6.65	57.3	14	17
Ruvuma	7.21	6.6	15	5
Iringa	8.46	2.6	16	2
Mwanza	8.54	20.7	17	15

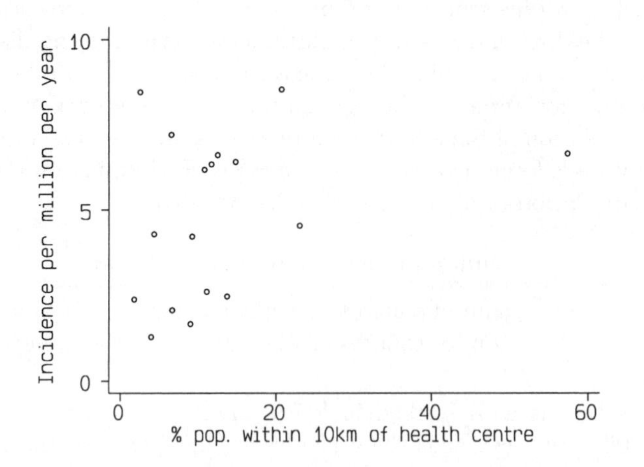

Fig. 12.1. Incidence of Kaposi's sarcoma per million per year and percentage of population within 10 km of a health centre, for 17 regions of mainland Tanzania

Table 12.7. Two sided 5% and 1% points of the distribution of Spearman's ρ

Sample size	Probability that ρ is as far or further from 0 than the tabulated value	
n	5%	1%
4	–	–
5	1.00	–
6	0.89	1.00
7	0.82	0.96
8	0.79	0.93
9	0.70	0.83
10	0.68	0.81

sample sizes up to 10. Note that, although the calculation is similar to that in §11.9–10, the distribution under the null hypothesis is not the same, and a different table is used. As n increases, so ρ tends to a Normal distribution when the null hypothesis is true, with expected value 0 and variance $1/(n-1)$. Thus $\rho/\sqrt{1/(n-1)} = \rho\sqrt{n-1}$ is from a Standard Normal distribution. The approximation is reasonable for $n > 10$.

For our data we have $0.38\sqrt{17-1} = 1.52$, which from Table 7.1 has two sided probability of 0.13. Hence we have not found any evidence of a relationship between the observed incidence of Kaposi's sarcoma and access to health centres. In this study there was no significant relationship with any of the possible explanatory variables and we concluded that the observed geographical distribution did not appear to be an artefact of population distribution or diagnostic provision.

We have ignored the problem of ties in the above. We treat observations with the same value as described in §12.2. We give them the average of the ranks they would have if they were separable and apply the rank correlation formula as described above. In this case the distribution of Table 12.7 is only approximate.

There are several ways of calculating this coefficient, resulting in formulae which appear quite different, though they give the same result (see Siegel 1956).

12.5 Kendall's rank correlation coefficient, τ

Spearman's rank correlation is quite satisfactory for testing the null hypothesis of no relationship, but is difficult to interpret as a measurement of the strength of the relationship. Kendall developed a different rank correlation coefficient, Kendall's τ, which has some advantages over Spearman's. (The Greek letter τ is pronounced 'tau'.) It is rather more tedious to calculate than Spearman's, but in the computer age this hardly matters. For each pair of subjects we observe whether the subjects are ordered in the same way by the two variables, a **concordant** pair, ordered in opposite ways, a **discordant** pair, or equal for one of the variables and so not ordered at all, a **tied** pair. **Kendall's** τ is the proportion of concordant pairs minus the proportion of discordant pairs. τ will be one if the rankings are identical, as all pairs will be ordered in the same way, and minus one if the rankings are exactly opposite, as all pairs will be ordered in the opposite way.

We shall denote the number of concordant pairs (ordered the same way) by n_c, the number of discordant pairs (ordered in opposite ways) by n_d, and the difference, $n_c - n_d$, by S. There are $n(n-1)/2$ pairs altogether, so

$$\tau = \frac{n_c - n_d}{\frac{1}{2}n(n-1)} = \frac{S}{\frac{1}{2}n(n-1)}$$

When there are no ties, $n_c + n_d = n(n-1)/2$.

The simplest way to calculate n_c is to order the observations by one of the variables, as in Table 12.6 which is ordered by disease incidence. Now consider the second ranking (% population within 10 km of a health centre). The first region, Coast, has 14 regions below it which have greater rank, so the pairs formed by the first region and these will be in the correct order. There are 2 regions below it which have lower rank, so the pairs formed by the first region and these will be in the opposite order. The second region, Shinyanga, has 10 regions below it with greater rank and so contributes 10 further pairs in the correct order. Note that the pair 'Coast and Shinyanga' has already been counted. There are 5 pairs in opposite order. The third region, Mbeya, has 10 regions below it in the same order and 4 in opposite orders, and so on. We add these numbers to get n_c and n_d:

$$n_c = 14+10+10+13+4+6+7+8+1+5+4+2+2+0+1+1+0 = 88$$

$$n_d = 2+5+4+0+8+5+3+1+7+2+2+3+2+3+1+0+0 = 48$$

The number of pairs is $n(n-1)/2 = 17 \times 16/2 = 136$. Because there are no ties, we could also calculate n_d by $n_d = n(n-1)/2 - n_c = 136 - 88 = 48$. $S = n_c - n_d = 88 - 48 = 40$. Hence $\tau = S/\left(n(n-1)/2\right) = 40/136 = 0.29$.

When there are ties, τ cannot be one. However, we could have perfect correlation if the ties were between the same subjects for both variables. To allow for this, we use a different version of τ, τ_b. Consider the denominator. There are $n(n-1)/2$ possible pairs. If there are t individuals tied at a particular rank for variable X, no pairs from these t individuals contribute to S. There are $t(t-1)/2$ such pairs. If we consider all the groups of tied individuals we have $\sum t(t-1)/2$ pairs which do not contribute to S, summing over all groups of tied ranks. Hence the total number of pairs which can contribute to S is $n(n-1) - \sum t(t-1)/2$, and S cannot be greater than $n(n-1)/2 - \sum t(t-1)/2$. The size of S is also limited by ties in the second ranking. If we denote the number of individuals with the same value of Y by u, then the number of pairs which can contribute to S is $n(n-1)/2 - \sum u(u-1)/2$. We now define τ_b by

$$\tau_b = \frac{S}{\sqrt{(n(n-1)/2 - \sum t(t-1)/2)\,(n(n-1)/2 - \sum u(u-1)/2)}}$$

Note that if there are no ties, $\sum t(t-1)/2 = 0 = \sum u(u-1)/2$, so $\tau_b = \tau$. When the rankings are identical $\tau_b = 1$, no matter how many ties there are. Kendall (1970) also discusses two other ways of dealing with ties, obtaining coefficients τ_a and τ_c, but their use is restricted.

We often want to test the null hypothesis that there is no relationship between the two variables in the population from which our sample was drawn. As usual, we are concerned with the probability of S being as or more extreme (i.e. far from zero) than the observed value. Table 12.8 was calculated in the same way as Tables 12.1 and 12.2. It shows the probability of being as extreme as the observed value of S for n up to 10. For convenience, S is tabulated rather than τ. When ties are present this is only an approximation.

When the sample size is greater than 10, S has an approximately Normal distribution under the null hypothesis, with mean zero. If there are no ties, the variance is

$$\mathrm{VAR}(S) = \frac{n(n-1)(2n+5)}{18}$$

When there are ties, the variance formula is very complicated (Kendall 1970). I shall omit it, as in practice these calculations will be done using computers anyway. If there are not many ties it will not make much difference if the simple form is used.

For the example, $S = 40$, $n = 17$ and there are no ties, so the Standard Normal variate is

$$\begin{aligned} \frac{S}{\sqrt{Var(S)}} &= \frac{S}{\sqrt{n(n-1)(2n+5)/18}} \\ &= \frac{40}{\sqrt{17 \times 18 \times 39/18}} \\ &= 1.55 \end{aligned}$$

From Table 7.1 of the Normal distribution we find that the two sided probability of a value as extreme as this is $0.06 \times 2 = 0.12$, which is very similar to that found using Spearman's ρ. The product moment correlation, r,

Table 12.8. Two sided 5% and 1% points of the distribution of S for Kendall's τ

Sample size	Probability that S is as far or further from the expected than the tabulated value	
n	5%	1%
4	–	–
5	10	–
6	13	15
7	15	19
8	18	22
9	20	26
10	23	29

gives $r = 0.30, \mathrm{P} = 0.24$, but of course the non-Normal distributions of the variables make this P invalid.

Why have two different rank correlation coefficients? Spearman's ρ is older than Kendall's τ, and can be thought of as a simple analogue of the product moment correlation coefficient, Pearson's r. τ is a part of a more general and consistent system of ranking methods, and has a direct interpretation, as the difference between the proportions of concordant and discordant pairs. In general, the numerical value of ρ is greater than that of τ. It is not possible to calculate τ from ρ or ρ from τ, they measure different sorts of correlation. ρ gives more weight to reversals of order when data are far apart in rank than when there is a reversal close together in rank, τ does not. However in terms of tests of significance both have the same power to reject a false null hypothesis, so for this purpose it does not matter which is used.

12.6 Continuity corrections

In this chapter, when samples were large we have used a continuous distribution, the Normal, to approximate to a discrete distribution, U, T or S. For example, Figure 12.2 shows the distribution of the Mann–Whitney U statistic for $n_1 = 4$, $n_2 = 4$ (Table 12.1) with the corresponding Normal curve. From the exact distribution, the probability that $U < 2$ is $0.014 + 0.014 + 0.029 = 0.057$. The corresponding Standard Normal deviate is

$$\frac{U - \frac{n_1 n_2}{2}}{\sqrt{\frac{n_1 n_2 (n_1 + n_2 + 1)}{12}}} = \frac{2 - \frac{4\times 4}{2}}{\sqrt{\frac{4\times 4\times 9}{12}}} = -1.732$$

This has a probability of 0.048, interpolating in Table 7.1. This is smaller than the exact probability. The disparity arises because the continuous distribution gives probability to values other than the integers 0, 1, 2, etc. The estimated probability for $U = 2$ can be found by the area under the curve between $U = 1.5$ and $U = 2.5$. The corresponding Normal

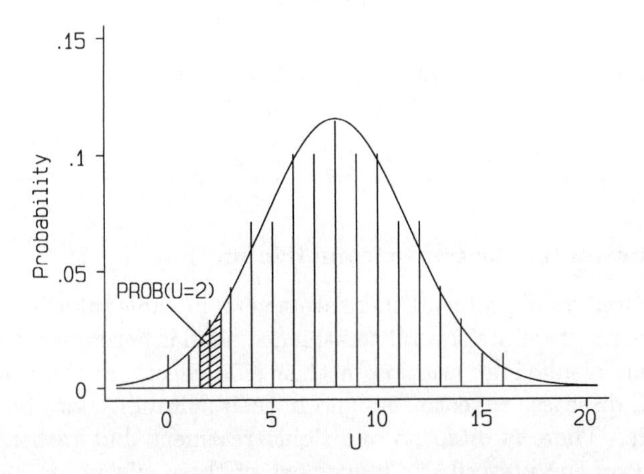

Fig. 12.2. Distribution of the Mann–Whitney U statistic, $n_1 = 4$, $n_2 = 4$, when the null hypothesis is true, with the corresponding Normal distribution and area estimating PROB($U = 2$)

deviates are -1.876 and -1.588, which have probabilities from Table 7.1 of 0.030 and 0.056. This gives the estimated probability for $U = 2$ to be $0.056 - 0.030 = 0.026$, which compares quite well with the exact figure of 0.029. Thus to estimate the probability that $U < 2$, we estimate the area below $U = 1.5$, not below $U = 2$. This gives us a Standard Normal deviate of -1.588, as already noted, and hence a probability of 0.056. This corresponds remarkably well with the exact probability of 0.057, especially when we consider how small n_1 and n_2 are.

We will get a better approximation from our Standard Normal deviate if we make U closer to its expected value by $\frac{1}{2}$. In general, we get a better fit if we make the observed value of the statistic closer to its expected value by half of the interval between adjacent discrete values. This is a **continuity correction.**

For S, the interval between adjacent values is 2, not 1, for $S = n_c - n_d = 2n_c - n(n-1)/2$, and n_c is an integer. A change of one unit in n_c produces a change of two units in S. The continuity correction is therefore half of 2, which is 1. We make S closer to the expected value of 0 by 1 before applying the Normal approximation. For the Kaposi's sarcoma data, we had $S = 40$, with $n = 17$. Using the continuity correction gives

$$\frac{S-1}{\sqrt{Var(S)}} = \frac{40-1}{\sqrt{17 \times 18 \times 39/18}} = \frac{39}{25.75} = 1.513$$

This gives a two sided probability of $0.066 \times 2 = 0.13$, slightly greater than the uncorrected value of 0.12.

Continuity corrections are important for small samples; for large samples they are negligible. We shall meet another in Chapter 13.

12.7 Parametric or non-parametric methods?

For many statistical problems there are several possible solutions, just as for many diseases there are several treatments, similar perhaps in their overall efficacy but displaying variation in their side effects, in their interactions with other diseases or treatments and in their suitability for different types of patients. There is often no one right treatment, but rather treatment is decided on the prescriber's judgement of these effects, past experience and plain prejudice. Many problems in statistical analysis are like this. In comparing the means of two small groups, for instance, we could use a t test, a t test with a transformation, a Mann–Whitney U test, or one of several others. Our choice of method depends on the plausibility of Normal assumptions, the importance of obtaining a confidence interval, the ease of calculation, and so on. It depends on plain prejudice, too. Some users of statistical methods are very concerned about the implications of Normal assumptions and will advocate non-parametric methods wherever possible, while others are too careless of the errors that may be introduced when assumptions are not met.

I sometimes meet people who tell me that they have used non-parametric methods throughout their analysis as if this is some kind of badge of statistical purity. It is nothing of the kind. It may mean that their significance tests have less power than they might have, and that results are left as 'not significant' when, for example, a confidence interval for a difference might be more informative.

On the other hand, such methods are very useful when the necessary assumptions of the t distribution method cannot be made, and it would be equally wrong to eschew their use. Rather, we should choose the method most suited to the problem, bearing in mind both the assumptions we are making and what we really want to know. We shall say more about what method to use when in Chapter 14.

There is a common misconception that when the number of observations is very small, usually said to be less than six, Normal distribution methods such as t tests and regression must not be used and that rank methods should be used instead. I have never seen any argument put forward in support of this, but inspection of Tables 12.2, 12.5, 12.7, and 12.8 will show that it is nonsense. For such small samples rank tests cannot produce any significance at the usual 5% level. Should one need statistical analysis of such small samples, Normal methods are required.

12M Multiple choice questions 62 to 66

(Each branch is either true or false)

62. For comparing the responses to a new treatment of a group of patients with the responses of a control group to a standard treatment, possible approaches include:

(a) the two sample t method;

(b) the sign test;

(c) the Mann–Whitney U test;

(d) the Wilcoxon matched pairs test;

(e) rank correlation between responses to the treatments.

63. Suitable methods for truly ordinal data include:

(a) the sign test;

(b) the Mann–Whitney U test;

(c) the Wilcoxon matched pairs test;

(d) the two sample t method;

(e) Kendall's rank correlation coefficient.

64. Kendall's rank correlation coefficient between two variables:

(a) depends on which variable is regarded as the predictor;

(b) is zero when there is no relationship;

(c) cannot have a valid significance test when there are tied observations;

(d) must lie between -1 and $+1$;

(e) is not affected by a log transformation of the variables.

65. Tests of significance based on ranks:

(a) are always to be preferred to methods which assume the data to be Normally distributed;

(b) are less powerful than methods based on the Normal distribution when data are Normally distributed;

(c) enable confidence intervals to be estimated easily;

(d) require no assumptions about the data;

(e) are often to be preferred when data cannot be assumed to follow any particular distribution.

66. Ten men with angina were given an active drug and a placebo on alternate days in random order. Patients were tested using the time in minutes for which they could exercise until angina or fatigue stopped them. The existence of an active drug effect could be examined by:

(a) paired t test;

(b) Mann–Whitney U test;

(c) sign test;

(d) Wilcoxon matched pairs test;

(e) Spearman's ρ.

12E Exercise: Application of rank methods

In this exercise we shall analyse the respiratory compliance data of §10E using non-parametric methods.

1. For the data of Table 10.17, use the sign test to test the null hypothesis that changing the waveform has no effect on static compliance.
2. Test the same null hypothesis using a test based on ranks.
3. Repeat step 1 using log transformed compliance. Does the transformation make any difference?
4. Repeat step 2 using log compliance. Why do you get a different answer?
5. What do you conclude about the effect of waveform from the non-parametric tests?
6. How do the conclusions of the parametric and non-parametric approaches differ?

13

The analysis of cross-tabulations

13.1 The chi-squared test for association

Table 13.1 shows the relationship between housing tenure for a sample of mothers and whether they had a preterm delivery. This kind of cross-tabulation of frequencies is also called a **contingency table** or **cross classification**. Each entry in the table is a frequency, the number of individuals having these characteristics. It can be quite difficult to measure the strength of the association between two qualitative variables, but it is easy to test the null hypothesis that there is no relationship or association between the two variables. If the sample is large, we do this by a chi-squared test.

The chi-squared test for association in a contingency table works like this. The null hypothesis is that there is no association between the two variables, the alternative being that there is an association of any kind. We find for each cell of the table the frequency which we would expect if the null hypothesis were true. To do this we use the row and column totals, so we are finding the expected frequencies for tables with these totals, called the **marginal** totals.

There are 1 443 women, of whom 899 were owner occupiers, a proportion 899/1 443. If there were no relationship between time of delivery and housing tenure, we would expect each column of the table to have the same proportion, 899/1 443, of its members in the first row. Thus the 99 patients in the first column would be expected to have $99 \times 899/1\,443 = 61.7$ in the first row. By 'expected' we mean the average frequency we would get in the

Table 13.1. Contingency table showing time of delivery by housing tenure

Housing tenure	Preterm	Term	Total
Owner-occupier	50	849	899
Council tenant	29	229	258
Private tenant	11	164	175
Lives with parents	6	66	72
Other	3	36	39
Total	99	1 344	1 443

Table 13.2. Expected frequencies under the null hypothesis for Table 13.1

Housing tenure	Preterm	Term	Total
Owner-occupier	61.7	837.3	899
Council tenant	17.7	240.3	258
Private tenant	12.0	163.0	175
Lives with parents	4.9	67.1	72
Other	2.7	36.3	39
Total	99	1 344	1 443

long run. We could not actually observe 61.7 subjects. The 1 344 patients in the second column would be expected to have $1\,344 \times 899/1\,443 = 837.3$ in the first row. The sum of these two expected frequencies is 899, the row total. Similarly, there are 258 patients in the second row and so we would expect $99 \times 258/1\,443 = 17.7$ in the second row, first column and $1\,344 \times 258/1\,443 = 240.3$ in the second row, second column. We calculate the expected frequency for each row and column combination, or cell. The 10 cells of Table 13.1 give us the expected frequencies shown in Table 13.2. Notice that the row and column totals are the same as in Table 13.1. In general, the expected frequency for a cell of the contingency table is found by

$$\frac{\text{row total} \times \text{column total}}{\text{grand total}}$$

It does not matter which variable is the row and which the column.

We now compare the observed and expected frequencies. If the two variables are not associated, the observed and expected frequencies should be close together, any discrepancy being due to random variation. We need a test statistic which measures this. The differences between observed and expected frequencies are a good place to start. We cannot simply sum them as the sum would be zero, both observed and expected frequencies having the same grand total, 1 443. We can resolve this as we resolved a similar problem with differences from the mean (§4.7), by squaring the differences. The size of the difference will also depend in some way on the number of patients. When the row and column totals are small, the difference between observed and expected is forced to be small. It turns out, for reasons discussed in §13A, that the best statistic is

$$\sum_{\text{all cells}} \frac{(\text{observed frequency} - \text{expected frequency})^2}{\text{expected frequency}}$$

This is often written as

$$\sum \frac{(O-E)^2}{E}$$

For Table 13.1 this is

$$\begin{aligned}\sum \frac{(O-E)^2}{E} &= \frac{(50-61.7)^2}{61.7} + \frac{(849-837.3)^2}{837.3} \\ &\quad + \frac{(29-17.7)^2}{17.7} + \frac{(229-240.3)^2}{240.3} \\ &\quad + \frac{(11-12.0)^2}{12.0} + \frac{(164-163.0)^2}{163.0} \\ &\quad + \frac{(6-4.9)^2}{4.9} + \frac{(66-67.1)^2}{67.1} \\ &\quad + \frac{(3-2.7)^2}{2.7} + \frac{(36-36.3)^2}{36.3} \\ &= 10.5\end{aligned}$$

As will be explained in §13A, the distribution of this test statistic when the null hypothesis is true and the sample is large enough is the Chi-squared distribution, with degrees of freedom given by

$$(\text{number of rows } - 1) \times (\text{number of columns } - 1)$$

We shall discuss what is meant by 'large enough' in §13.3.

For Table 13.1 we have $(5-1)\times(2-1) = 4$ degrees of freedom. Table 13.3 shows some percentage points of the Chi-squared distribution for selected degrees of freedom. These are the upper percentage points, as shown in Figure 13.1. We see that for 4 degrees of freedom the 5% point is 9.49 and 1% point is 13.28, so our observed value of 10.5 has probability between 1% and 5%, or 0.01 and 0.05. If we use a computer program which prints out the actual probability, we find P = 0.03. The data are not consistent with the null hypothesis and we can conclude that there is good evidence of a relationship between housing and time of delivery.

The chi-squared statistic is not an index of the strength of the association. If we double the frequencies in Table 13.1, this will double chi-squared, but the strength of the association is unchanged. We can only use the chi-squared test when the numbers in the cells are frequencies, not when they are percentages, proportions or measurements.

13.2 Tests for 2 by 2 tables

Consider the data on cough symptom and history of bronchitis discussed in §9.8. We had 273 children with a history of bronchitis of whom 26 were reported to have day or night cough, and 1 046 children without history of bronchitis, of whom 44 were reported to have day or night cough. We can set these data out as a contingency table, as in Table 13.4.

Let us use the chi-squared test to test the null hypothesis of no association between cough and history. The expected values are shown in Table 13.5. The test statistic is

Table 13.3. Percentage points of the Chi-squared distribution

Degrees of freedom	Probability that the tabulated value is exceeded (Figure 13.1)			
	10%	5%	1%	0.1%
1	2.71	3.84	6.63	10.83
2	4.61	5.99	9.21	13.82
3	6.25	7.81	11.34	16.27
4	7.78	9.49	13.28	18.47
5	9.24	11.07	15.09	20.52
6	10.64	12.59	16.81	22.46
7	12.02	14.07	18.48	24.32
8	13.36	15.51	20.09	26.13
9	14.68	16.92	21.67	27.88
10	15.99	18.31	23.21	29.59
11	17.28	19.68	24.73	31.26
12	18.55	21.03	26.22	32.91
13	19.81	22.36	27.69	34.53
14	21.06	23.68	29.14	36.12
15	22.31	25.00	30.58	37.70
16	23.54	26.30	32.00	39.25
17	24.77	27.59	33.41	40.79
18	25.99	28.87	34.81	42.31
19	27.20	30.14	36.19	43.82
20	28.41	31.41	37.57	45.32

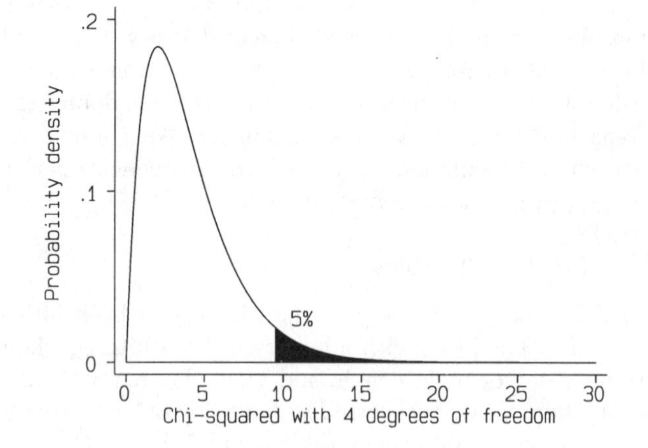

Fig. 13.1. Percentage point of the Chi-squared distribution

Table 13.4. Cough during the day or at night at age 14 for children with and without a history of bronchitis before age 5 (Holland *et al.* 1978)

	Bronchitis	No Bronchitis	Total
Cough	26	44	70
No cough	247	1 002	1 249
Total	273	1 046	1 319

Table 13.5. Expected frequencies for Table 13.4

	Bronchitis	No Bronchitis	Total
Cough	14.49	55.51	70
No cough	258.51	990.49	1 249
Total	273	1 046	1 319

$$\begin{aligned}\sum \frac{(O-E)^2}{E} &= \frac{(26-14.49)^2}{14.49} + \frac{(44-55.51)^2}{55.51} \\ &\quad + \frac{(247-258.51)^2}{258.51} + \frac{(1\,002-990.49)^2}{990.49} \\ &= 12.2\end{aligned}$$

We have $r = 2$ rows and $c = 2$ columns, so there are $(r-1)(c-1) = (2-1)\times(2-1) = 1$ degree of freedom. We see from Table 13.3 that the 5% point is 3.84, and the 1% point is 6.63, so we have observed something very unlikely if the null hypothesis were true. Hence we reject the null hypothesis of no association and conclude that there is a relationship between present cough and history of bronchitis.

Now the null hypothesis 'no association between cough and bronchitis' is the same as the null hypothesis 'no difference between the proportions with cough in the bronchitis and no bronchitis groups'. If there were a difference, the variables would be associated. Thus we have tested the same null hypothesis in two different ways. In fact these tests are exactly equivalent. If we take the Normal deviate from §9.8, which was 3.49, and square it, we get 12.2, the chi-squared value. The method of §9.8 and §8.6 has the advantage that it can also give us a confidence interval for the size of the difference, which the chi-squared method does not.

13.3 The chi-squared test for small samples

When the null hypothesis is true, the test statistic $\sum(O-E)^2/E$, which we can call the **chi-squared statistic**, follows the Chi-squared distribution provided the expected values are large enough. This is a large sample test, like those of §9.7 and §9.8. The smaller the expected values become, the more dubious will be the test.

Table 13.6. Observed and expected frequencies of categories of radiological appearance at six months as compared with appearance on admission in the MRC streptomycin trial, patients with an initial temperature of 100–100.9°F

Radiological	Streptomycin		Control		Total
assessment	observed	expected	observed	expected	
Improvement	13	8.4	5	9.6	18
Deterioration	2	4.2	7	4.8	9
Death	0	2.3	5	2.7	5
Total	15	15	17	17	32

The conventional criterion for the test to be valid is usually attributed to the great statistician W. G. Cochran. The rule is this: the chi-squared test is valid if at least 80% of the expected frequencies exceed 5 and all the expected frequencies exceed 1. We can see that Table 13.2 satisfies this requirement, since only 2 out of 10 expected frequencies, 20%, are less than 5 and none is less than 1. Note that this condition applies to the expected frequencies, not the observed frequencies. It is quite acceptable for an observed frequency to be 0, provided the expected frequencies meet the criterion.

This criterion is open to question. Simulation studies appear to suggest that the condition may be too conservative and that the chi-squared approximation works for smaller expected values, especially for larger numbers of rows and columns. At the time of writing the analysis of tables based on small sample sizes, particularly 2 by 2 tables, is the subject of hot dispute among statisticians. As yet, no-one has succeeded in devising a better rule than Cochran's, so I would recommend keeping to it until the theoretical questions are resolved. Any chi-squared test which does not satisfy the criterion is always open to the charge that its validity is in doubt.

If the criterion is not satisfied we can usually combine or delete rows and columns to give bigger expected values. Of course, this cannot be done for 2 by 2 tables, which we consider in more detail below. For example, Table 13.6 shows data from the MRC streptomycin trial (§2.2), the results of radiological assessment for a subgroup of patients defined by a prognostic variable. We want to know whether there is evidence of a streptomycin effect within this subgroup, so we want to test the null hypothesis of no effect using a chi-squared test. There are 4 out of 6 expected values less than 5, so the test on this table would not be valid. We can combine the rows so as to raise the expected values. Since the small expected frequencies are in the 'deterioration' and 'death' rows, it makes sense to combine these to give a 'deterioration or death' row. The expected values are then all greater than 5 and we can do the chi-squared test with 1 degree of freedom. This

Table 13.7. Reduction of Table 13.6 to a 2 by 2 table

Radiological	Streptomycin		Control		Total
assessment	observed	expected	observed	expected	
Improvement	13	8.4	5	9.6	18
Deterioration or death	2	6.6	12	7.4	14
Total	15	15.0	17	17.0	32

editing must be done with regard to the meaning of the various categories. In Table 13.6, there would be no point in combining rows 1 and 3 to give a new category of 'considerable improvement or death' to be compared to the remainder, as the comparison would be absurd. The new table is shown in Table 13.7. We have

$$\sum \frac{(O-E)^2}{E} = \frac{(13-8.4)^2}{8.4} + \frac{(5-9.6)^2}{9.6} + \frac{(2-6.6)^2}{6.6} + \frac{(12-7.4)^2}{7.4} = 10.8$$

Under the null hypothesis this is from a Chi-squared distribution with one degree of freedom, and from Table 13.3 we can see that the probability of getting a value as extreme as 10.8 is less than 1%. We have data inconsistent with the null hypothesis and we can conclude that the evidence suggests a treatment effect in this subgroup.

If the table does not meet the criterion even after reduction to a 2 by 2 table, we can apply either a continuity correction to improve the approximation to the Chi-squared distribution (§13.5), or an exact test based on a discrete distribution (§13.4).

13.4 Fisher's exact test

The chi-squared test described in §13.1 is a large sample test. When the sample is not large and expected values are less than 5, we can turn to an exact distribution like that for the Mann Whitney U statistic (§12.2). This method is called **Fisher's exact test.**

The exact probability distribution for the table can only be found when the row and column totals are given. Just as with the large sample chi-squared test, we restrict our attention to tables with these totals. This difficulty has led to much controversy about the use of this test. I shall show how the test works, then discuss its applicability.

Consider the following artificial example. In an experiment, we randomly allocate 4 patients to treatment A and 4 to treatment B, and get the outcome shown in Table 13.8. We want to know the probability of so large a difference in mortality between the two groups if the treatment have the same effect (the null hypothesis). We could have randomized the subjects into two groups in many ways, but if the null hypothesis is true the same three would have died. The row and column totals would therefore

Table 13.8. Artificial data to illustrate Fisher's exact test

	Survived	Died	Total
Treatment A	3	1	4
Treatment B	2	2	4
Total	5	3	8

Table 13.9. Possible tables for the totals of Table 13.8

i.

	S	D	T
A	4	0	4
B	1	3	4
T	5	3	8

ii.

	S	D	T
A	3	1	4
B	2	2	4
T	5	3	8

iii.

	S	D	T
A	2	2	4
B	3	1	4
T	5	3	8

iv.

	S	D	T
A	1	3	4
B	4	0	4
T	5	3	8

be the same for all these possible allocations. If we keep the row and column totals constant, there are only 4 possible tables, shown in Table 13.9. These tables are found by putting the values 0, 1, 2, 3 in the 'Died in group A' cell. Any other values would make the D total greater than 3.

Now, let us label our subjects a to h. The survivors we will call a to e, and the deaths f, g, h. How many ways can these patients be arranged in two groups of 4 to give tables i, ii, iii and iv? Table i can arise in 5 ways. Patients f, g, and h would have to be in group B, to give 3 deaths, and the remaining member of B could be a, b, c, d or e. Table ii can arise in 30 ways. The 3 survivors in group A can be abc, abd, abe, acd, ace, ade, bcd, bce, bde, cde, 10 ways. The death in A can be f, g or h, 3 ways. Hence the group can be made up in $10 \times 3 = 30$ ways. Table iii is the same as table ii, with A and B reversed, so arises in 30 ways. Table iv is the same as table i with A and B reversed, so arises in 5 ways. Hence we can arrange the 8 patients into 2 groups of 4 in $5 + 30 + 30 + 5 = 70$ ways. Now, the probability of any one arrangement arising by chance is $1/70$, since they are all equally likely if the null hypothesis is true. If there are 3 deaths, table i arises from 5 of the 70 arrangements, so had probability $5/70 = 0.071$. Table ii arises from 30 out of 70 arrangements, so has probability $30/70 = 0.429$. Similarly, table iii has probability $30/70 = 0.429$, and table iv has probability $5/70 = 0.071$.

Hence, under the null hypothesis that there is no association between treatment and survival, table ii, which we observed, has a probability of 0.429. It could easily have arisen by chance and so it is consistent with the null hypothesis. As in §9.2, we must also consider tables more extreme than the observed. In this case, there is one more extreme table in the direction of the observed difference, table i. In the direction of the observed difference, the probability of the observed table or a more extreme one is $0.071 + 0.429 = 0.5$. This is the P value for a one sided test (§9.5).

Fisher's exact test is essentially one sided. It is not clear what the corresponding deviations in the other direction would be, especially when all the marginal totals are different. This is because the distribution is asymmetrical, unlike those of §12.2–5. One solution is to double the one sided probability to get a two sided test when this is required. I follow Armitage and Berry (1987) in preferring this option. Another solution is to calculate probabilities for every possible table and sum all probabities less than or equal to the probability for the observed table to give the P value. This may give a smaller P value than the doubling method.

There is no need to enumerate all the possible tables, as above. The probability can be found from a simple formula (§13B). The probability of observing a set of frequencies $f_{11}, f_{12}, f_{21}, f_{22}$, when the row and column totals are r_1, r_2, c_1, and c_2 and the grand total is n, is

$$\frac{r_1!r_2!c_1!c_2!}{n!f_{11}!f_{12}!f_{21}!f_{22}!}$$

(See §6A for the meaning of $n!$.) We can calculate this for each possible table and so find the probability for the observed table and each more extreme one. For the example:

$$\text{table i: } \frac{5!3!4!4!}{8!4!0!1!3!} = 0.071$$

$$\text{table ii: } \frac{5!3!4!4!}{8!3!1!2!2!} = 0.429$$

giving a total of 0.50 as before.

Unlike the exact distributions for the rank statistics, this distribution is fairly easy to calculate but difficult to tabulate. A good table of this distribution required a small book (Finney *et al.* 1963).

We can apply this test to Table 13.7. The 2 by 2 tables to be tested and their probabilities are:

Table:		Probability
13	5	0.001 378 2
2	12	
14	4	0.000 075 7
1	13	
15	3	0.000 001 4
0	14	

The total one sided probability is 0.001 455 3, which doubled for a two sided test gives 0.002 9. The method using all smaller probabilities gives P=0.001 59. Either is larger than the probability for the χ^2 value of 10.6, which is 0.001 1.

13.5 Yates' continuity correction for the 2 by 2 table

The discrepancy in probabilities between the chi-squared test and Fisher's exact test arises because we are estimating the discrete distribution of the test statistic by the continuous Chi-squared distribution. A continuity correction like those of §12.6, called **Yates' correction**, can be used to improve the fit. The observed frequencies change in units of one, so we make them closer to their expected values by one half. Hence the formula for the corrected chi-squared statistic for a 2 by 2 table is

$$\sum \frac{(|O-E|-\frac{1}{2})^2}{E}$$

where $|O-E|$ means the absolute value or modulus of the difference, without sign. For Table 13.7 we have:

$$\begin{aligned}\sum \frac{(|O-E|-\frac{1}{2})^2}{E} &= \frac{(|13-8.4|-\frac{1}{2})^2}{8.4} + \frac{(|5-9.6|-\frac{1}{2})^2}{9.6} \\ &\quad + \frac{(|2-6.6|-\frac{1}{2})^2}{6.6} + \frac{(|12-7.4|-\frac{1}{2})^2}{7.4} \\ &= \frac{(4.6-\frac{1}{2})^2}{8.4} + \frac{(4.6-\frac{1}{2})^2}{9.6} \\ &\quad + \frac{(4.6-\frac{1}{2})^2}{6.6} + \frac{(4.6-\frac{1}{2})^2}{7.4} \\ &= 8.6\end{aligned}$$

This has probability 0.0037, which is closer to the exact probability, though there is still a considerable discrepancy. At such extremely low values any approximate probability model is liable to break down. In the critical area between 0.10 and 0.01, the continuity correction usually gives a very good fit to the exact probability.

13.6 The validity of Fisher's and Yates' methods

There has been much dispute among statisticians about the validity of the exact test and the continuity correction which approximates to it. Among the more argumentative of the founding fathers of statistical inference, such as Fisher and Neyman, this was quite acrimonious. The problem is still unresolved, and generating almost as much heat as light.

Note that the 2 by 2 tables Tables 13.4 and 13.7 arose in different ways. In Table 13.7, the column totals were fixed by the design of the experiment and only the row totals are from a random variable. In Table 13.4 neither row nor column totals were set in advance. Both are from the Binomial distribution, depending on the incidence of bronchitis and prevalence of chronic cough in the population. There is a third possibility, that both the row and column totals are fixed. This is rare in practice, but it can be achieved by the following experimental design. We want to know whether a subject can distinguish an active treatment from a placebo. We present him with 10 tablets, 5 of each, and ask him to sort the tablets into the 5 active and 5 placebo. This would give a 2 by 2 table, subject's choice versus truth, in which all row and column totals are preset to 5. There are several variations on these types of table, too. It can be shown that the same chi-squared test applies to all these cases when samples are large. When samples are small, this is not necessarily so. A discussion of the problem is well beyond the scope of this book. For some of these cases, Fisher's exact test and Yates' correction may be conservative, that is, give rather larger probabilities than they should, though this is a matter of debate. My own opinion is that Yates' correction and Fisher's exact test should be used. If we must err, it seems better to err on the side of caution.

13.7 The odds ratio

If the probability of an event is p then the odds of that event is $o = p/(1-p)$. The probability that a coin shows a head is 0.5, the odds is $0.5/(1-0.5) = 1$. The odds has advantages for some types of analysis, as it is not constrained to lie between 0 and 1, but can take any value from zero to infinity. We often use the logarithm to the base e of the odds, the **log odds** or **logit**:

$$\log_e(o) = \log_e\left(\frac{p}{1-p}\right)$$

This can vary from minus infinity to plus infinity and thus is very useful in fitting regression type models (§17.8). The logit is zero when $p = 1/2$ and the logit of $1-p$ is minus the logit of p:

$$\log_e(o_p) = \log_e\left(\frac{p}{1-p}\right) = -\log_e\left(\frac{1-p}{p}\right) = -\log_e\left(o_{1-p}\right)$$

Consider Table 13.4. The probability of cough for children with a history of bronchitis is $26/273 = 0.095\,24$. The odds of cough for children with a history of bronchitis is $26/247 = 0.105\,26$. The probability of cough for children without a history of bronchitis is $44/1\,046 = 0.042\,07$. The odds of cough for children without a history of bronchitis is $44/1\,002 = 0.043\,91$.

Table 13.10. 2 by 2 table in symbolic notation

			Total
	a	b	$a+b$
	c	d	$c+d$
Total	$a+c$	$b+d$	$a+b+c+d$

One way to compare children with and without bronchitis is to find the ratio of the proportions of children with cough in the two groups (the relative risk, §8.6). Another is to find the **odds ratio**, the ratio of the odds of cough in children with bronchitis and children without bronchitis. This is $(26/247)/(44/1\,002) = 0.105\,26/0.043\,91 = 2.397\,18$. Thus the odds of cough in children with a history of bronchitis is 2.397 18 times the odds of cough in children without a history of bronchitis.

If we denote the frequencies in the table by a, b, c, and d, as in Table 13.10, the odds ratio is given by

$$or = \frac{a/c}{b/d} = \frac{ad}{bc}$$

This is symmetrical; we get the same thing by

$$or = \frac{a/b}{c/d} = \frac{ad}{bc}$$

We can estimate the standard error and confidence interval using the log of the odds ratio (§13C). The standard error of the log odds ratio is:

$$\text{SE}\left(\log_e(or)\right) = \sqrt{\frac{1}{a} + \frac{1}{b} + \frac{1}{c} + \frac{1}{d}}$$

Hence we can find the 95% confidence interval. For Table 13.4, the log odds ratio is $\log_e(2.397\,18) = 0.874\,29$, with standard error

$$\text{SE}\left(\log_e(or)\right) = \sqrt{\frac{1}{26} + \frac{1}{44} + \frac{1}{247} + \frac{1}{1\,002}} = \sqrt{0.066\,24} = 0.257\,36$$

Provided the sample is large enough, we can assume that the log odds ratio comes from a Normal distribution and hence the approximate 95% confidence interval is

$$0.874\,29 - 1.96 \times 0.257\,36 \text{ to } 0.874\,29 + 1.96 \times 0.257\,36 = 0.369\,86 \text{ to } 1.378\,72$$

To get a confidence interval for the odds ratio itself we must antilog:

$$e^{0.369\,86} \text{ to } e^{1.378\,72} = 1.45 \text{ to } 3.97$$

The odds ratio can be used to estimate the relative risk in a case control study. The calculation of relative risk in §8.6 depended on the fact that we could estimate the risks. We could do this because we had a prospective study and so knew how many of the risk group developed the symptom. This cannot be done if we start with the outcome, in this case cough at age 14, and try to work back to the risk factor, bronchitis, as in a case control study.

Table 13.11 shows data from a case control study of smoking and lung cancer (see §3.8). We start with a group of cases, patients with lung cancer and a group of controls, here hospital patients without cancer. We cannot calculate risks (the column totals would meaningless and have been omitted), but we can still estimate the relative risk.

Suppose the prevalence of lung cancer is p, which must be a small number, and the table is as Table 13.10. Then we can estimate the probability of lung cancer and being a smoker by $pa/(a+b)$ and the probability of being a smoker without lung cancer is $(1-p)c/(c+d)$. The probability of being a smoker is therefore $pa/(a+b)+(1-p)c/(c+d)$, the probability of being a smoker with lung cancer plus the probability of being a smoker without lung cancer. Because p is much smaller than $1-p$, the first term can be ignored and the probability of being a smoker is approximately $(1-p)c/(c+d)$. The risk of lung cancer for smokers is found by dividing the probability of being a smoker with lung cancer by the probability being a smoker:

$$\frac{pa/(a+b)}{(1-p)c/(c+d)}$$

Similarly, the probability of both being a non-smoker and having lung cancer is $pb/(a+b)$ and the probability of being a non-smoker without lung cancer is $(1-p)d/(c+d)$. The probability of being a non-smoker is therefore $pb/(a+b)+(1-p)d/(c+d)$, and since p is much smaller than $1-p$, the first term can be ignored and the probability of being a non-smoker is approximately $(1-p)d/(c+d)$. This gives a risk of lung cancer among non-smokers of approximately

$$\frac{pb/(a+b)}{(1-p)d/(c+d)}$$

The relative risk of lung cancer for smokers is thus, approximately,

$$\frac{pa/(a+b)}{(1-p)c/(c+d)} \bigg/ \frac{pb/(a+b)}{(1-p)d/(c+d)} = \frac{ad}{bc}$$

This is, of course, the odds ratio. Thus for case control studies the relative risk is approximated by the odds ratio. This actually works whether the disease is rare or not, see Rodrigues and Kirkwood (1990).

Table 13.11. Smokers and non-smokers among male cancer patients and controls (Doll and Hill 1950)

	Smokers	Non-smokers	Total
Lung cancer	647	2	649
Controls	622	27	649

For Table 13.11 we have

$$\frac{ad}{bc} = \frac{647 \times 27}{2 \times 622} = 14.04$$

Thus the risk of lung cancer in smokers is about 14 times that of non-smokers. This is a surprising result from a table with so few non-smokers, but a direct estimate from the cohort study (Table 3.1) is 0.90/0.07=12.9, which is very similar. The log odds ratio is 2.64210 and its standard error is

$$\text{SE}\,(\log_e(or)) = \sqrt{\frac{1}{647} + \frac{1}{2} + \frac{1}{622} + \frac{1}{27}} = \sqrt{0.540\,19} = 0.734\,98$$

Hence the approximate 95% confidence interval is

$$2.642\,10 - 1.96 \times 0.734\,98 \text{ to } 2.642\,10 + 1.96 \times 0.734\,98 = 1.201\,54 \text{ to } 4.082\,65$$

To get a confidence interval for the odds ratio itself we must antilog:

$$e^{1.201\,54} \text{ to } e^{4.082\,65} = 3.3 \text{ to } 59.3$$

The very wide confidence interval is because the numbers of non-smokers, particularly for lung cancer cases, are so small.

13.8 The chi-squared test for trend

Consider the data of Table 13.12. Using the chi-squared test described in §13.1, we can test the null hypothesis that there is no relationship between reported cough and smoking against the alternative that there is a relationship of some sort. The chi-squared statistic is 64.25, with 2 degrees of freedom, $P < 0.001$. The data are not consistent with the null hypothesis.

Now, we would have got the same value of chi-squared whatever the order of the columns. The test ignores the natural ordering of the columns, but we might expect that if there were a relationship between reported cough and smoking, the prevalence of cough would be greater for greater amounts of smoking. In other words, we look for a trend in cough prevalence from one end of the table to the other. We can test for this using the **chi-squared test for trend**.

Table 13.12. Cough during the day or at night and cigarette smoking by 12 year old boys (Bland *et al.* 1978)

	Boys' smoking						Total
	Non-smoker		Occasional		Regular		
Cough	266	20.4%	395	28.8%	80	46.5%	741
No cough	1 037	79.6%	977	71.2%	92	53.5%	2 106
Total	1 303	100.0%	1 372	100.0%	172	100.0%	2 847

First, we define two variables, X and Y, whose values depend on the categories of the row and column variables. For example, we could put X=1 for non-smokers, $X = 2$ for occasional smokers and $X = 3$ for regular smokers, and put $Y = 1$ for 'cough' and $Y = 2$ for 'no cough'. Then for a non-smoker who coughs, the value of X is 1 and the value of Y is 1. Both X and Y may have more than two categories, provided both are ordered. If there are n individuals, we have n pairs of observations (x_i, y_i). If there is a linear trend across the table, there will be linear regression of Y on X which has non-zero slope. We fit the usual least squares regression line, $Y = a + bX$, where

$$b = \frac{\sum(y_i - \bar{y})(x_i - \bar{x})}{\sum(x_i - \bar{x})^2}$$

$$\text{SE}(b) = \sqrt{\frac{s^2}{(x_i - \bar{x})^2}}$$

where s^2 is the estimated variance of Y. In simple linear regression, as described in Chapter 11, we are usually concerned with estimating b and making statements about its precision. Here we are only going to test the null hypothesis that in the population $b = 0$. Under the null hypothesis, the variance about the line is equal to the total variance of Y, since the line has zero slope. We use the estimate

$$s^2 = \frac{1}{n}\sum (y_i - \bar{y})^2$$

(We use n as the denominator here, not $n-1$, because the test is conditional on the row and column totals as described in §13A. There is a good reason for it, but it is not worth going into here.) As in §11.5, the standard error of b is

$$\text{SE}(b) = \sqrt{\frac{s^2}{\sum(x_i - \bar{x})^2}} = \sqrt{\frac{\sum(y_i - \bar{y})^2}{n\sum(x_i - \bar{x})^2}}$$

As in §11.5, b is the sum of many independent, identically distributed random variables $\sum(y_i - \bar{y})(x_i - \bar{x})$, and so follows a Normal distribution by the central limit theorem (§7.2). As n is large, SE(b) should be a good estimate of the standard deviation of this distribution. Hence, if the null hypothesis is true and $E(b) = 0$, $b/\mathrm{SE}(b)$ is an observation from a Standard Normal distribution. Hence the square of this, $b^2/\mathrm{SE}(b)^2$, is from a Chi-squared distribution with one degree of freedom.

$$\begin{aligned}\frac{b^2}{\mathrm{SE}(b)^2} &= \left(\frac{\sum(y_i-\bar{y})(x_i-\bar{x})}{\sum(x_i-\bar{x})^2}\right)^2 \Big/ \frac{\sum(y_i-\bar{y})^2}{n\sum(x_i-\bar{x})^2} \\ &= \frac{n\left(\sum(y_i-\bar{y})(x_i-\bar{x})\right)^2}{\sum(x_i-\bar{x})^2\sum(y_i-\bar{y})^2}\end{aligned}$$

For practical calculations we use the alternative forms of the sums of squares and products:

$$\chi_1^2 = \frac{n\left(\sum y_i x_i - \frac{(\sum y_i)(\sum x_i)}{n}\right)^2}{\left(\sum x_i - \frac{(\sum x_i)^2}{n}\right)\left(\sum y_i^2 - \frac{(\sum y_i)^2}{n}\right)}$$

Note that is does not matter which variable is X and which is Y. The sums of squares and products are easy to work out. For example, for the column variable, X, we have 1 303 individuals with X=1, 1 372 with X=2 and 172 with X=3. For our data we have

$$\sum x_i^2 = 1^2 \times 1\,303 + 2^2 \times 1\,372 + 3^2 \times 172 = 8\,339$$

$$\sum x_i = 1 \times 1\,303 + 2 \times 1\,372 + 3 \times 172 = 4\,563$$

$$\begin{aligned}\sum x_i y_i &= 1 \times 1 \times 266 + 2 \times 1 \times 395 + 3 \times 1 \times 80 \\ &+ 1 \times 2 \times 1\,037 + 2 \times 2 \times 977 + 3 \times 2 \times 92 \\ &= 7\,830\end{aligned}$$

Similarly, $\sum y_i^2 = 9\,165$ and $\sum y_i = 4\,953$.

$$\chi_1^2 = \frac{2\,847 \times \left(7\,830 - \frac{4\,563 \times 4\,953}{2\,847}\right)^2}{\left(8\,339 - \frac{4\,563^2}{2\,847}\right)\left(9\,165 - \frac{4\,953^2}{2\,847}\right)} = 59.47$$

If the null hypothesis is true, χ_1^2 is an observation from the Chi-squared distribution with 1 degree of freedom. The value 59.47 is highly unlikely from this distribution and the trend is significant.

There are several points to note about this method. The choice of values for X and Y is arbitrary. By putting $X = 1$, 2 or 3 we assumed that the difference between non-smokers and occasional smokers is the same as that between occasional smokers and smokers. This need not be so and a different choice of X would give a different chi-squared for trend statistic. The choice is not critical, however. For example, putting $X = 1$, 2 or 4, so making regular smokers more different from occasional smokers than occasional smokers are from non-smokers, we get χ^2 for trend to be 64.22. The fit to the data is rather better, but the conclusions are unchanged.

The trend may be significant even if the overall contingency table chi-squared is not. This is because the test for trend has greater power for detecting trends than has the ordinary chi-squared test. On the other hand, if we had an association where those who were occasional smokers had far more symptoms than either non-smokers or regular smokers, the trend test would not detect it. If the hypothesis we wish to test involves the order of the categories, we should use the trend test, if it does not we should use the contingency table test of §13.1. Note that the trend test statistic is always less than the overall chi-squared statistic.

The distribution of the trend chi-squared statistic depends on a large sample regression model, not on the theory given in §13A. The table does not have to meet Cochran's rule (§13.3) for the trend test to be valid. As long as there are at least 30 observations the approximation should be valid.

Some computer programs offer a slightly different test, the Mantel Haenzsel trend test (not to be confused with the Mantel Haenzsel method for combining 2 by 2 tables, §17.11). This is almost identical to the method described here. As an alternative to the chi-squared test for trend, we could calculate Kendall's rank correlation coefficient, τ_b, between X and Y (§12.5). For Table 13.12 we get $\tau_b = -0.136$ with standard error 0.018. We get a χ^2_1 statistic by $(\tau_b/\mathrm{SE}(\tau_b))^2 = 57.09$. This is very similar to the χ^2 for trend value 59.47.

13.9 McNemar's test for matched samples

The chi-squared test described above enables us, among other things, to test the null hypothesis that binomial proportions estimated from two independent samples are the same. We can do this for the one sample or matched sample problem also. For example, Holland *et al.* (1978) obtained respiratory symptom questionnaires for 1 319 Kent schoolchildren at ages 12 and 14. One question we asked was whether the prevalence of reported symptoms was different at the two ages. At age 12, 347 (26%) children were reported to have had severe colds in the past 12 months compared to 454 (34%) at age 14. Was there evidence of a real increase?

Just as in the one sample or paired t test (§10.2) we would hope to improve our analysis by taking into account the fact that this is the same sample. We might expect, for instance, that symptoms on the two occasions

Table 13.13. Severe colds reported at two ages for Kent schoolchildren (Holland *et al.* 1978)

Severe colds at age 12	Severe colds at age 14		Total
	Yes	No	
Yes	212	144	356
No	256	707	963
Total	468	851	1 319

will be related.

The method which enables us to do this is McNemar's test, another version of the sign test. We need to know that 212 children were reported to have colds on both occasions, 144 to have colds at 12 but not at 14, 256 to have colds at 14 but not at 12 and 707 to have colds at neither age. Table 13.13 shows the data in tabular form.

The null hypothesis is that the proportions saying yes on the first and second occasions are the same, the alternative being that one exceeds the other. This is a hypothesis about the row and column totals, quite different from that for the contingency table chi-squared test. If the null hypothesis were true we would expect the frequencies for 'yes, no' and 'no, yes' to be equal. In other words, as many should go up as down. (Compare this with the sign test, §9.2.) If we denote these frequencies by f_{yn} and f_{ny}, then the expected frequencies will be $(f_{yn} + f_{ny})/2$. We get the test statistic:

$$\sum \frac{(O-E)^2}{E} = \frac{(f_{yn} - \frac{f_{yn}+f_{ny}}{2})^2}{\frac{f_{yn}+f_{ny}}{2}} + \frac{(f_{ny} - \frac{f_{yn}+f_{ny}}{2})^2}{\frac{f_{yn}+f_{ny}}{2}}$$

which follows a Chi-squared distribution provided the expected values are large enough. There are two observed frequencies and one constraint, that the sum of the observed frequencies = the sum of the expected frequencies. Hence there is one degree of freedom. The test statistic can be simplified considerably, to:

$$\chi^2 = \frac{(f_{yn} - f_{ny})^2}{f_{yn} + f_{ny}}$$

For Table 13.13, we have

$$\chi^2 = \frac{(f_{yn} - f_{ny})^2}{f_{yn} + f_{ny}} = \frac{(144-256)^2}{144+256} = 31.4$$

This can be referred to Table 13.3 with one degree of freedom and is clearly highly significant. There was a difference between the two ages. As there was no change in any of the other symptoms studied, we thought that this was possibly due to an epidemic of upper respiratory tract infection just before the second questionnaire.

There is a continuity correction, again due to Yates. If the observed frequency f_{yn} increases by 1, f_{ny} decreases by 1 and $f_{yn} - f_{ny}$ increases by 2. Thus half the difference between adjacent possible values is 1 and we make the observed difference nearer to the expected difference (zero) by 1. Thus the continuity corrected test statistic is

$$\chi^2 = \frac{(|f_{yn} - f_{ny}| - 1)^2}{f_{yn} + f_{ny}}$$

where $|f_{yn} - f_{ny}|$ is the absolute value, without sign. For Table 13.13:

$$\chi^2 = \frac{(|f_{yn} - f_{ny}| - 1)^2}{f_{yn} + f_{ny}} = \frac{(|144 - 256| - 1)^2}{144 + 256} = \frac{(112 - 1)^2}{400} = 30.8$$

There is very little difference because the expected values are so large but if the expected values are small, say less than 20, the correction is advisable. For small samples, we can also take f_{ny} as an observation from the Binomial distribution with $p = \frac{1}{2}$ and $n = f_{yn} + f_{ny}$ and proceed as for the sign test (§9.2).

We can also find a large sample confidence interval for the difference between the proportions. The estimated difference is $p_1 - p_2 = (f_{yn} - f_{ny})/n$ and the standard error is

$$\text{SE}(p_1 - p_2) = \sqrt{\frac{f_{yn} + f_{ny}}{n^2} - \frac{(f_{yn} - f_{ny})^2}{n^3}}$$

For the example, $p_1 - p_2 = (144 - 256)/1\,319 = -0.085$ and the standard error is

$$\text{SE}(p_1 - p_2) = \sqrt{\frac{144 + 256}{1\,319^2} - \frac{(144 - 256)^2}{1\,319^3}} = 0.015$$

The 95% confidence interval is $-0.085 - 1.96 \times 0.015 = -0.11$ to $-0.085 - 1.96 \times 0.015 = -0.06$. We estimate that the proportion of colds on the first equation was less than that on the second by between 0.06 and 0.11.

We may wish to compare the distribution of a variable with three or more categories in matched samples. If the categories are ordered, like smoking experience in Table 13.12, we are usually looking for a shift from one end of the distribution to the other, and we can use the sign test (§9.2), counting positives when smoking increased, negative when it decreased, and zero if the category was the same. When the categories are not ordered, as Table 13.1 there is a test due to Stuart (1955), described by Maxwell (1970). The test is difficult to do and the situation is very unusual, so I shall omit details.

Table 13.14. Parity of 125 women attending antenatal clinics at St. George's Hospital, with the calculation of the chi-squared goodness of fit test

Parity	Frequency	Poisson probability	Expected frequency		$\frac{(O-E)^2}{E}$
0	59	0.442 20	55.275		0.251
1	44	0.360 83	45.104		0.027
2	14	0.147 22	18.402		1.053
3	3	0.040 04	5.005		
4	4	0.008 17	1.021		
5	1	0.001 33	0.167	6.219	1.666
> 5	0	0.000 21	0.026		
Total	125	1.000 00	125.000		2.997

13.10 The chi-squared goodness of fit test

Another use of the Chi-squared distribution is the goodness of fit test. Here we test the null hypothesis that a frequency distribution follows some theoretical distribution such as the Poisson or Normal. Table 13.14 shows a frequency distribution. We shall test the null hypothesis that it is from a Poisson distribution, i.e. that conception is a random event among fertile women.

First we estimate the parameter of the Poisson distribution, its mean, μ, in this case 0.816. We then calculate the probability for each value of the variable, using the Poisson formula of §6.7:

$$\frac{e^{-\mu}\mu^r}{r!}$$

where r is the number of events. The probabilities are shown in Table 13.14. The probability that the variable exceeds five is found by subtracting the probabilities for 0, 1, 2, 3, 4, and 5 from 1.0. We then multiply these by the number of observations, 125, to give the frequencies we would expect from 125 observations from a Poisson distribution with mean 0.816.

We now have a set of observed and expected frequencies and can compute a chi-squared statistic in the usual way. We want all the expected frequencies to be greater than 5 if possible. We achieve this here by combining all the categories for parity greater than 3. We then add $(O-E)^2/E$ for the categories to give a χ^2 statistic. We now find the degrees of freedom. This is the number of categories minus the number of parameters fitted from the data (one in the example) minus one. Thus we have $4-1-1=2$ degrees of freedom. From Table 13.3 the observed χ^2 value of 2.99 has P > 0.10 and the deviation from the Poisson distribution is clearly not significant.

The same test can be used for testing the fit of any distribution. For example, Wroe *et al.* (1992) studied diurnal variation in onset of strokes. Table 13.15 shows the frequency distribution of times of onset. If the null

Table 13.15. Time of onset of 554 strokes (Wroe *et al.* 1992)

Time	Frequency	Time	Frequency
00.01–02.00	21	12.01–14.00	34
02.01–04.00	16	14.01–16.00	59
04.01–06.00	22	16.01–18.00	44
06.01–08.00	104	18.01–20.00	51
08.01–10.00	95	20.01–22.00	32
10.01–12.00	66	22.01–24.00	10

hypothesis that there is no diurnal variation were true, the time at which strokes occurred would follow a Uniform distribution (§7.2). The expected frequency in each time interval would be the same. There were 554 cases altogether, so the expected frequency for each time is $554/12 = 46.167$. We then work out $(O-E)^2/E$ for each interval and add to give the chi-squared statistic, in this case equal to 218.8. There is only one constraint, that the frequencies total 554, as no parameters have been estimated. Hence if the null hypothesis were true we would have an observation from the Chi-squared distribution with $12 - 1 = 11$ degrees of freedom. The calculated value of 218.8 is very unlikely, $P < 0.001$ from Table 13.3, and the data are not consistent with the null hypothesis.

13A Appendix: Why the chi-squared test works

We noted some of the properties of the Chi-squared distribution in §7A. In particular, it is the sum of the squares of a set of independent Standard Normal variables, and if we look at a subset of values defined by independent linear relationships between these variables we lose one degree of freedom for each constraint. It is on these two properties that the chi-squared test depends.

Suppose we did not have a fixed size to the birth study of Table 13.1, but observed subjects as they delivered. Then in any given time interval the number in a given cell of the table would be from a Poisson distribution and the set of Poisson variables corresponding to the cell frequency would be independent of one another. Our table is one set of samples from these Poisson distributions. However, we do not know the expected values of these distributions under the null hypothesis; we only know their expected values if the table has the row and column totals we observed. We can only consider the subset of outcomes of these variables which has the observed row and column totals. The test is said to be **conditional** on these row and column totals.

The mean and variance of a Poisson variable are equal (§6.7). If the null hypothesis is true, the means of these variables will be equal to the expected frequency calculated in §13.1. Thus O, the observed cell frequency,

Table 13.16. Symbolic representation of a 2 × 2 table

			Total
	f_{11}	f_{12}	r_1
	f_{21}	f_{22}	r_2
Total	c_1	c_2	n

is from a Poisson distribution with mean E, the expected cell frequency, and standard deviation $\sqrt{E}$. Provided E is large enough, this Poisson distribution will be approximately Normal. Hence $(O - E)/\sqrt{E}$ is from a Normal distribution mean 0 and variance 1. Hence if we find

$$\sum\left(\frac{O-E}{\sqrt{E}}\right)^2 = \sum\frac{(O-E)^2}{E}$$

this is the sum of the squares of a set of Normally distributed random variables with mean 0 and variance 1, and so is from a Chi-squared distribution (§7A).

We will now find the degrees of freedom. Although the underlying variables are independent, we are only considering a subset defined by the row and column totals. Consider the table as in Table 13.16. Here, f_{11} to f_{22} are the observed frequencies, r_1, r_2 the row totals, c_1, c_2 the column totals, and n the grand total. Denote the corresponding expected values by e_{11} to e_{22}. There are three linear constraints on the frequencies:

$$f_{11} + f_{12} + f_{21} + f_{22} = n$$

$$f_{11} + f_{12} = r_1$$

$$f_{11} + f_{21} = c_1$$

Any other constraint can be made up of these. For example, we must have

$$f_{21} + f_{22} = r_2$$

This can be found by subtracting the second equation from the first. Each of these linear constraints on f_{11} to f_{22} is also a linear constraint on $(f_{11} - e_{11})/\sqrt{e_{11}}$ to $(f_{22} - e_{22})/\sqrt{e_{22}}$. This is because e_{11} is fixed and so $(f_{11} - e_{11})/\sqrt{e_{11}}$ is a linear function of f_{11}. There are four observed frequencies and so four $(O - E)/\sqrt{E}$ variables, with three constraints. We lose one degree of freedom for each constraint and so have $4 - 3 = 1$ degree of freedom.

If we have r rows and c columns, then we have one constraint then the sum of the frequencies is n. Each row must add up, but when we reach the last row the constraint can be obtained by subtracting the first $r - 1$ rows from the grand total. The rows contribute only $r-1$ further constraints. Similarly the columns contribute $c-1$ constraints. Hence, there being rc frequencies, the degrees of freedom are

$$rc-1-(r-1)-(c-1) = rc-1-r+1-c+1 = rc-r-c+1 = (r-1)(c-1)$$

So we have degrees of freedom given by the number of rows minus one times the number of columns minus one.

13B Appendix: The formula for Fisher's exact test

The derivation of Fisher's formula is strictly for the algebraically minded. Remember that the number of ways of choosing r things out of n things (§6A) is $n!/r!(n-r)!$. Now, suppose we have a 2 by 2 table made up of n as shown in Table 13.16. First, we ask how many ways n individuals can be arranged to give marginal totals, r_1, r_2, c_1 and c_2. They can be arranged in columns in $n!/c_1!c_2!$ ways, since we are choosing c_1 objects out of n, and in rows $n!/r_1!r_2!$ ways. (Remember $n-c_1 = c_2$ and $n-r_1 = r_2$.) Hence they can be arranged in

$$\frac{n!}{c_1!c_2!} \times \frac{n!}{r_1!r_2!} = \frac{n!n!}{c_1!c_2!r_1!r_2!}$$

ways. For example, the table with totals

		4
		4
5	3	8

can happen in

$$\frac{8!}{5! \times 3!} \times \frac{8!}{4! \times 4!} = 56 \times 70 = 3620 \text{ ways.}$$

As we saw in §13.4, the columns can be arranged in 70 ways. Now we ask, of these ways how many make up a particular table? We are now dividing the n into four groups of sizes f_{11}, f_{12}, f_{21} and f_{12}. We can choose the first group in $n!/f_{11}!(n-f_{11})!$ ways, as before. We are now left with $n-f_{11}$ individuals, so we can choose f_{12} in $(n-f_{11})!/f_{12}!(n-f_{11}-f_{12})!$. We are now left with $n-f_{11}-f_{12}$, and so we choose f_{21} in $(n-f_{11}-f_{12})!/f_{21}!$ ways. This leaves $n-f_{11}-f_{12}-f_{21}$, which is, of course, equal to f_{22} and so f_{22} can only be chosen in one way. Hence we have altogether:

$$\frac{n!}{f_{11}! \times (n-f_{11})!} \times \frac{(n-f_{11})!}{f_{12}! \times (n-f_{11}-f_{12})!} \times \frac{(n-f_{11}-f_{12})!}{f_{21}! \times (n-f_{11}-f_{12}-f_{12})!}$$

$$= \frac{n!}{f_{11}! \times f_{12}! \times f_{21}! \times (n - f_{11} - f_{12} - f_{12})!}$$
$$= \frac{n!}{f_{11}! \times f_{12}! \times f_{21}! \times f_{22}!}$$

because $n - f_{11} - f_{12} - f_{12} = f_{22}$. So out of the

$$\frac{n! \times n!}{c_1! \times c_2! \times r_1! \times r_2!}$$

possible tables, the given tables arises in

$$\frac{n!}{f_{11}! \times f_{12}! \times f_{21}! \times f_{22}!}$$

ways. The probability of this table arising by chance is

$$\frac{n!}{f_{11}! \times f_{12}! \times f_{21}! \times f_{22}!} \Big/ \frac{c_1! \times c_2! \times r_1! \times r_2!}{n! \times n!} =$$
$$\frac{n! \times f_{11}! \times f_{12}! \times f_{21}! \times f_{22}!}{c_1! \times c_2! \times r_1! \times r_2!}$$

13C Appendix: Standard error for the log odds ratio

This is for the mathematical reader. We start with a general result concerning log transformations. If X is a random variable with mean μ, the approximate variance of $\log_e(X)$ is given by

$$\text{VAR}(\log_e(X)) = \frac{\text{VAR}(X)}{\mu^2}$$

This is why when the standard deviation of a variable is proportional to its mean, and hence the variance is proportional to the mean squared, a log transformation makes the variance independent of the mean. Two particular cases are important here. The approximate variance of the sampling distribution of the log of a binomial proportion estimate $\hat{p}$ is given by

$$\text{VAR}\left(\log_e(\hat{p})\right) = \frac{\text{VAR}(\hat{p})}{p^2} = \frac{p(1-p)/n}{p^2} = \frac{1-p}{np}$$

Thus the standard error of the log proportion estimate is

$$\text{SE}\left(\log_e(\hat{p})\right) = \sqrt{\frac{1-p}{np}}$$

For the estimate $\hat{\mu}$ of a Poisson rate μ, the approximate variance is given by

$$\text{VAR}\left(\log_e(\hat{\mu})\right) = \frac{\text{VAR}(\hat{\mu})}{\mu^2} = \frac{\mu}{\mu^2} = \frac{1}{\mu}$$

If an event happens a times and does not happen b times, the log odds is $\log_e(a/b) = \log_e(a) - \log_e(b)$. The frequencies a and b are from independent

Poisson distributions with means estimated by a and b respectively. Hence their variances are estimated by $1/a$ and $1/b$ respectively. The variance of the log odds is given by

$$\mathrm{VAR}\,(\log_e(o)) = \mathrm{VAR}\,(\log_e(a/b)) = \mathrm{VAR}\,(\log_e(a) - \log_e(b))$$

$$= \mathrm{VAR}\,(\log_e(a)) + \mathrm{VAR}\,(\log_e(b)) = \frac{1}{a} + \frac{1}{b}$$

The standard error of the log odds is thus given by

$$\mathrm{SE}\,(\log_e(o)) = \sqrt{\frac{1}{a} + \frac{1}{b}}$$

The log odds ratio is the difference between the log odds:

$$\log_e(o_1/o_2) = \log_e(o_1) - \log_e(o_2)$$

The variance of the log odds ratio is the sum of the variances of the log odds and for table 2 we have

$$\mathrm{VAR}\,(\log_e(or)) = \left(\frac{1}{a} + \frac{1}{b} + \frac{1}{c} + \frac{1}{d}\right)$$

The standard error is the square root of this:

$$\mathrm{SE}\,(\log_e(or)) = \sqrt{\frac{1}{a} + \frac{1}{b} + \frac{1}{c} + \frac{1}{d}}$$

13M Multiple choice questions 67 to 74

(Each branch is either true or false)

67. In a chi-squared test for a 5 by 3 contingency table:

(a) variables must be quantitative;

(b) observed frequencies are compared to expected frequencies;

(c) there are 15 degrees of freedom;

(d) at least 12 cells must have expected values greater than 5;

(e) all the observed values must be greater than 1.

Table 13.17. Cough first thing in the morning in a group of schoolchildren, as reported by the child and by the child's parents (Bland *et al.* 1979)

Parents' Report	Child's Report Yes	No	Total
Yes	29	104	133
No	172	5097	5269
Total	201	5201	5402

68. In Table 13.17:

(a) the association between reports by parents and children can be tested by a chi-squared test;

(b) the difference between symptom prevalence as reported by children and parents can be tested by McNemar's test;

(c) if McNemar's test is significant, the contingency chi-squared test is not valid;

(d) the contingency chi-squared test has one degree of freedom;

(e) it would be important to use the continuity correction in the contingency chi-squared test.

69. Table 13.18 appeared in the report of a case control study of infection with *Campylobacter jejuni*:

(a) A chi-squared test for trend could be used to test the null hypothesis that risk of disease does not increase with the number of bird attacks;

(b) 'OR' means the odds ratio;

(c) A significant chi-squared test would show that risk of disease increases with increasing numbers of bird attacks;

(d) 'OR' provides an estimate of the relative risk of *Campylobacter jejuni* infection;

(e) Kendall's rank correlation coefficient, τ_b, could be used to test the null hypothesis that risk of disease does not increase with the number of bird attacks.

Table 13.18. Bird attacks on milk bottles reported by cases of *Campylobacter jejuni* infection and controls (Southern *et al.* 1990)

Number of days of week when attacks took place	Number of cases	Number of controls	OR
0	3	42	1
1-3	11	3	51
4-5	5	1	70
6-7	10	1	140

70. Fisher's exact test for a contingency table:

(a) applies to 2 by 2 tables;

(b) usually gives a larger probability than the ordinary chi-squared test;

(c) usually gives about the same probability as the chi-squared test with Yates' continuity correction;

(d) is suitable when expected frequencies are small;

(e) is difficult to calculate when the expected frequencies are large.

71. The standard chi-squared test for a 2 by 2 contingency table is not valid unless:

(a) all the expected frequencies are greater than five;

(b) both variables are continuous;

(c) at least one variable is from a Normal distribution;

(d) all the observed frequencies are greater than five;

(e) the sample is very large.

72. McNemar's test could be used:

(a) to compare the numbers of cigarette smokers among cancer cases and age and sex matched healthy controls;

(b) to examine the change in respiratory symptom prevalence in a group of asthmatics from winter to summer;

(c) to look at the relationship between cigarette smoking and respiratory symptoms in a group of asthmatics;

(d) to examine the change in PEFR in a group of asthmatics from winter to summer;

(e) to compare the number of cigarette smokers among a group of cancer cases and a random sample of the general population.

73. In a study of boxers, computer tomography revealed brain atrophy in 3 of 6 professionals and 1 of 8 amateurs (Kaste *et al.* 1982). These groups could be compared using:

(a) Fisher's exact test;

(b) the chi-squared test;

(c) the chi-squared test with Yates' correction;

(d) McNemar's test;

(e) the two sample t test.

74. When an odds ratio is calculated from a two by two table:

(a) the odds ratio is a measure of the strength of the relationship between the row and column variables;

(b) if the order of the rows and the order of the columns is reversed, the odds ratio will be unchanged;

(c) the ratio may take any positive value;

(d) the odds ratio will be changed to its reciprocal if the order of the columns is changed;

(e) the odds ratio is the ratio of the proportions of observations in the first row for the two columns.

13E Exercise: Admissions to hospital in a heatwave

In this exercise we shall look at some data assembled to test the hypothesis that there is a considerable increase in the number of admissions to geriatric wards during heatwaves. Table 13.19 shows the number of admissions to geriatric wards in a health district for each week during the summers of 1982, which was cold, and 1983, which was hot. Also shown are the average of the daily peak temperatures for each week.

1. When do you think the heatwave began and ended?

2. How many admissions were there during the heatwave and in the corresponding period of 1982? Would this be sufficient evidence to conclude that heatwaves produce an increase in admissions?

3. We can use the periods before and after the heatwave weeks as controls for changes in other factors between the years. Divide the years into three periods, before, during, and after the heatwave and set up a two way table showing numbers of admissions by period and year.

4. We can use this table to test for a heatwave effect. State the null hypothesis and calculate the frequencies expected if the null hypothesis were true.

5. Test the null hypothesis. What conclusions can you draw?

6. What other information could be used to test the relationship between heatwaves and geriatric admissions?

Table 13.19. Mean peak daily temperatures for each week from May to September of 1982 and 1983, with geriatric admissions in Wandsworth (Fish 1985)

Week	Mean peak temp, °C		Admissions		Week	Mean peak temp, °C		Admissions	
	1982	1983	1982	1983		1982	1983	1982	1983
1	12.4	15.3	24	20	12	21.7	25.0	11	25
2	18.2	14.4	22	17	13	22.5	27.3	6	22
3	20.4	15.5	21	21	14	25.7	22.9	10	26
4	18.8	15.6	22	17	15	23.6	24.3	13	12
5	25.3	19.6	24	22	16	20.4	26.5	19	33
6	23.2	21.6	15	23	17	19.6	25.0	13	19
7	18.6	18.9	23	20	18	20.2	21.2	17	21
8	19.4	22.0	21	16	19	22.2	19.7	10	28
9	20.6	21.0	18	24	20	23.3	16.6	16	19
10	23.4	26.5	21	21	21	18.1	18.4	24	13
11	22.8	30.4	17	20	22	17.3	20.7	15	29

14

Choosing the statistical method

14.1 Method oriented and problem oriented teaching

The choice of method of analysis for a problem depends on the comparison to be made and the data to be used. In Chapters 8 to 13, statistical methods have been arranged by type of data, large samples, Normal, ordinal, categorical, etc. In this chapter we look at how the appropriate method is chosen for the three most common problems in statistical inference:

- comparison of two independent groups, for example, groups of patients given different treatments;
- comparison of the response of one group under different conditions, as in a cross-over trial, or of matched pairs of subjects, as in some case-control studies;
- investigation of the relationship between two variables measured on the same sample of subjects.

This chapter acts as a map of the methods described in Chapters 8 to 13. Subsequent chapters describe methods for special problems in clinical medicine, population study, dealing with several factors at once, and the choice of sample size.

As was discussed in §12.7, there are often several different approaches to even a simple statistical problem. The methods described here and recommended for particular types of question may not be the only methods, and may not always be universally agreed as the best method. Statisticians are at least as prone to disagree as clinicians. However, these would usually be considered as valid and satisfactory methods for the purposes for which they are suggested here.

14.2 Types of data

The study design is one factor which determines the method of analysis, the variable being analysed is another. We can classify variables into the following types:

Ratio scales. The ratio of two quantities has a meaning, so we can say that one observation is twice another. Human height is a ratio scale.

Interval scales. The interval or distance between points on the scale has precise meaning, a change in one unit at one scale point is the same as a change in one unit at another. For example, temperature in ^{O}C is an interval scale, whereas anxiety score calculated from a questionnaire is not. Centigrade temperature is not a ratio scale, because the zero is arbitrary. We can add and subtract on an interval scale. All ratio scales are also interval scales.

Ordinal scale. The scale enables us to order the subjects, from that with the lowest value to that with the highest. Any ties which cannot be ordered are assumed to be because the measurement is not sufficiently precise.

Ordered nominal scale. We can group subjects into several categories, which have an order. For example, we can ask patients if their condition is much improved, improved a little, no change, a little worse, much worse.

Nominal scale. We can group subjects into categories which need not be ordered in any way. Eye colour is measured on a nominal scale.

Dichotomous scales. Subjects are grouped into only two categories, for example: survived or died. This is a special case of the nominal scale.

Clearly these classes are not mutually exclusive, and an interval scale is also ordinal. Sometimes it is useful to apply methods appropriate to a lower level of measurement, ignoring some of the information.

Ratio and interval scales allow us to calculate means and variances, and to find standard errors and confidence intervals for these. For example, in comparing two groups we can find the difference in mean between them, and estimate limits within which this difference should lie in the population from which the sample was drawn. For large samples, the estimation of confidence intervals presents no problem, as the means will follow Normal distributions and the variances reasonably good estimates of their population values. For small samples, we must assume that the observations themselves are from a Normal distribution. Many interval scales do follow a Normal distribution, and if not they can often be made to do so by a suitable transformation. Provided the assumption of a Normal distribution is valid, methods based on this are the most powerful available. If Normal assumptions do not apply, methods based on ranks can be used. For the ordinal and lower levels of measurement, most simple analyses produce tests of significance only, which are less satisfactory.

14.3 Comparing two groups

The methods used for comparing two groups are summarized in Table 14.1.

Table 14.1. Methods for comparing two samples

Type of data	Size of sample	Method
Interval	large, > 50 each sample	Normal distribution for means (§8.5, §9.7)
	small, < 50 each sample, with Normal distribution and uniform variance	two sample t method (§10.3)
	small, < 50 each sample, non-Normal	Mann–Whitney U test (§12.2)
Ordinal	any	Mann–Whitney U test (§12.2)
Nominal, ordered	large, $n > 30$	chi-squared for trend (§13.8)
Nominal, not ordered	large, most expected frequencies > 5	chi-squared test (§13.1)
	small, more than 20% expected frequencies < 5	reduce number of categories by combining or excluding as appropriate (§13.3)
Dichotomous	large, all expected frequencies > 5	comparison of two proportions (§8.6, §9.8), chi-squared test (§13.1), odds ratio (§13.7)
	small, at least one expected frequency < 5	chi-squared test with Yates' correction (§13.5), Fisher's exact test (§13.4)

Interval data. For large samples, say more than 50 in each group, confidence intervals for the mean can be found by the Normal approximation (§8.5). For smaller samples, confidence intervals for the mean can be found using the t distribution provided the data follow or can be transformed to a Normal distribution (§10.3, §10.4). If not, a significance test of the null hypothesis that the means are equal can be carried out using the Mann–Whitney U test (§12.2). This can be useful when the data are censored, that is, there are values too small or too large to measure. This happens, for example, when concentrations are too small to measure and labelled 'not detectable'. Provided that data are from Normal distributions, it is possible to compare the variances of the groups using the F test (§10.8).

Ordinal data. The tendency for one group to exceed members of the other is tested by the Mann–Whitney U test (§12.2).

Ordered nominal data. First the data is set out as a two way table, one variable being group and the other the ordered nominal data. A chi-

Table 14.2. Methods for differences in one or paired sample

Type of data	Size of sample	Method
Interval	large, > 100	Normal distribution (§8.3)
	small, < 100, Normal differences	Paired t method (§10.2)
	small, < 100, non-Normal differences	Wilcoxon matched pairs test (§12.3)
Ordinal	any	sign test (§9.2)
Nominal, ordered	any	sign test (§9.2)
Nominal	any	Stuart test (§13.9)
Dichotomous	any	McNemar's test (§13.9)

squared test (§13.1) will test the null hypothesis that there is no relationship between group and variable, but takes no account of the ordering. This is done by using the chi-squared test for trend, which takes the ordering into account and provides a much more powerful test (§13.8).

Nominal data. Set the data out as a two way table as described above. The chi-squared test for a two way table is the appropriate test (§13.1). The condition for validity of the test, that at least 80% of the expected frequencies should be greater than 5, must be met by combining or deleting categories as appropriate (§13.3). If the table reduces to a 2 by 2 table without the condition being met, use Fisher's exact test.

Dichotomous data. For large samples, either present the data as two proportions and use the Normal approximation to find the confidence interval for the difference (§8.6), or set the data up as a 2 by 2 table and do a chi-squared test (§13.1). These are equivalent methods. An odds ratio can also be calculated (§13.7). If the sample is small, the fit to the Chi-squared distribution can be improved by using Yates' correction (§13.5). Alternatively, use Fisher's exact test (§13.4).

14.4 One sample and paired samples

Methods of analysis for paired samples are summarized in Table 14.2.

Interval data. Inferences are on differences between the variable as observed on the two conditions. For large samples, say $n > 100$, the confidence interval for the mean difference is found using the Normal approximation (§8.3). For small samples, provided the differences are from a Normal distribution, use the paired t test (§10.2). This assumption is often very reasonable, as most of the variation between individuals is removed and random error is largely made up of measurement error. Furthermore, the error is the result of two added measurement errors and so tends to follow a Normal distribution anyway. If not, transformation of the original data will often make differences Normal (§10.4). If no assumption of a Normal distribution can be made, use the Wilcoxon signed-rank matched pairs test (§12.3).

It is rarely asked whether there is a difference in variability in paired data. This can be tested by finding the differences between the two conditions and their sum. Then if there is no change in variance the correlation between difference and sum has expected value zero (Pitman's test). This is not obvious but it is true.

Ordinal data. If the data do not form an interval scale, as noted in §14.2 the difference between conditions is not meaningful. However, we can say what direction the difference is in, and this can be examined by the sign test (§9.2).

Ordered nominal data. Use the sign test, with changes in one direction being positive, in the other negative, no change as zero (§9.2).

Nominal data. With more than two categories, this is difficult. Use Stuart's generalization to more than two categories of McNemar's test (§13.9).

Dichotomous data. Here we are comparing the proportions of individuals in a given state under the two conditions. The appropriate test is McNemar's test (§13.9).

14.5 Relationship between two variables

The methods for studying relationships between variables are summarized in Table 14.3. Relationships with dichotomous variables can be studied as the difference between two groups (§14.3), the groups being defined by the two states of the dichotomous variable. Dichotomous data have been excluded from the text of this section, but are included in Table 14.3.

Interval and interval data. Two methods are used: regression and correlation. Regression (§11.2, §11.5) is usually preferred, as it gives information about the nature of the relationship as well as about its existence. Correlation (§11.9) measures the strength of the relationship. For regression, residuals about the line must follow a Normal distribution with uniform variance. For estimation, the correlation coefficient requires an assumption that both variables follow a Normal distribution, but to test the null hypothesis only one variable needs to follow a Normal distribution. If neither variable can be assumed to follow a Normal distribution or be transformed to it (§11.8), use rank correlation (§12.4, §12.5).

Interval and ordinal data. Rank correlation coefficient (§12.4, §12.5).

Interval and ordered nominal data. This can be approached by rank correlation, using Kendall's τ (§12.5) because it copes with the large num-

Table 14.3. Methods for relationships between variables

	Interval, Normal	Interval, non-Normal	Ordinal
Interval Normal	regression (§11.2) correlation (§11.9)	regression (§11.2) rank correlation (§12.4, §12.5)	rank correlation (§12.4, §12.5)
Interval, non-Normal	regression (§11.2) rank correlation (§12.4, 12.5)	rank correlation (§12.4, §12.5)	rank correlation (§12.4, §12.5)
Ordinal	rank correlation (§12.4, §12.5)	rank correlation (§12.4, §12.5)	rank correlation (§12.4, 12.5)
Nominal, ordered	Kendall's rank correlation (§12.5)	Kendall's rank correlation (§12.5)	Kendall's rank correlation (§12.5)
Nominal	analysis of variance (§10.9)	Kruskal Wallis test (§12.2)	Kruskal Wallis test (§12.2)
Dichotomous	t-test (§10.3) Normal test (§8.5, §9.7)	large sample Normal test (§8.5, §9.7), Mann–Whitney U test (§12.2)	Mann–Whitney U test (§12.2)

	Nominal, ordered	Nominal	Dichotomous
Interval Normal	rank correlation (§12.4, 12.5)	analysis of variance (§10.9)	t-test (§10.3) Normal test (§8.5, §9.7)
Interval, non-Normal	Kendall's rank correlation (§12.5)	Kruskal Wallis test (§12.2)	large sample Normal test (§8.5, §9.7), Mann–Whitney U test (§12.2)
Ordinal	Kendall's rank correlation (§12.5)	Kruskal Wallis test (§12.2)	Mann–Whitney U test (§12.2)
Nominal, ordered	chi-squared test for trend (§13.8)	chi-squared test (§13.1)	chi-squared test for trend (§13.8)
Nominal	chi-squared test (§13.1)	chi-squared test (§13.1)	chi-squared test (§13.1)
Dichotomous	chi-squared test for trend (§13.8)	chi-squared test (§13.1)	chi-squared (§13.1, §13.5) Fisher's exact test (§13.4)

ber of ties better than does Spearman's ρ, or by analysis of variance as described for interval and nominal data. The latter requires an assumption of Normal distribution and uniform variance for the interval variable. These two approaches are not equivalent.

Interval and nominal data. If the interval scale follows a Normal distribution, use one-way analysis of variance (§10.9). The assumption is that within categories the interval variable is from Normal distributions with uniform variance. If this assumption is not reasonable, use Kruskal Wallis analysis of variance by ranks (§12.2).

Ordinal and ordinal data. Use a rank correlation coefficient, Spearman's ρ (§12.4) or Kendall's τ (§12.5). Both will give very similar answers for testing the null hypothesis of no relationship in the absence of ties. For data with many ties and for comparing the strengths of different relationships, Kendall's τ is preferable.

Ordinal and ordered nominal data. Use Kendall's rank correlation coefficient, τ (§12.5).

Ordinal and nominal data. Kruskal Wallis one-way analysis of variance by ranks (§12.2).

Ordered nominal and ordered nominal data. Use chi-squared for trend (§13.8).

Ordered nominal and nominal data. Use the chi-squared test for a two-way table (§13.1).

Nominal and nominal data. Use the chi-squared test for a two-way table (§13.1), provided the expected values are large enough. Otherwise use Yates' correction (§13.5) or Fisher's exact test (§13.4).

14M Multiple choice questions 75 to 80

(Each branch is either true or false)

75. The following variables have interval scales of measurement:

(a) height;

(b) presence or absence of asthma;

(c) Apgar score;

(d) age;

(e) Forced Expiratory Volume.

Table 14.4. Number of rejection episodes over 16 weeks following heart transplant in two groups of patients

Episodes	Group A	Group B	Total
0	10	8	18
1	15	6	21
2	4	0	4
3	3	0	3
Total patients	32	14	46

76. Table 14.4 shows the number of rejection episodes following heart transplant in two groups of patients:

(a) The rejection rates in the two populations could be compared by a Mann Whitney U test;

(b) The rejection rates in the two populations could be compared by a two sample t test;

(c) The rejection rates in the two populations could be compared by a chi-squared test for trend;

(d) The chi-squared test for a 4 by 2 table would not be valid;

(e) The hypothesis that the number of episodes follows a Poisson distribution could be investigated using a chi-squared test for goodness of fit.

77. Twenty arthritis patients were given either a new analgesic or aspirin on successive days in random order. The grip strength of the patients was measured. Methods which could be used to investigate the existence of a treatment effect include:

(a) Mann Whitney U test;

(b) paired t method;

(c) sign test;

(d) Normal confidence intervals for the mean difference;

(e) Wilcoxon matched pairs signed rank test.

78. The following methods may be used to investigate a relationship between two continuous variables:

(a) paired t test;

(b) the correlation coefficient, r;

(c) simple linear regression;

(d) Kendall's τ;

(e) Spearman's ρ.

79. When analysing categorical variables the following statistical methods may be used:

(a) simple linear regression;

(b) correlation coefficient, r;

(c) paired t test;

(d) Kendall's τ;

(e) chi-squared test.

80. To compare levels of a continuous variable in two groups, possible methods include:

(a) the Mann Whitney U test;

(b) Fisher's exact test;

(c) a t test;

(d) Wilcoxon matched pairs signed rank test;

(e) the sign test.

14E Exercise: Choosing a statistical method

1. In a crossover trial to compare two appliances for ileostomy patients, of 14 patients who received system A first, 5 expressed a preference for A, 9 for system B and none had no preference. Of the patients who received system B first, 7 preferred A, 5 preferred B and 4 had no preference. How would you decide whether one treatment was preferable? How would you decide whether the order of treatment influenced the choice?

2. Burr *et al.* (1976) tested a procedure to remove house-dust mites from the bedding of adult asthmatics in attempt to improve subjects' lung function, which they measured by PEFR. The trial was a two period crossover design, the control or placebo treatment being thorough dust removal from the living room. The means and standard errors for PEFR in the 32 subjects were:

active treatment:	335 litres/min, SE = 19.6 litres/min
placebo treatment:	329 litres/min, SE = 20.8 litres/min
differences within subjects: (treatment-placebo)	6.45 litres/min, SE = 5.05 litres/min

How would you decide whether the treatment improves PEFR?

3. In a trial of screening and treatment for mild hypertension (Reader *et al.* 1980), 1138 patients completed the trial on active treatment, with 9 deaths, and 1080 completed on placebo, with 19 deaths. A further 583 patients allocated to active treatment withdrew, of whom 6 died, and 626 allocated to placebo withdrew, of whom 16 died during the trial period. How would

Table 14.5. Gastric pH and urinary nitrite concentrations in 26 subjects (Hall and Northfield, private communication)

pH	Nitrite	pH	Nitrite	pH	Nitrite	pH	Nitrite
1.72	1.64	2.64	2.33	5.29	50.6	5.77	48.9
1.93	7.13	2.73	52.0	5.31	43.9	5.86	3.26
1.94	12.1	2.94	6.53	5.50	35.2	5.90	63.4
2.03	15.7	4.07	22.7	5.55	83.8	5.91	81.2
2.11	0.19	4.91	17.8	5.59	52.5	6.03	19.5
2.17	1.48	4.94	55.6	5.59	81.8		
2.17	9.36	5.18	0.0	5.71	21.9		

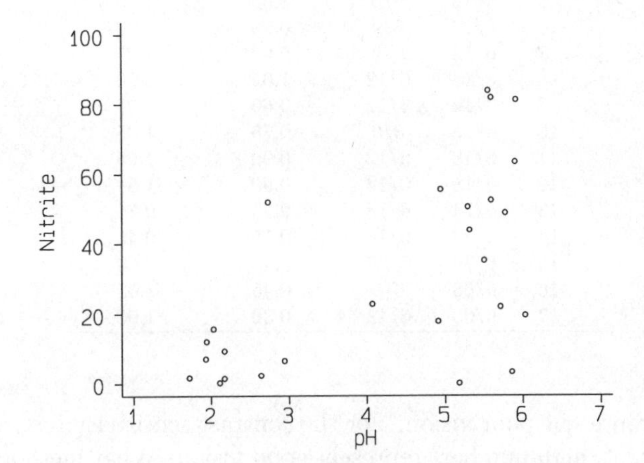

Fig. 14.1. Gastric pH and urinary nitrite

you decide whether screening and treatment for mild hypertension reduces the risk of dying?

4. Table 14.5 shows the pH and nitrite concentrations in samples of gastric fluid from 26 patients. A scatter diagram is shown in Figure 14.1. How would you assess the evidence of a relationship between pH and nitrite concentration?

5. The lung function of 79 children with a history of hospitalization for whooping cough and 178 children without a history of whooping cough, taken from the same school classes, was measured. The mean transit time for the whooping cough cases was 0.49 s (s.d. = 0.14 s) and for the controls 0.47 s (s.d. = 0.11 s), (Johnston *et al.* 1983). How could you analyse the difference in lung function between children who had had whooping cough and those who had not? Each case had two matched controls. If you had all the data, how could you use this information?

6. Table 14.6 shows some data from a pre- and post-treatment study of cataract patients. The second number in the visual acuity score represents the size of letter which can be read at a distance of six metres, so high

Table 14.6. Visual acuity and results of a contrast sensitivity vision test before and after cataract surgery (Wilkins, personal communication)

Case	Visual acuity		Contrast sensitivity test	
	before	after	before	after
1	6/9	6/9	1.35	1.50
2	6/9	6/9	0.75	1.05
3	6/9	6/9	1.05	1.35
4	6/9	6/9	0.45	0.90
5	6/12	6/6	1.05	1.35
6	6/12	6/9	0.90	1.20
7	6/12	6/9	0.90	1.05
8	6/12	6/12	1.05	1.20
9	6/12	6/12	0.60	1.05
10	6/18	6/6	0.75	1.05
11	6/18	6/12	0.90	1.05
12	6/18	6/12	0.90	1.50
13	6/24	6/18	0.45	0.75
14	6/36	6/18	0.15	0.45
15	6/36	6/36	0.45	0.60
16	6/60	6/9	0.45	1.05
17	6/60	6/12	0.30	1.05

numbers represent poor vision. For the contrast sensitivity test, which is a measurement, high numbers represent good vision. What method could be used to test the difference in visual acuity and in the contrast sensitivity test pre- and post-operation? What method could be used to investigate the relationship between visual acuity and the contrast sensitivity test post-operation?

7. Table 14.7 shows the relationship between age of onset of asthma in children and maternal age at the child's birth. How would you test whether these were related? The children were all born in one week in March, 1958. Apart from the possibility that young mothers in general tend to have children prone to asthma, what other possible explanations are there for this finding?

Table 14.7. Asthma or wheeze by maternal age (Anderson *et al.* 1986)

Asthma or wheeze	mother's age at child's birth		
reported	15–19	20–29	30+
never	261	4 017	2 146
onset by age 7	103	984	487
onset from 8 to 11	27	189	95
onset from 12 to 16	20	157	67

15

Clinical measurement

15.1 Making measurements

In this chapter we shall look at a number of problems associated with clinical measurement. These include how precisely we can measure, how different methods of measurement can be compared, how measurements can be used in diagnosis and how to deal with incomplete measurements of survival.

When we make a measurement, particularly a biological measurement, the number we obtain is the result of several things: the true value of the quantity we want to measure, biological variation, the measurement instrument itself, the position of the subject, the skill, experience and expectations of the observer, and even the relationship between observer and subject. Some of these factors, such as the variation within the subject, are outside the control of the observer. Others, such as position, are not, and it is important to standardize these. One which is most under our control is the precision with which we read scales and record the result. When blood pressure is measured, for example, some observers record to the nearest 5 mm Hg, others to the nearest 10 mm Hg. Some observers may record diastolic pressure at Korotkov sound four, others at five. Observers may think that as blood pressure is such a variable quantity, errors in recording of this magnitude are unimportant. In the monitoring of the individual patient, such lack of uniformity may make apparent changes difficult to interpret. In research, imprecise measurement can have serious consequences for the interpretation of data and can lead to false conclusions.

How precisely should we record data? While this must depend to some extent on the purpose for which the data are to be recorded, any data which are to be subjected to statistical analysis should be recorded as precisely as possible. A study can only be as good as the data, and data are often very costly and time-consuming to collect. The precision to which data are to be recorded and all other procedures to be used in measurement should be decided in advance and stated in the protocol, the written statement of how the study is to be carried out. We should bear in mind that the precision of recording depends on the number of significant figures (§5.2) recorded, not the number of decimal places. The observations 0.15 and 1.66 from Table 15.6, for example, are both recorded to one decimal place, but 0.15 has two significant figures and 1.66 has three. The second observation is recorded more precisely. This becomes very important when we come to analyse the data, for the data of Table 15.6 have a skew distribution which we wish to log transform. The greater imprecision of recording at the lower end of the scale is magnified by the transformation.

In measurement there is usually uncertainty in the last digit. Observers will often have some values for this last digit which they record more often than others. Many observers are more likely to record a terminal zero than a nine or a one, for example. This is known as **digit preference**. The tendency to read blood pressure to the nearest five or ten mm Hg mentioned above is an example of this. Observer training and awareness of the problem help to minimize digit preference, but if possible readings should be taken to sufficient significant figures for the last digit to be unimportant. Digit preference is particularly important when differences in the last digit are of importance to the outcome, as it might be in Table 15.1, where we are dealing with the difference between two similar numbers. Because of this it is a mistake to have one measurer take readings under one set of conditions and a second under another, as their degree of digit preference may differ. It is also important to agree the precision to which data are to be recorded and to ensure that instruments have sufficiently fine scales for the job in hand.

15.2 Repeatability and measurement error

I have already discussed some factors which may produce bias in measurements (§2.7, §2.8, §3.6). I have not yet considered the natural biological variability, in subject and in measurement method, which may lead to measurement error. 'Error' comes from a Latin root meaning 'to wander', and its use in statistics in closely related to this, as in §11.2, for example. Thus error in measurement may include the natural continual variation of a biological quantity, when a single observation will be used to characterize the individual. For example, in the measurement of blood pressure we are dealing with a quantity that varies continuously, not only from heart beat to heart beat but from day to day, season to season, and even with the sex of the measurer. The measurer, too, will show variation in the perception of the sound and reading of the manometer. Because of this, most clinical measurements cannot be taken at face value without some consideration being given to their error.

The quantification of measurement error is not difficult in principle. To do it we need a set of replicate readings, obtained by measuring each member of a sample of subjects more than once. We can then estimate

Table 15.1. Pairs of readings made with a Wright Peak Flow Meter on 17 healthy volunteers

Subject	PEFR (litres/min)		Subject	PEFR (litres/min)	
	First	Second		First	Second
1	494	490	10	433	429
2	395	397	11	417	420
3	516	512	12	656	633
4	434	401	13	267	275
5	476	470	14	478	492
6	557	611	15	178	165
7	413	415	16	423	372
8	442	431	17	427	421
9	650	638			

Table 15.2. Analysis of variance by subject for the PEFR data of Table 15.1

Source of variation	Degrees of freedom	Sum of squares	Mean square	Variance ratio (F)	Probability
Total	33	445 581.5			
Between subjects	16	441 598.5	27 599.9	117.8	P < 0.000 1
Residual (within subjects)	17	3 983.0	234.3		

the standard deviation of repeated measurements on the same subject. Table 15.1 shows some replicated measurements of peak expiratory flow rate, made with a Wright Peak Flow Meter (see §10.2). We can find the standard deviation within subjects by one way analysis of variance (§10.9), the subjects being the 'groups' of §10.9 (Table 15.2). The variance within subjects is the residual mean square in the analysis of variance table, and the standard deviation, s_w, is the square root of this. From Table 15.2, this is $s_w = \sqrt{234.3} = 15.3$ litres/min.

There are a number of ways in which the measurement error may be presented to the user of the measurement. It may be as the standard deviation calculated above, or it may be, as recommended by the British Standards Institution (1979), the value below which the difference between two measurements will lie with probability 0.95. Provided the measurement errors are from a Normal distribution, this is estimated by $1.96 \times \sqrt{2s_w^2} = 2.77s_w$. BSI (1979) recommend $2 \times \sqrt{2s_w^2} = 2.83s_w$. Clearly, 2.77 is more exact, but it makes no practical difference. Throughout this chapter I have used 2 standard deviations on either side of the mean rather than 1.96, for simplicity.

Measurement error may also be reported as the **within subjects coefficient of variation,** which is the standard deviation divided by the mean, often multiplied by 100 to give a percentage. For our data the mean PEFR is 447.9 litres/min, so the coefficient of variation is $15.3/447.9 = 0.034$ or 3.4%. The difference between the observed value, with measurement er-

Table 15.3. Analysis of variance by subject for the log (base e) transformed PEFR data of Table 15.1

Source of variation	Degrees of freedom	Sum of squares	Mean square	Variance ratio (F)	Probability
Total	33	3.160 104			
Subjects	16	3.139 249	0.196 203	159.9	P < 0.000 1
Residual (within subjects)	17	0.020 855	0.001 227		

ror, and the subject's true value will be at most two standard deviations with probability 0.95. By 'true value' here, I mean the average value which would be obtained over many measurements. The precision may be quoted as being within two standard deviations, $2 \times 15.3 = 30$ litres/min, or within twice the coefficient of variation, $2 \times 15.3/447.9 = 0.068$, or 7%.

The trouble with quoting the error as a percentage in this example is that 7% of the smallest observation, 165 litres, is only 12 litres/min, compared to 7% of the largest, 656, which is 46 litres/min. This is not a good method if the interval is great compared to the size of the smallest observations and the error does not depend on the value of the measurement. It is a good method if the standard deviation is proportional to the mean. In that case a logarithmic transformation (§10.4) can be used. Although there is no reason to apply the log transformation to the data of Table 15.1, I shall do it for illustration. Table 15.3 gives the within subjects standard deviation on the log scale as $s_w = \sqrt{0.001\,227} = 0.035\,0$. This standard deviation does not have the same units as the original data, but is a pure number. If we back-transform by taking the antilog to give $\exp(0.035\,0) = 1.036$, we do not get a standard deviation on the PEFR scale. This is because to get s_w we subtracted the log of one number from the log of another, the mean on the log scale from the observations on the log scale. Now, the difference between the logs of two numbers is the log of their ratio. By subtracting on the log scale we divide one PEFR by the other to get a dimensionless ratio. Thus the antilog of s_w is the ratio of the mean plus one standard deviation to the mean. If we subtract one from this ratio, we get the ratio of the standard deviation to the mean, which is the coefficient of variation. For the example this is $1.036 - 1 = 0.036$, or 3.6%, actually very similar to the 3.4% by the crude method above. When the standard deviation is proportional to the mean we therefore can have a valid use and method of estimation of the coefficient of variation. From it, we can estimate the standard deviation of repeated measurements at any point within the interval of measurement.

We should check to see whether the error does depend on the value of the measurement, usually being larger for larger values. We can do this by plotting a scatter diagram of the absolute value of the difference (i.e. ignoring the sign) and the mean of the two observations (Figure 15.1).

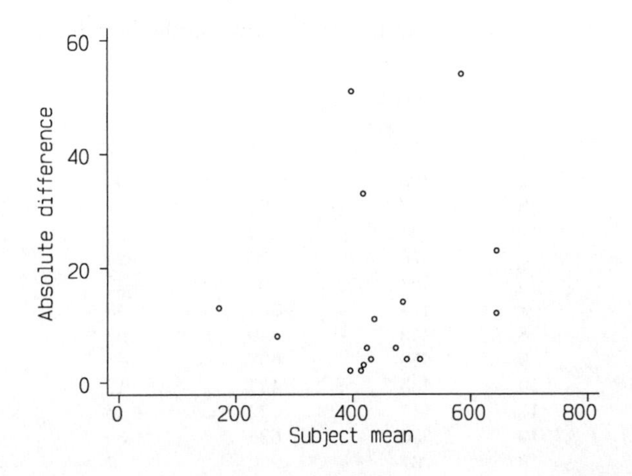

Fig. 15.1. Absolute difference versus sum for 17 pairs of Wright Peak Flow Meter measurements

For the PEFR data, there is no obvious relationship. We can check this by calculating a correlation (§11.9) or rank correlation coefficient (§12.4, §12.5). For Figure 15.1 we have $\tau = 0.17$, $P = 0.3$, so there is little to suggest that the measurement error is related to the size of the PEFR. Hence the coefficient of variation is not as appropriate as the within subjects standard deviation as a representation of the measurement error. For most medical measurements, the standard deviation is either independent of or proportional to the measurement and so one of these two approaches can be used.

Measurement error may also be presented as the correlation coefficient between pairs of readings. This is sometimes called the **reliability** of the measurement, and is often used for psychological measurements using questionnaire scales. However, the correlation depends on the amount of variation between subjects. If we deliberately choose subjects to have a wide spread of possible values, the correlation will be bigger than if we take a random sample of subjects. Thus this method should only be used if we have a representative sample of the subjects in whom we are interested. The **intra-class correlation coefficient**, a special form which does not take into account the order in which observations were taken, and which can be used with more than two observations per subject, is preferred for this application.

15.3 Comparing two methods of measurement

In clinical measurement, most of the things we want to measure, hearts, lungs, livers and so on, are deep within living bodies and out of reach.

Table 15.4. Comparison of two methods of measuring PEFR

Subject number	PEFR (litres/min) Wright Meter	Mini-meter	Difference Wright−mini
1	494	512	−18
2	395	430	−35
3	516	520	−4
4	434	428	6
5	476	500	−24
6	557	600	−43
7	413	364	49
8	442	380	62
9	650	658	−8
10	433	445	−12
11	417	432	−15
12	656	626	30
13	267	260	7
14	478	477	1
15	178	259	−81
16	423	350	73
17	427	451	−24
Total			−36
Mean			2.1
S.d.			38.8

This means that many of the methods we use to measure them are indirect and we cannot be sure how closely they are related to what we really want to know. When a new method of measurement is developed, rather than compare its outcome to a set of known values we must often compare it to another method just as indirect. This is a common type of study, and one which is often badly done (Altman and Bland 1983, Bland and Altman 1986).

Table 15.4 shows measurements of PEFR by two different methods, the Wright meter data coming from Table 15.1. For simplicity, I shall use only one measurement by each method here. We could make use of the duplicate data by using the average of each pair first, but this introduces an extra stage in the calculation. Bland and Altman (1986) give details.

The first step in the analysis is to plot the data as a scatter diagram (Figure 15.2). If we draw the line of equality, along which the two measurements would be exactly equal, this gives us an idea of the extent to which the two methods agree. This is not the best way of looking at data of this type, because much of the graph is empty space and the interesting information is clustered along the line. A better approach is to plot the difference between the methods against the sum or average. The sign of the difference is important, as there is a possibility that one method may give higher values than the other and this may be related to the true value we are trying to measure. This plot is also shown in Figure 15.2.

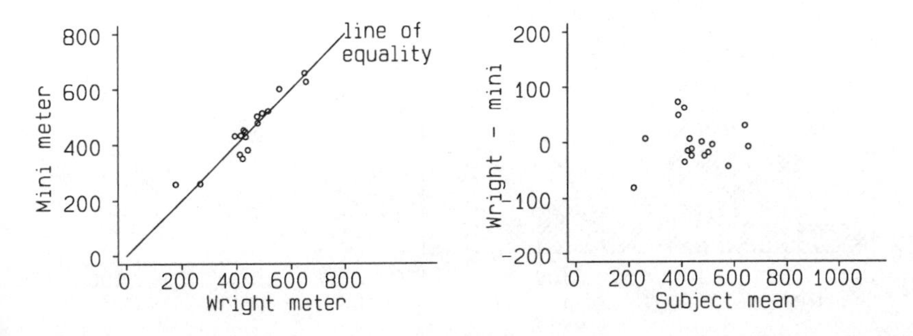

Fig. 15.2. PEFR measured by two different instruments, mini meter vs Wright meter and difference versus mean of mini and Wright meters

Two methods of measurement agree if the difference between observations on the same subject using both methods is small enough for us to use the methods interchangeably. How small this this difference has to be depends on the measurement and the use to which it is to be put. It is a clinical, not a statistical, decision. We quantify the differences by estimating the bias, which is the mean difference, and the limits within which most differences will lie. We estimate these limits from the mean and standard deviation of the differences. If we are to estimate these quantities, we want them to be the same for high values and for low values of the measurement. We can check this from the plot. There is no clear evidence of a relationship between difference and mean in Figure 15.4, and we can check this by a test of significance using the correlation coefficient. We get $r = 0.19, P = 0.5$.

The mean difference is close to zero, so there is little evidence of overall bias. We can find a confidence interval for the mean difference as described in §10.2. The differences have a mean of –2.1 litres/min, and a standard deviation of 38.8. The standard error of the mean is thus $s/\sqrt{n} = 38.8/\sqrt{17} = 9.41$ litres/min and the corresponding value of t with 16 degrees of freedom is 2.12. The 95% confidence interval for the bias is thus $-2.1 \pm 2.12 \times 9.41 = -22$ to $+18$ litres/min. Thus on the basis of these data we could have a bias of as much as 22 litres/min, which could be clinically important. The original comparison of these instruments used a much larger sample and found that any bias was very small (Oldham *et al.* 1979).

The standard deviation of the differences between measurements made by the two methods provides a good index of the comparability of the methods. If we can estimate the mean and standard deviation reliably, with small standard errors, we can then say that the difference between methods will be at most two standard deviations on either side of the mean for 95% of observations. These $\bar{x} \pm 2s$ limits for the difference are called the **95%**

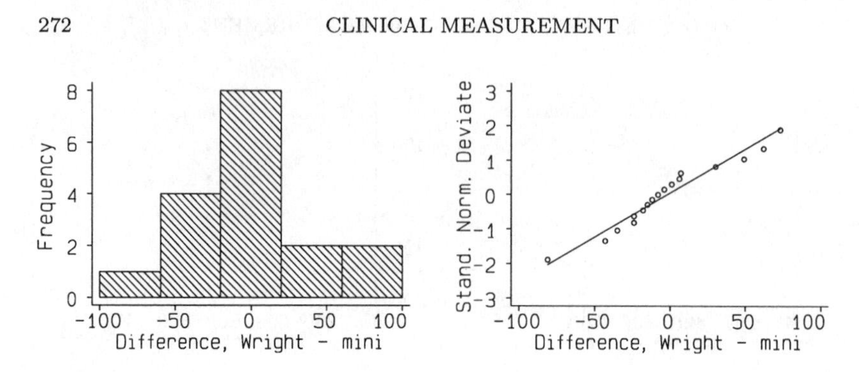

Fig. 15.3. Distribution of differences between PEFR measured by two methods

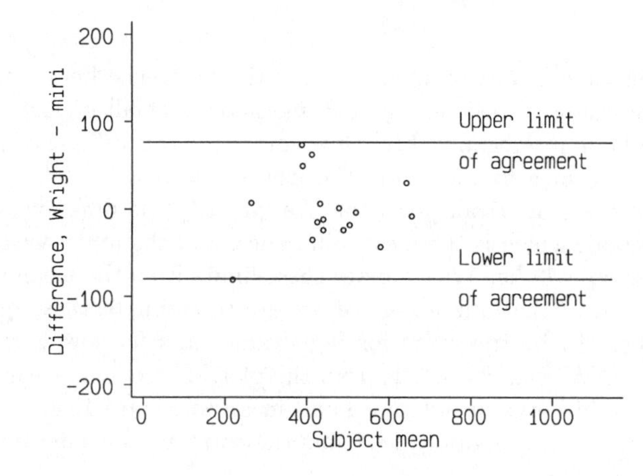

Fig. 15.4. Difference versus sum for PEFR measured by two methods

limits of agreement. For the PEFR data, the standard deviation of the differences is estimated to be 38.8 litres/min and the mean is –2 litres/min. Two standard deviations is therefore 78 litres/min. The reading with the mini-meter is expected to be 80 litres below to 76 litres above for most subjects. These limits are shown as horizontal lines in Figure 15.4. The limits depend on the assumption that the distribution of the differences is approximately Normal, which can be checked by histogram and Normal plot (§7.5) (Figure 15.3).

On the basis of these data we would not conclude that the two methods are comparable or that the mini-meter could reliably replace the Wright peak flow meter. As remarked in §10.2, this meter had received considerable wear.

When there is a relationship between the difference and the mean, we can try to remove it by a transformation. This is usually accomplished by the logarithm, and leads to an interpretation of the limits similar to that described in §15.2. Bland and Altman (1986) give details.

15.4 Sensitivity and specificity

One of the main reasons for making clinical measurements is to aid in diagnosis. This may be to identify one of several possible diagnoses in a patient, or to find people with a particular disease in an apparently healthy population. The latter is known as screening. In either case the measurement provides a test which enables us to classify subjects into two groups, one group whom we think are likely to have disease in which we are interested, and another group unlikely to have the disease. When developing such a test, we need to compare the test result with a true diagnosis. The test may be based on a continuous variable and the disease indicated if it is above or below a given level, or it may be a qualitative observation such as carcinoma *in situ* cells on a cervical smear. In either case I shall call the test positive if it indicates the disease and negative if not, and the disease positive if the disease is later confirmed, negative if not.

How do we measure the effectiveness of the test? Table 15.5 shows three artificial sets of test and disease data. We could take as an index of test effectiveness the proportion giving the correct diagnosis from the test. For Test 1 in the example it is 94%. Now consider Test 2, which always gives a negative result. Test 2 will never detect any cases of the disease. We are now right for 95% of the subjects! However, the first test is useful, in that it detects some cases of the disease, and the second is not, so this is clearly a poor index.

There is no one simple index which enables us to compare different tests in all the ways we would like. This is because there are two things we need to measure: how good the test is at finding disease positives, i.e. those with the condition, and how good the test is at excluding disease negatives, i.e. those who do not have the condition. The indices conventionally employed to do this are:

$$\textbf{sensitivity} = \frac{\text{Number who are both disease positive and test positive}}{\text{Number who are disease positive}}$$

$$\textbf{specificity} = \frac{\text{Number who are both disease negative and test negative}}{\text{Number who are disease negative}}$$

In other words, the sensitivity is a proportion of disease positives who are test positive, and the specificity is the proportion of disease negatives who are test negatives. For our three tests these are:

Table 15.5. Some artificial test and diagnosis data

Test 1	Disease positive	Disease negative	Total
positive	4	5	9
negative	1	90	91
Total	5	95	100

Test 2	Disease positive	Disease negative	Total
positive	0	0	0
negative	5	95	100
Total	5	95	100

Test 3	Disease positive	Disease negative	Total
positive	2	0	2
negative	3	95	98
Total	5	95	100

	Sensitivity	Specificity
Test 1	0.80	0.95
Test 2	0.00	1.00
Test 3	0.40	1.00

Test 2, of course, misses all the disease positives and finds all the disease negatives, by saying all are negative. The difference between Tests 1 and 3 is brought out by the greater sensitivity of 1 and the greater specificity of 3. We are comparing tests in two dimensions. We can see that Test 3 is better than Test 2, because its sensitivity is higher and specificity the same. However, it is more difficult to see whether Test 3 is better than Test 1. We must come to a judgement based on the relative importance of sensitivity and specificity in the particular case. Sensitivity and specificity are often multiplied by 100 to give percentages. They are both binomial proportions, so their standard errors and confidence intervals are found as described in §8.4. The sample size required for their reliable estimation can be calculated as described in §18.2.

When the test is based on a continuous variable, we can alter the sensitivity and specificity by changing the cut-off point. If high values indicate the disease, raising the cut-off point will mean fewer cases will be detected and so the sensitivity will be decreased. However, there will be fewer false positives, positives on test but who do not in fact have the disease, and the specificity will be increased. On the other hand, if we lower the cut-off point we will detect more cases and the sensitivity will be increased, but we will have more false positives and the specificity will be decreased.

For a practical example, Maxwell *et al.* (1983) observed that a remarkable number of alcoholics had evidence at x-ray of past rib fractures. We asked whether this would be of any value in the detection of alcoholism in patients. Among 74 patients with alcoholic liver disease, 20 had evidence of at least one past fracture on chest x-ray and 11 had evidence of bilateral or multiple fractures. In a control group of 181 patients with non-alcoholic liver disease or gastro-intestinal disorders, 6 had evidence of at least one fracture and 2 of bilateral or multiple fractures.

For any fractures as a test for alcoholism, the sensitivity was $20/74 = 0.27$, and the specificity $(181 - 6)/181 = 0.97$. For bilateral or multiple fractures the sensitivity was $11/74 = 0.15$ and the specificity was $(181 - 2)/181 = 0.99$. Hence both tests were very specific; few non-alcoholics would be indicated as alcoholics. On the other hand, neither was very sensitive; many alcoholics would be missed. As we should expect, the more stringent test of bilateral or multiple fractures was more specific and less sensitive than the test of any fracture.

We can also estimate the **positive predictive value**, the probability that a subject who is test positive will be a **true positive** (i.e. has the disease and is correctly classified), and the **negative predictive value**, the probability that a subject who is test negative will be a **true negative** (i.e. does not have the disease and is correctly classified). These depend on the prevalence of the condition, p_{prev}, as well as the sensitivity, p_{sens}, and the specificity, p_{spec}. The probability of being disease positive and test positive is $p_{prev} \times p_{sens}$ and the probability of being disease negative and test positive is $(1 - p_{prev}) \times (1 - p_{spec})$ (see §6.2), so the probability of being test positive is $p_{prev} \times p_{sens} + (1 - p_{prev}) \times (1 - p_{spec})$ and the positive predictive value is

$$\text{PPV} = \frac{p_{prev}p_{sens}}{p_{prev}p_{sens} + (1 - p_{prev})(1 - p_{spec})}$$

Similarly, the negative predictive value is

$$\text{NPV} = \frac{(1 - p_{prev})p_{spec}}{(1 - p_{prev})p_{spec} + p_{prev}(1 - p_{sens})}$$

In screening situations the prevalence is almost always small and the PPV is low. Suppose we have a fairly sensitive and specific test, $p_{sens} = 0.95$ and $p_{spec} = 0.90$, and the disease has prevalence $p_{prev} = 0.01$ (1%). Then

$$\text{PPV} = \frac{0.01 \times 0.95}{0.01 \times 0.95 + (1 - 0.01) \times (1 - 0.90)} = 0.088$$

$$\text{NPV} = \frac{(1 - 0.01) \times 0.90}{(1 - 0.01) \times 0.90 + 0.01 \times (1 - 0.95)} = 0.999$$

so only 8.8% of test positives would be true positives, but almost all test negatives would be true negatives. Most screening tests are dealing with much smaller prevalences than this, so most test positives are false positives.

15.5 Normal range or reference interval

In §15.4 we were concerned with the diagnosis of particular diseases. In this section we look at it the other way round and ask what values measurements on normal, healthy people are likely to have. There are difficulties in doing this. Who is 'normal' anyway? In the UK population almost everyone has hard fatty deposits in their coronary arteries, which result in death for many of them. Very few Africans have this; they die from other causes. So it is normal in the UK to have an abnormality. We usually say that normal people are the apparently healthy members of the local population. We can draw a sample of these as described in Chapter 3 and make the measurement on them.

The next problem is to estimate the set of values. If we use the range of the observations, the difference between the two most extreme values, we can be fairly confident that if we carry on sampling we will eventually find observations outside it, and the range will get bigger and bigger (§4.7). To avoid this we use a range between two quantiles (§4.7), usually the 2.5 centile and the 97.5 centile, which is called the **normal range, 95% reference range** or **95% reference interval**. This leaves 5% of normals outside the 'normal range', which is the set of values within which 95% of measurements from apparently healthy individuals will lie.

A third difficulty comes from confusion between 'normal' as used in medicine and 'Normal distribution' as used in statistics. This has led some people to develop approaches which say that all data which do not fit under a Normal curve are abnormal! Such methods are simply absurd, there is no reason to suppose that all variables follow a Normal distribution (§7.4, §7.5). The term 'reference interval', which is becoming widely used, has the advantage of avoiding this confusion. However, the most commonly used method of calculation rests on the assumption that the variable follows a Normal distribution.

We have already seen that in general most observations fall within two standard deviations of the mean, and that for a Normal distribution 95% are within these limits with 2.5% below and 2.5% above. If we estimate the mean and standard deviation of data from a Normal population we can estimate the reference interval as $\bar{x} - 2s$ to $\bar{x} + 2s$.

Consider the FEV1 data of Table 4.5. We will estimate the reference interval for FEV1 in male medical students. We have 57 observations, mean 4.06 and standard deviation 0.67 litres. The reference interval is thus 2.7 to 5.4 litres. From Table 4.4 we see that in fact only one student (2%) is outside these limits, although the sample is rather small.

As the observations are assumed to be from a Normal distribution, standard errors and confidence intervals for these limits are easy to find. The estimates $\bar{x}$ and s are independent (§7A) with standard errors $\sqrt{s^2/n}$ and $\sqrt{s^2/2(n-1)}$ (§8.2, §8.7). $\bar{x}$ follows a Normal distribution and s a distribution which is approximately Normal. Hence $\bar{x}-2s$ is from a Normal distribution with variance:

$$\begin{aligned}\text{VAR}(\bar{x}-2s) &= \text{VAR}(\bar{x}) + \text{VAR}(2s) = \text{VAR}(\bar{x}) + 4\text{VAR}(s) \\ &= \frac{s^2}{n} + 4 \times \frac{s^2}{2(n-1)} = s^2\left(\frac{1}{n} + \frac{2}{n-1}\right)\end{aligned}$$

Hence, provided Normal assumptions hold, the standard error of the limit of the reference interval is

$$\sqrt{s^2\left(\frac{1}{n} + \frac{2}{n-1}\right)}$$

If n is large, this is approximately $\sqrt{3s^2/n}$. For the FEV1 data, this is $\sqrt{3 \times 0.67^2/57} = 0.15$. Hence the 95% confidence intervals for these limits are $2.7 \pm 1.96 \times 0.15$ and $5.4 \pm 1.96 \times 0.15$, i.e. from 2.4 to 3.0 and 5.1 to 5.7 litres.

Compare the serum triglyceride measurements of Table 15.6. As already noted (§7.4), the data are highly skewed, and we cannot use the Normal method directly. If we did, the lower limit would be 0.07, well below any of the observations, and the upper limit would be 0.94, greater than which are 5% of the observations. It is possible for such data to give a negative lower limit.

Figure 7.11 shows the $\log_{10}$ transformed data, which give a breathtakingly symmetrical distribution ($\bar{x} = -0.331$, $s = 0.171$). The lower limit in the transformed data is –0.67, corresponding to a triglyceride level of 0.21, below which are 2.1% of observations. The upper limit is 0.01, corresponding to 1.02, above which are 2.5% of observations. The fit to the log transformed data is excellent. For the standard error of the reference limit we have $\sqrt{3 \times 0.171^2/282} = 0.0176$. The 95% confidence intervals are thus $-0.673 \pm 1.96 \times 0.0176$ and $0.011 \pm 1.96 \times 0.0176$, i.e. –0.707 to –0.639 and –0.023 to 0.045. In the untransformed data this gives 0.196 to 0.230 and 0.948 to 1.109, found by taking the antilogs. These confidence limits can be transformed back to the original scale, unlike those in §10.4, because no subtraction of means has taken place.

Because of the obviously unsatisfactory nature of the Normal method for some data, some authors have advocated the estimation of the percentiles directly (§4.5), without any distributional assumptions. This is

Table 15.6. Serum triglyceride measurements in cord blood from 282 babies

0.15	0.29	0.34	0.38	0.41	0.46	0.52	0.56	0.64	0.80
0.16	0.30	0.34	0.38	0.41	0.46	0.52	0.56	0.64	0.80
0.20	0.30	0.34	0.38	0.41	0.46	0.52	0.56	0.65	0.82
0.20	0.30	0.34	0.39	0.42	0.46	0.52	0.57	0.66	0.82
0.20	0.30	0.34	0.39	0.42	0.47	0.52	0.57	0.66	0.82
0.20	0.30	0.34	0.39	0.42	0.47	0.52	0.58	0.66	0.82
0.21	0.30	0.34	0.39	0.42	0.47	0.52	0.58	0.66	0.83
0.22	0.30	0.35	0.39	0.42	0.47	0.53	0.58	0.66	0.84
0.24	0.30	0.35	0.40	0.42	0.47	0.54	0.58	0.67	0.84
0.25	0.30	0.35	0.40	0.44	0.48	0.54	0.59	0.67	0.84
0.26	0.31	0.35	0.40	0.44	0.48	0.54	0.59	0.68	0.86
0.26	0.31	0.35	0.40	0.44	0.48	0.54	0.59	0.70	0.87
0.26	0.32	0.35	0.40	0.44	0.48	0.54	0.59	0.70	0.88
0.27	0.32	0.36	0.40	0.44	0.48	0.54	0.60	0.70	0.88
0.27	0.32	0.36	0.40	0.44	0.48	0.55	0.60	0.70	0.95
0.27	0.32	0.36	0.40	0.44	0.48	0.55	0.60	0.72	0.96
0.28	0.32	0.36	0.40	0.44	0.48	0.55	0.60	0.72	0.96
0.28	0.32	0.36	0.40	0.44	0.48	0.55	0.60	0.74	0.99
0.28	0.32	0.36	0.40	0.45	0.48	0.55	0.60	0.75	1.01
0.28	0.32	0.36	0.40	0.45	0.48	0.55	0.60	0.75	1.02
0.28	0.33	0.36	0.40	0.45	0.48	0.55	0.60	0.76	1.02
0.28	0.33	0.36	0.40	0.45	0.49	0.55	0.61	0.76	1.04
0.28	0.33	0.37	0.40	0.45	0.49	0.56	0.62	0.78	1.08
0.28	0.33	0.37	0.40	0.45	0.49	0.56	0.62	0.78	1.11
0.29	0.33	0.37	0.41	0.46	0.50	0.56	0.63	0.78	1.20
0.29	0.33	0.37	0.41	0.46	0.50	0.56	0.64	0.78	1.28
0.29	0.33	0.38	0.41	0.46	0.50	0.56	0.64	0.78	1.64
0.29	0.33	0.38	0.41	0.46	0.50	0.56	0.64	0.78	1.66
0.29	0.34								

an attractive idea. We want to know the point below which 2.5% of values will fall. Let us simply rank the observations and find the point below which 2.5% of the observations fall. For the 282 triglycerides, the 2.5 and 97.5 centiles are found as follows. For the 2.5 centile, we find $i = q(n+1) = 0.025 \times (282+1) = 7.08$. The required quantile will be between the 7th and 8th observation. The 7th is 0.21, the 8th is 0.22 so the 2.5 centile would be estimated by $0.21 + (0.22 - 0.21) \times (7.08 - 7) = 0.211$. Similarly the 97.5 centile is 1.039.

This approach gives an unbiassed estimate whatever the distribution. The log transformed triglyceride would give exactly the same results. Note that the Normal theory limits from the log transformed data are very similar. We now look at the confidence interval. The 95% confidence interval for the q quantile, here q being 0.025 or 0.975, estimated directly from the data is found by an application of the Binomial distribution (§6.4, §6.6) (see Conover 1980). The number of observations less than the q quantile will be an observation from a Binomial distribution with parameters n and

q, and hence has mean nq and standard deviation $\sqrt{nq(1-q)}$. Calculate j and k:

$$j = nq - 1.96\sqrt{nq(1-q)}$$

$$k = nq + 1.96\sqrt{nq(1-q)}$$

We round j and k up to the next integer. Then the 95% confidence interval is between the jth and the kth observations in the ordered data. For the triglyceride data, $n = 282$ and so for the lower limit, $q = 0.025$, we have

$$j = 282 \times 0.025 - 1.96\sqrt{282 \times 0.025 \times 0.975}$$

$$k = 282 \times 0.025 + 1.96\sqrt{282 \times 0.025 \times 0.975}$$

This gives $j = 1.9$ and $k = 12.2$, which we round up to $j = 2$ and $k = 13$. In the triglyceride data the second observation, corresponding to $j = 2$, is 0.16 and the 13th is 0.26. Thus the 95% confidence interval for the lower reference limit is 0.16 to 0.26. The corresponding calculation for $q = 0.975$ gives $j = 270$ and $k = 281$. The 270th observation is 0.96 and the 281st is 1.64, giving a 95% confidence interval for the upper reference limit of 0.96 to 1.64. These are wider confidence intervals than those found by the Normal method, those for the long tail particularly so. This method of estimating percentiles in long tails is relatively imprecise.

15.6 Survival data

We often have data which represent the time from some event to death, such as time from diagnosis or from entry to a clinical trial, but survival analysis does not have to be about death. In cancer studies we can use survival analysis for the time to metastasis or to local recurrence of a tumour, in a study of medical care we can use it to analyse the time to readmission to hospital, in a study of breast-feeding we could look at the age at which breast-feeding ceased or at which bottle feeding was first introduced, and in a study of the treatment of infertility we can treat the time from treatment to conception as survival data. We usually refer to the terminal event, death, conception, etc., as the **endpoint**.

Problems arise in the measurement of survival because often we do not know the exact survival times of all subjects. This is because some will still be surviving when we want to analyse the data. When cases have entered the study at different times, some of the recent entrants may be surviving, but only have been observed for a short time. Their observed survival time may be less than those cases admitted early in the study and who have since died. The method of calculating survival curves described below takes this into account. Observations which are known only to be greater than some value are **right censored**, often shortened to **censored**.

Table 15.7. Survival time in years of patients after diagnosis of parathyroid cancer

Alive	Deaths
<1	<1
<1	2
1	6
1	6
4	7
5	9
6	9
8	11
10	14
10	
17	

(We get **left censored** data when the measurement method cannot detect anything below some cut-off value, and observations are recorded as 'none dectectable'. The rank methods in Chapter 12 are useful for such data).

Table 15.7 shows some survival data, for patients with parathyroid cancer. The survival times are recorded in completed years. A patient who survived for 6 years and then died can be taken as having lived for 6 years and then died in the seventh. In the first year from diagnosis, one patient died. Three patients were observed for only part of this year and 17 survived into the next year. The 2 who have only been observed for part of the year are censored, also called **lost to follow-up** or **withdrawn from follow-up**. (These are rather misleading names, often wrongly interpreted as meaning that these subjects have dropped out of the study. This may be the case, but most of these subjects are simply still alive and their further survival is unknown.) There is no information about the survival of these subjects after the first year, because it has not happened yet. These patients are only at risk of dying for part of the year and we cannot say that 1 out of 20 died as they may yet contribute another death in the first year. We can say that such patients will contribute half a year of risk, on average, so the number of patient years at risk in the first year is 18 (17 who survived and 1 who died) plus 2 halves for those withdrawn from follow-up, giving 19 altogether. We get an estimate of the probability of dying in the first year of 1/19, and an estimated probability of surviving of $1-\frac{1}{19}$. We can do this for each year until the limits of the data are reached. We thus trace the survival of these patients estimating the probability of death or survival at each year and the cumulative probability of survival to each year. This set of probabilities is called a **life table**.

To carry out the calculation, we first set out for each year, x, the number alive at the start, n_x, the number withdrawn during the year, w_x, and the number at risk, r_x, and the number dying, d_x (Table 15.8). Thus in year

Table 15.8. Life table calculation for parathyroid cancer survival

Year	Number at start	Withdrawn during year	At risk	Deaths	Prob of death	Prob of surviving year x	Cumulative prob of surviving x years
x	n_x	w_x	r_x	d_x	q_x	p_x	P_x
1	20	2	19	1	0.0526	0.9474	0.9474
2	17	2	16	0	0	1	0.9474
3	15	0	15	1	0.0667	0.9333	0.8842
4	14	0	14	0	0	1	0.8842
5	14	1	13.5	0	0	1	0.8842
6	13	1	12.5	0	0	1	0.8842
7	12	1	11.5	2	0.1739	0.8261	0.7304
8	9	0	9	1	0.1111	0.8889	0.6493
9	8	1	7.5	0	0	1	0.6493
10	7	0	7	2	0.2857	0.7143	0.4638
11	5	2	4	0	0	1	0.4638
12	3	0	3	1	0.3333	0.6667	0.3092
13	2	0	2	0	0	1	0.3092
14	2	0	2	0	0	1	0.3092
15	2	0	2	1	0.5000	0.5000	0.1546
16	1	0	1	0	0	1	0.1546
17	1	0	1	0	0	1	0.1546
18	1	1	0.5	0	0	1	0.1546

$r_x = n_x - \frac{1}{2}w_x$, $q_x = d_x/r_x$, $p_x = 1 - q_x$, $P_x = p_x P_{x-1}$

1 the number at the start is 20, the number withdrawn is 2, the number at risk $r_1 = n_1 - \frac{1}{2}w_1 = 20 - \frac{1}{2} \times 2 = 19$ and the number of deaths is 1. As there were 2 withdrawals and 1 death the number at the start of year 2 is 17. For each year we calculate the probability of dying in that year for patients who have reached the beginning of it, $q_x = d_x/r_x$, and hence the probability of surviving to the next year, $p_x = 1 - q_x$. Finally we calculate the cumulative survival probability. For the first year, this is the probability of surviving that year, $P_1 = p_1$. For the second year, it is the probability of surviving up to the start of the second year, P_1, times the probability of surviving that year, p_2, to give $P_2 = p_2 P_1$. The probability of surviving for 3 years is similarly $P_3 = p_3 P_2$, and so on. From this life table we can estimate the **five year survival rate**, a useful measure of prognosis in cancer. For the parathyroid cancer, the 5 year survival rate 0.8842, or 88%. We can see that the prognosis for this cancer is quite good. If we know the exact time of death or withdrawal for each subject, then instead of using fixed time intervals we use x as the exact time, with a row of the table for each time when either an endpoint or a withdrawal occurs. Then $r_x = n_x$ and we can omit the $r_x = n_x - \frac{1}{2}w_x$ step.

We can draw a graph of the cumulative survival probability, the **survival curve**. This is usually drawn in steps, with abrupt changes in probability (Figure 15.5). This convention emphasizes the relatively poor esti-

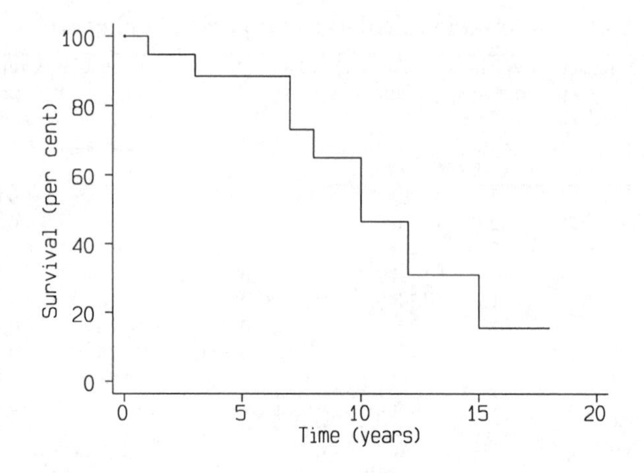

Fig. 15.5. Survival curve for parathyroid cancer patients

mation at the long survival end of the curve, where the small numbers at risk produced large steps. When the exact times of death and censoring are known, this is called a **Kaplan–Meier survival curve**. The times at which observations are censored may be marked by small vertical lines above the survival curve (Figure 15.6), and the number remaining at risk may be written at suitable intervals below the time axis.

The standard error and confidence interval for the survival probabilities can be found (see Armitage and Berry 1987). These are useful for estimates such as five year survival rate. They do not provide a good method for comparing survival curves, as they do not include all the data, only using those up to the chosen time. Survival curves start off together at 100% survival, possibly diverge, but eventually come together at zero survival. Thus the comparison would depend on the time chosen. Survival curves can be compared by several significance tests, of which the best known is the **logrank** test. This is a non-parametric test which makes use of the full survival data without making any assumption about the shape of the survival curve.

Table 15.9 shows the time to recurrence of gallstones following dissolution by bile acid treatment or lithotrypsy. Here we shall compare the two groups defined by having single or multiple gallstones, using the logrank test. We shall look at the quantitative variables diameter of gallstone and months to dissolve in §17.9. Figure 15.6 show the time to recurrence for subjects with single primary gall stones and multiple primary gall stones. The null hypothesis is that there is no difference in recurrence-free survival time, the alternative that there is such a difference. The calculation of the logrank test is set out in Table 15.10. For each time at which a recurrence

Table 15.9. Time to recurrence of gallstones following dissolution, whether previous gall stones were multiple, maximum diameter of previous gall stones, and months previous gall stones took to dissolve

time	rec.	mult	diam	dis
3	no	yes	4	10
3	no	no	18	3
3	no	yes	5	27
4	no	yes	4	4
5	no	no	19	20
6	no	yes	3	10
6	no	yes	4	6
6	no	yes	4	20
6	yes	yes	5	8
6	yes	yes	3	18
6	yes	yes	7	9
6	no	no	25	9
6	no	yes	4	6
6	yes	yes	10	38
6	yes	yes	8	15
6	no	yes	4	13
7	yes	yes	4	15
7	no	yes	3	7
7	yes	yes	10	48
8	yes	yes	14	29
8	yes	no	18	14
8	yes	yes	6	6
8	no	no	15	1
8	no	yes	1	12
8	no	yes	5	6
9	no	yes	2	15
9	yes	yes	7	6
9	no	no	19	8
10	yes	yes	14	8
11	no	yes	8	12
11	no	no	15	15
11	yes	no	5	8
11	no	yes	3	6
11	yes	yes	5	12
11	no	yes	4	6
11	no	yes	4	3
11	no	yes	13	18
11	yes	no	7	8
12	yes	yes	5	7
12	yes	yes	8	12
12	no	yes	4	6
12	no	yes	4	8
12	yes	yes	7	19
12	yes	no	7	3
12	no	yes	5	22
12	yes	no	8	1
12	no	no	6	6
12	no	no	26	4
13	no	yes	5	6
13	no	no	13	6
13	no	no	11	6
13	no	no	22	33
13	no	no	13	9
13	yes	yes	8	12
14	no	yes	6	6
14	no	no	23	15
14	no	no	15	10
16	yes	yes	5	6
16	yes	yes	6	8
16	no	no	18	4
17	no	no	7	10
17	no	yes	4	3
17	no	yes	7	6
17	yes	no	8	8
17	no	yes	5	6
18	yes	no	10	9
18	yes	yes	8	38
18	no	yes	11	11
19	no	no	26	6
19	no	yes	11	16
19	yes	yes	5	7
20	no	no	11	2
20	no	no	13	9
20	no	no	6	7
21	no	yes	11	1
21	no	yes	13	24
21	no	yes	4	11
22	no	no	10	4
22	no	no	20	20
23	no	no	16	6
24	no	no	15	4
24	no	yes	3	6
24	no	no	15	2
24	yes	yes	7	6
25	no	no	13	10
25	yes	yes	6	3
25	no	no	4	11
26	no	no	17	5
26	no	yes	6	12
26	yes	no	16	8
28	no	no	20	3
28	yes	no	30	4
29	no	no	16	3
29	yes	no	12	15
29	yes	yes	10	7
29	no	yes	7	6
30	no	yes	4	4
30	no	no	9	12
30	yes	yes	22	10
30	yes	yes	6	3

Table 15.9. continued

time	rec.	mult	diam	dis	time	rec.	mult	diam	dis
31	no	yes	5	6	38	no	no	10	18
31	no	no	26	3	38	yes	yes	5	10
31	no	no	7	24	38	no	no	7	4
32	yes	yes	10	12	40	no	no	23	1
32	no	yes	5	6	41	no	no	16	2
32	no	no	4	6	41	no	no	4	14
32	no	no	18	10	42	no	no	15	43
33	no	no	13	9	42	no	yes	16	6
34	no	no	15	8	42	no	yes	9	11
34	no	no	20	30	42	no	yes	14	9
34	no	yes	15	8	43	yes	no	4	17
34	no	no	27	8	44	no	yes	7	6
35	no	no	6	12	44	no	yes	10	8
36	no	no	18	5	45	no	no	12	17
36	no	yes	6	16	47	no	yes	4	3
36	no	yes	5	6	48	no	no	21	11
36	no	yes	8	17	48	no	no	9	10
36	no	no	5	4	53	no	yes	6	9
37	no	yes	5	7	60	yes	no	15	15
37	no	no	19	4	61	no	no	10	11
37	no	yes	4	4	65	no	yes	5	3
37	no	yes	4	12	70	no	yes	7	12

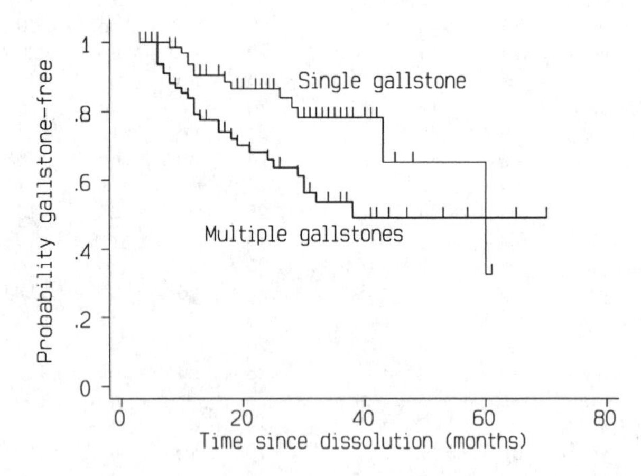

Fig. 15.6. Gallstone-free survival after the dissolution of single and multiple gallstones

Table 15.10. Calculation for the logrank test

time	n_1	d_1	w_1	n_2	d_2	w_2	p_d	e_1	e_2
3	65	0	1	79	0	2	0.000	0.000	0.000
4	64	0	0	77	0	1	0.000	0.000	0.000
5	64	0	1	76	0	0	0.000	0.000	0.000
6	63	0	1	76	5	5	0.036	2.266	2.734
7	62	0	0	66	2	1	0.016	0.969	1.031
8	62	1	1	63	2	2	0.024	1.488	1.512
9	60	0	1	59	1	1	0.008	0.504	0.496
10	59	0	0	57	1	0	0.009	0.509	0.491
11	59	2	1	56	1	5	0.026	1.539	1.461
12	56	2	2	50	3	3	0.047	2.642	2.358
13	52	0	4	44	1	1	0.010	0.542	0.458
14	48	0	2	42	0	1	0.000	0.000	0.000
16	46	0	1	41	2	0	0.023	1.057	0.943
17	45	1	1	39	0	3	0.012	0.536	0.464
18	43	1	0	36	1	1	0.025	1.089	0.911
19	42	0	1	34	1	1	0.013	0.553	0.447
20	41	0	3	32	0	0	0.000	0.000	0.000
21	38	0	0	32	0	3	0.000	0.000	0.000
22	38	0	2	29	0	0	0.000	0.000	0.000
23	36	0	1	29	0	0	0.000	0.000	0.000
24	35	0	2	29	1	1	0.016	0.547	0.453
25	33	0	2	27	1	0	0.017	0.550	0.450
26	31	1	1	26	0	1	0.018	0.544	0.456
28	29	1	1	25	0	0	0.019	0.537	0.463
29	27	1	1	25	1	1	0.038	1.038	0.962
30	25	0	1	23	2	1	0.042	1.042	0.958
31	24	0	2	20	0	1	0.000	0.000	0.000
32	22	0	2	19	1	1	0.024	0.537	0.463
33	20	0	1	17	0	0	0.000	0.000	0.000
34	19	0	3	17	0	1	0.000	0.000	0.000
35	16	0	1	16	0	0	0.000	0.000	0.000
36	15	0	2	16	0	3	0.000	0.000	0.000
37	13	0	1	13	0	3	0.000	0.000	0.000
38	12	0	2	10	1	0	0.045	0.545	0.455
40	10	0	1	9	0	0	0.000	0.000	0.000
41	9	0	2	9	0	0	0.000	0.000	0.000
42	7	0	1	9	0	3	0.000	0.000	0.000
43	6	1	0	6	0	0	0.083	0.500	0.500
44	5	0	0	4	0	2	0.000	0.000	0.000
45	5	0	1	4	0	0	0.000	0.000	0.000
47	4	0	0	4	0	1	0.000	0.000	0.000
48	4	0	2	3	0	0	0.000	0.000	0.000
53	2	0	0	3	0	1	0.000	0.000	0.000
60	2	1	0	2	0	0	0.250	0.500	0.500
61	1	0	1	2	0	0	0.000	0.000	0.000
65	0	0	0	2	0	1	0.000	0.000	0.000
70	0	0	0	1	0	1	0.000	0.000	0.000
Total		12			27			20.032	18.968

$p_d = (d_1 + d_2)/(n_1 + n_2)$, $e_1 = p_d n_1$, $e_2 = p_d n_2$

or a censoring occurred, we have the numbers under observation in each group, n_1 and n_2, the number of recurrences, d_1 and d_2 (d for death), and the number of censorings, w_1 and w_2 (w for withdrawal). For each time, we calculate the probability of recurrence, $p_d = (d_1 + d_2)/(n_1 + n_2)$, which each subject would have if the null hypothesis were true. For each group, we calculate the expected number of recurrences, $e_1 = p_d \times n_1$ and $e_2 = p_d \times n_2$. We then calculate the numbers at risk at the next time, $n_1 - d_1 - w_1$ and $n_2 - d_2 - w_2$. We do this for each time. We then add the d_1 and d_2 columns to get the observed numbers of recurrences, and the e_1 and e_2 columns to get the numbers of recurrences expected if the null hypothesis were true.

We have observed frequencies of recurrence d_1 and d_2, and expected frequencies e_1, and e_2. Of course, $d_1 + d_2 = e_1 + e_2$, so we only need to calculate e_1 as in Table 15.10, and hence e_2 by subtraction. This only works for two groups, however, and the method of Table 15.10 works for any number of groups. We can test the null hypothesis that the risk of recurrence in any month is equal for the two populations by a chi-squared test:

$$\sum \frac{(d_i - e_i)^2}{e_i} = \frac{(12 - 20.032)^2}{20.032} + \frac{(27 - 18.968)^2}{18.968} = 6.62$$

There is one constraint, that the two frequencies add to the sum of the expected (i.e. the total number of recurrences), so we lose one degree of freedom, giving $2 - 1 = 1$ degree of freedom. From Table 13.3, this has a probability of 0.01.

Some texts describe this test differently, saying that under the null hypothesis d_1 is from a Normal distribution with mean e_1 and variance $e_1 e_2/(e_1 + e_2)$. This is algebraically identical to the chi-squared method, but only works for two groups.

The logrank test is nonparametric, because we make no assumptions about either the distribution of survival time or any difference in recurrence rates. It requires the survival or censoring times to be exact. A similar method for grouped data as in Table 15.8 is given by Mantel (1966).

15.7 Computer aided diagnosis

Reference intervals (§15.5) are one area where statistical methods are involved directly in diagnosis, computer aided diagnosis is another. The 'aided' is put in to persuade clinicians that the main purpose is not to do them out of a job, but, naturally, they have their doubts. Computer aided diagnosis is partly a statistical exercise. There are two types of computer aided diagnosis: statistical methods, where diagnosis is based on a set of data obtained from past cases, and decision tree methods, which try to imitate the thought processes of an expert in the field. We shall look briefly at each approach.

There are several methods of statistical computer aided diagnosis. One uses **discriminant analysis**. In this we start with a set of data on subjects whose diagnosis was subsequently confirmed, and calculate one or more discriminant functions. A discriminant function has the form:

$$constant_1 \times variable_1 + constant_2 \times variable_2 + \ldots + constant_k \times variable_k$$

The constants are calculated so that the values of the functions are as similar as possible for members of the same group and as different as possible for members of different groups. In the case of only two groups, we have one discriminant function and all the subjects in one group will have high values of the function and all subjects in the other will have low values. For each new subject we evaluate the discriminant function and use it to allocate the subject to a group or diagnosis. We can estimate the probability of the subject falling in that group, and in any other. Many forms of discriminant analysis have been developed to try and improve this form of computer diagnosis, but it does not seem to make much difference which is used. Logistic regression (§17.8) can also be used.

A different approach uses **Bayesian** analysis. This is based on **Bayes' Theorem**, a result about probability which may be stated in terms of the probability of diagnosis A being true if we have observed data B, as:

$$\text{PROB(diag } A \text{ if data } B) = \frac{\text{PROB(data } B \text{ if diag } A) \times \text{PROB(diag } A)}{\text{PROB(data } B)}$$

If we have a large data set of known diagnoses and their associated symptoms and signs, we can determine PROB(diagnosis A) easily. It is simply the proportion of times A has been diagnosed. The problem of finding the probability of a particular combination of symptoms and signs is more difficult. If they are all independent, we can say that the probability of a given symptom is the proportion of times it occurs, and the probability of the symptom for each diagnosis is found in the same way. The probability of any combination of symptoms can be found by multiplying their individual probabilities together, as described in §6.2. In practice the assumption that signs and symptoms are independent is most unlikely to be met and a more complicated analysis would be required to deal with this. However, some systems of computer aided diagnosis have been found to work quite well with the simple approach.

Expert or knowledge-based systems work in a different way. Here the knowledge of a human expert or group of experts in the field is converted into a series of decision rules, e.g. 'if the patient has post bilateral rib fractures then the patient is an alcoholic, if not then on to the next decision'.

These systems can be modified by asking further experts to test the system with cases from their own experience and to suggest further decision rules if the program fails. They also have the advantage that the program can 'explain' the reason for its 'decision' by listing the series of steps which led to it. Most of Chapter 14 consists of rules of just this type and could be turned into an expert system for statistical analysis.

Although there have been some impressive achievements in the field of computer diagnosis, it has to date made little progress towards acceptance in routine medical practice. As computers become more familiar to clinicians, more common in their surgeries and more powerful in terms of data storage and processing speed, we may expect computer aided diagnosis to become as well established as computer aided statistical analysis is today.

15.M Multiple choice questions 81 to 86

(Each answer is true or false)

81. The repeatability or precision of measurements may be measured by:

(a) the coefficient of variation of repeated measurements;

(b) the standard deviation of measurements between subjects;

(c) the standard deviation of the difference between pairs of measurements;

(d) the standard deviation of repeated measurements within subjects;

(e) the difference between the means of two sets of measurements on the same set of subjects.

82. The specificity of a test for a disease:

(a) has a standard error derived from the Binomial distribution;

(b) measures how well the test detects cases of the disease;

(c) measures how well the test excludes subjects without the disease;

(d) measures how often a correct diagnosis is obtained from the test;

(e) is all we need to tell us how good the test is.

83. The level of an enzyme measured in blood is used as a diagnostic test for a disease, the test being positive if the enzyme concentration is above a critical value. The sensitivity of the diagnostic test:

(a) is one minus the specificity;

(b) is a measure of how well the test detects cases of the disease;

(c) is the proportion of people with the disease who are positive on the test;

(d) increases if the critical value is lowered;

(e) measures how well people without the disease are excluded.

84. A 95% reference interval, 95% reference range, or normal range:

(a) may be calculated as two standard deviations on either side of the mean;

(b) may be calculated directly from the frequency distribution;

(c) can only be calculated if the observations follow a Normal distribution;

(d) gets wider as the sample size increases;

(e) may be calculated from the mean and its standard error.

85. If the 95% reference interval for haematocrit in men is 43.2 to 49.2:

(a) any man with haematocrit outside these limits is abnormal;

(b) haematocrits outside these limits are proof of disease;

(c) a man with a haematocrit of 46 must be very healthy;

(d) a woman with a haematocrit of 48 has a haematocrit within normal limits;

(e) a man with a haematocrit of 42 may be ill.

86. When a survival curve is calculated from censored survival times:

(a) the estimated proportion surviving becomes less reliable as survival time increases;

(b) individuals withdrawn during the first time interval are excluded from the analysis;

(c) survival estimates depend on the assumption that survival rates remain constant over the study period;

(d) it may be that the survival curve will not reach zero survival;

(e) the five year survival rate can be calculated even if some of the subjects were identified less than five years ago.

15.E Exercise: A reference interval

In this exercise we shall estimate a reference interval. Mather *et al.* (1979) measured plasma magnesium in 140 apparently healthy people, to compare with a sample of diabetics. The normal sample was chosen from blood donors and people attending day centres for the elderly in the area of St. George's Hospital, to give 10 male and 10 female subjects in each age decade from 15–24 to 75 years and over. Questionnaires were used to exclude any subject with persistent diarrhoea, excessive alcohol intake or who were on regular drug therapy other than hypnotics and mild analgesics in the elderly. The distribution of plasma magnesium is shown in Figure 15.7. The mean was 0.810 mmol/litre and the standard deviation 0.057 mmol/litre.

1. What do you think of the sampling method? Why use blood donors and elderly people attending day centres?

2. Why were some potential subjects excluded? Was this a good idea? Why were certain drugs allowed for the elderly?

3. Does plasma magnesium appear to follow a Normal distribution?

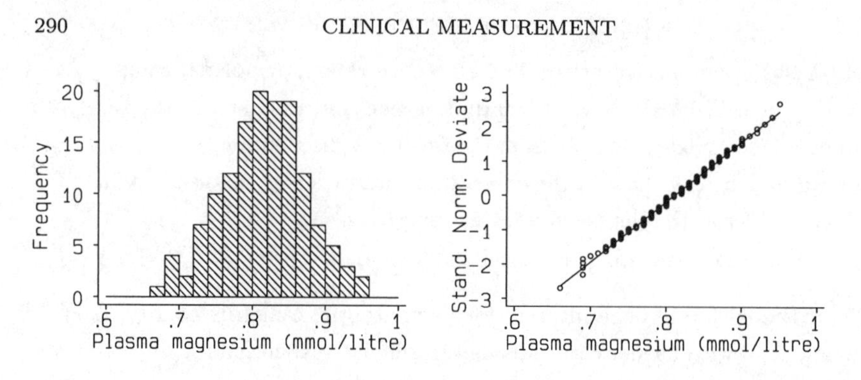

Fig. 15.7. Distribution of plasma magnesium in 140 apparently healthy people

4. What is the reference interval for plasma magnesium, using the Normal distribution method?
5. Find confidence intervals for the reference limits.
6. Would it matter if mean plasma magnesium in normal people increased with age? What method might be used to improve the estimate of the reference interval in this case?

16

Mortality statistics and population structure

16.1 Mortality rates

Mortality statistics are one of our principal sources of information about the changing pattern of disease within a country and the differences in disease between countries. In most developed countries, any death must be certified by a doctor, who records the cause, date and place of death and some data about the deceased. In Britain, these include the date of birth, area of residence and last known occupation. These death certificates form the raw material from which mortality statistics are compiled by a national bureau of censuses, in Britain the Office of Population Censuses and Surveys. The numbers of deaths can be tabulated by cause, sex, age, types of occupation, area of residence, and marital status. Table 5.1 shows one such tabulation, of deaths by cause and sex.

For purposes of comparison we must relate the number of deaths to the number in the population in which they occur. We have this information fairly reliably at 10 year intervals from the decennial census of the country. We can estimate the size and age and sex structure of the population between censuses using registration of births and deaths. Each birth or death is notified to an official registrar, and so we can keep some track of changes in the population. There are other, less well documented changes taking place, such as immigration and emigration, which mean that population size estimates between the census years are only approximations. Some estimates, such as the numbers in different occupations, are so unreliable that mortality data is only tabulated by them for census years.

If we take the number of deaths over a given period of time and divide it by the number in the population and the time period, we get a mortality rate, the number of deaths per unit time per person. We usually take the number of deaths over one calendar year, although when the number of deaths is small we may take deaths over several years, to increase the precision of the numerator. The number in the population is changing continually, and we take as the denominator the estimated population at the mid-point of the time period. Mortality rates are often very small numbers, so we usually multiply them by a constant, such as 1 000 or 100 000, to avoid strings of zeros after the decimal point.

When we are dealing with deaths in the whole population, irrespective of age, the rate we obtain is called the **crude mortality rate** or **crude death rate**. The terms 'death rate' and 'mortality rate' are used interchangeably. We calculate the crude mortality rate for a population as:

$$\frac{\text{deaths occurring over given period}}{\text{number in population at mid-point of period} \times \text{length of period}} \times 1\,000$$

If the period is in years, this gives the crude mortality rate as deaths per 1 000 population per year.

The crude mortality rate is so called because no allowance is made for the age distribution of the population, and comparisons between populations with different age structures. For example, in 1901 the crude mortality rate among adult males (aged over 15 years) in England and Wales was 15.7 per 1 000 per year, and in 1981 it was 15.6 per 1 000 per year. It seems strange that with all the improvements in medicine, housing and nutrition between these times there has been so little improvement in the crude mortality rate. To see why we must look at the **age specific mortality rates**, the mortality rates within narrow age groups. Age specific mortality rates are usually calculated for one, five or ten year age groups. In 1901 the age specific mortality rate for men aged 15 to 19 was 3.5 deaths per 1 000 per year, whereas in 1981 it was only 0.8. As Table 16.1 shows, the age specific mortality rate in 1901 was greater than that in 1981 for every age group. However in 1901 there was a much greater proportion of the population in the younger age groups, where mortality was low, than there was in 1981. Correspondingly, there was a smaller proportion of the 1901 population than the 1981 population in the higher mortality older age groups. Although mortality was lower at any given age in 1981, the greater proportion of older people meant that there were as many deaths as in 1901.

To eliminate the effects of different age structures in the populations which we want to compare, we can look at the age specific death rates. But if we are comparing several populations, this is a rather cumbersome procedure, and it is often more convenient to calculate a single summary figure from the age specific rates. There are many ways of doing this, of which three are frequently used: the direct and indirect methods of age standardization and the life table.

Table 16.1. Age specific mortality rates and age distribution in adult males, England and Wales, 1901 and 1981

Age group (years)	Age-specific death rate per 1 000 per year		% adult population in age group	
	1901	1981	1901	1981
15–19	3.5	0.8	15.36	11.09
20–24	4.7	0.8	14.07	9.75
25–34	6.2	0.9	23.76	18.81
35–44	10.6	1.8	18.46	15.99
45–54	18.0	6.1	13.34	14.75
55–64	33.5	17.7	8.68	14.04
65–74	67.8	45.6	4.57	10.65
75–84	139.8	105.2	1.58	4.28
85+	276.5	226.2	0.17	0.64

16.2 Age standardization using the direct method

I shall describe the **direct method** first. We use a standard population structure, i.e. a standard age distribution or set of proportions of people in each age group. We then calculate the overall mortality rate which a population with the standard age structure would have if it experienced the age specific mortality rates of the observed population, the population whose mortality rate is to be adjusted. We shall take the 1901 population as the standard and calculate the mortality rate the 1981 population would have experienced if the age distribution were the same as in 1901. We do this by multiplying each 1981 age specific mortality rate by the proportion in that age group in the standard 1901 population, and adding. This then gives us an average mortality rate for the whole population, the **age standardized mortality rate.** For example, the 1981 mortality rate in age group 15–19 was 0.8 per 1 000 per year and the proportion in the standard population in this age group is 15.36% or 0.153 6. The contribution of this age group is $0.8 \times 0.1536 = 0.1229$. The calculation is set out in Table 16.2.

If we used the population's own proportions in each age group in this calculation we would get the crude mortality rate. Since 1901 has been chosen as the standard population, its crude mortality rate of 15.7 is also the age standardized mortality rate. The age standardized mortality rate for 1981 was 7.3 per 1 000 men per year. We can see that there was a much higher age standardized mortality in 1901 than 1981, reflecting the difference in age specific mortality rates.

16.3 Age standardization by the indirect method

The direct method relies upon age specific mortality rates for the observed population. If we have very few deaths, these age specific rates will be very poorly estimated. This will be particularly so in the younger age

Table 16.2. Calculation of the age standardized mortality rate by the direct method

Age group (years)	Standard proportion in age group (a)	Observed mortality rate per 1 000 (b)	$a \times b$
15–19	0.153 6	0.8	0.122 9
20–24	0.140 7	0.8	0.112 6
25–34	0.237 6	0.9	0.213 8
35–44	0.184 6	1.8	0.332 3
45–54	0.133 4	6.1	0.813 7
55–64	0.086 8	17.7	1.536 4
65–74	0.045 7	45.6	2.083 9
75–84	0.015 8	105.2	1.662 2
85+	0.001 7	226.2	0.384 5
Sum			7.262 3

groups, where we may even have no deaths at all. Such situations arise when considering mortality due to particular conditions or in relatively small groups, such as those defined by occupation. The indirect method of standardization is used for such data. We calculate the number of deaths we would expect in the observed population if it experienced the age specific mortality rates of a standard population. We then compare the expected number of deaths with that actually observed.

I shall take as an example the deaths due to cirrhosis of the liver among male qualified medical practitioners in England and Wales, recorded around the 1971 census. There were 14 deaths among 43 570 doctors aged below 65, a crude mortality rate of $14/43\,570 = 321$ per million, compared to 1 423 out of 15 247 980 adult males (aged 15–64), or 93 per million. The mortality among doctors appears high, but the medical population may be older than the population of men as a whole, as it will contain relatively few below the age of 25. Also the actual number of deaths among doctors is small and any difference not explained by the age effect may be due to chance. The indirect method enables us to test this. Table 16.3 shows the age specific mortality rates for cirrhosis of the liver among all men aged 15 to 65, and the number of men estimated in each ten-year age group, for all men and for doctors. We can see that the two age distributions do appear to be different.

The calculation of the expected number of deaths is similar to the direct method, but different populations and rates are used. For each age group, we take the number in the observed population, and multiply it by the standard age specific mortality rate, which would be the probability of dying if mortality in the observed population were the same as that in the standard population. This give us the number we would expect to die in this age group in the observed population. We add these over the age groups and obtain the expected number of deaths. The calculation is set

Table 16.3. Age specific mortality rates due to cirrhosis of the liver and age distributions of all men and medical practitioners, England and Wales, 1971

Age group (years)	Mortality per million men per year	Number of men	Number of doctors
15–24	5.859	3 584 320	1 080
25–34	13.050	3 065 100	12 860
35–44	46.937	2 876 170	11 510
45–54	161.503	2 965 880	10 330
55–64	271.358	2 756 510	7 790

Table 16.4. Calculation of the expected number of deaths due to cirrhosis of the liver among practitioners, using the indirect method

Age group (years)	Standard mortality rate per 1 000 (*a*)	Observed population: number of doctors (*b*)	$a \times b$
15–24	0.000 005 859	1 080	0.006 3
25–34	0.000 013 050	12 860	0.167 8
35–44	0.000 046 937	11 510	0.540 2
45–54	0.000 161 503	10 330	1.668 3
55–64	0.000 271 358	7 790	2.113 9
Total			4.496 5

out in Table 16.4.

The expected number of deaths is 4.4965, which is considerably less than the 14 observed. We usually express the result of the calculation as the ratio of observed to expected deaths, called the **standardized mortality ratio** or **SMR**. Thus the SMR for cirrhosis among doctors is

$$\text{SMR} = \frac{14}{4.4965} = 3.11$$

We usually multiply the SMR by 100 to get rid of the decimal point, and report the SMR as 311. If we do not adjust for age at all, the ratio of the crude death rates is 3.44, compared to the age adjusted figure of 3.11, so the adjustment has made some, but not much, difference in this case.

We can calculate a confidence interval for the SMR quite easily. Denote the observed deaths by O and expected by E. It is reasonable to suppose that the deaths are independent of one another and happening randomly in time, so the observed number of deaths is from a Poisson distribution (§6.7). The standard deviation of this Poisson distribution is the square root of its mean and so can be estimated by the square root of the observed deaths, $\sqrt{O}$. The expected number is calculated from a very much larger sample and is so well estimated it can be treated as a constant, so the standard deviation of $100 \times O/E$, which is the standard error of the SMR, is estimated by $100 \times \sqrt{O}/E$. Provided the number of deaths is large enough, say more than 10, an approximate 95% confidence interval is given by

$$100 \times \frac{O}{E} - 1.96 \times 100 \times \frac{\sqrt{O}}{E} \text{ to } 100 \times \frac{O}{E} + 1.96 \times 100 \times \frac{\sqrt{O}}{E}$$

For small observed frequencies tables based on the exact probabilities of the Poisson distribution are available (Pearson and Hartley 1970). For the cirrhosis data the formula gives

$$311 - 1.96 \times 100 \times \frac{\sqrt{14}}{4.496\,5} \text{ to } 311 + 1.96 \times 100 \times \frac{\sqrt{14}}{4.496\,5}$$

$$= 311 - 163 \text{ to } 311 + 163$$

$$= 148 \text{ to } 474$$

The confidence interval clearly excludes 100 and the high mortality cannot be ascribed to chance.

We can also test the null hypothesis that in the population the SMR = 100. If the null hypothesis is true, O is from a Poisson distribution with mean E and hence standard deviation $\sqrt{E}$, provided the sample is large enough, say $E > 10$. Then $(O - E)/\sqrt{E}$ would be an observation from the Standard Normal distribution if the null hypothesis were true. The sample of doctors is too small for this test to be reliable, but if it were, we would have $(O-E)/\sqrt{E} = (14-4.496\,5)/\sqrt{4.496\,5} = 4.48$, P = 0.000 1. The news is not all bad for medical practitioners, however. Their SMR for cancer of the trachea, bronchus and lung is only 32. Doctors may drink, but they don't smoke!

Better approximation and exact methods of calculating confidence intervals are described by Morris and Gardner (1989) and Breslow and Day (1987).

16.4 Demographic life tables

We have already discussed a use of the life table technique for the analysis of clinical survival data (§15.6). The life table was found by following the survival of a group of subjects from some starting point to death. In **demography**, which means the study of human populations, life tables are generated in a different way. Rather than charting the progress of a group from birth to death, we start with the present age specific mortality rates. We then calculate what would happen to a cohort of people from birth if these age specific mortality rates applied unchanged throughout their lives. We denote the probability of dying between ages x and $x + 1$ years (the age specific mortality rate at age x) by q_x. As in Table 15.8, the

Table 16.5. Extract from English Life Table Number 11, 1950–52, Males

Age in years	Expected number alive at age x	Probability an individual dies between ages x and $x+1$	Expected life at age x years
x	l_x	q_x	e_x
0	100 000	0.032 66	66.42
1	96 734	0.002 41	67.66
2	96 501	0.001 41	66.82
3	96 395	0.001 02	65.91
4	96 267	0.000 84	64.98
.	.	.	.
.	.	.	.
.	.	.	.
100	23	0.440 45	1.67
101	13	0.450 72	1.62
102	7	0.460 11	1.58
103	4	0.468 64	1.53
104	2	0.476 36	1.50

probability of surviving from age x to $x+1$ is $p_x = 1 - q_x$. We now suppose that we have a cohort of size l_0 at age 0, i.e. at birth. l_0 is usually 100 000 or 10 000. The number who would still be alive after x years is l_x. We can see that the number alive after $x+1$ years is $l_{x+1} = p_x \times l_x$, so given all the p_x from $x = 0$ onwards we can calculate the l_x. The cumulative survival probability to age x is then $P_x = l_x/l_0$.

Table 16.5 shows an extract from Life Table Number 11, 1950–52, for England and Wales. With the exception of 1941, a life table like this has been produced every 10 years since 1871, based on the decennial census year. The life table is based on the census year because only then do we have a good measure of the number of people at each age, the denominator in the calculation of q_x. A three year period is used to increase the number of deaths for a year of age and so improve the estimation of q_x. Separate tables are produced for males and females because the mortality of the two sexes is very different. Age specific deaths rates are higher in males than females at every age. Between census years life tables are still produced but are only published in an abridged form, giving l_x at five year intervals only after age five (Table 16.6).

The final column in Tables 16.5 and 16.6 is the **expected life, expectation of life** or **life expectancy**, e_x. This is the average life still to be lived by those reaching age x. We have already calculated this as the expected value of the probability distribution of year of death (§6E). We can do the calculation in a number of other ways. For example, if we add l_{x+1}, l_{x+2}, l_{x+3}, etc. we will get the total number of years to be lived, because the l_{x+1} who survive to $x+1$ will have added l_{x+1} years to the total, the

Table 16.6. Abridged Life Table 1988–90, England and Wales

Age	Males		Females	
x	l_x	e_x	l_x	e_x
0	10 000	73.0	10 000	78.5
1	9 904	72.7	9 928	78.0
2	9 898	71.7	9 922	77.1
3	9 893	70.8	9 919	76.1
4	9 890	69.8	9 916	75.1
5	9 888	68.8	9 914	74.2
10	9 877	63.9	9 907	69.2
15	9 866	58.9	9 899	64.3
20	9 832	54.1	9 885	59.4
25	9 790	49.3	9 870	54.4
30	9 749	44.5	9 852	49.5
35	9 702	39.7	9 826	44.6
40	9 638	35.0	9 784	39.8
45	9 542	30.3	9 718	35.1
50	9 375	25.8	9 607	30.5
55	9 097	21.5	9 431	26.0
60	8 624	17.5	9 135	21.7
65	7 836	14.0	8 645	17.8
70	6 689	11.0	7 918	14.2
75	5 177	8.4	6 869	11.0
80	3 451	6.4	5 446	8.2
85	1 852	4.9	3 659	5.9

l_{x+2} of these who survive from $x+1$ to $x+2$ will add a further l_{x+2} years, and so on. If we divide this sum by l_x we get the average number of whole years to be lived. If we then remember that people do not die on their birthdays, but scattered throughout the year, we can add half to allow for the average of half year lived in the year of death. We thus get

$$e_x = \frac{1}{2} + \frac{1}{l_x} \sum_{i=x+1}^{\infty} l_i$$

i.e. summing the l_i from age $x+1$ to the end of the life table.

If many people die in early life, with high age specific death rates for children, this has a great effect on expectation of life at birth. Table 16.7 shows expectation of life at selected ages from four English Life Tables (OPCS 1992). In 1981, for example, expectation of life at birth for males was 71 years, compared to only 40 years in 1841, an improvement of 31 years. However expectation of life at age 45 in 1981 was 29 years compared to 23 years in 1841, an improvement of only 6 years. At age 65, male expectation of life was 11 years in 1841 and 13 years in 1981, an even smaller change. Hence the change in life expectancy at birth was due to changes in mortality in early life, not late life.

Table 16.7. Life expectancy in 1841, 1901, 1951, and 1981, England and Wales

Age	Sex	Expectation of life in years			
		1841	1901	1951	1981
Birth	Males	40	49	66	71
	Females	42	52	72	77
15 yrs	Males	43	47	54	57
	Females	44	50	59	63
45 yrs	Males	23	23	27	29
	Females	24	26	31	34
65 yrs	Males	11	11	12	13
	Females	12	12	14	17

There is a common misconception that a life expectancy at birth of 40 years, as in 1841, meant that most people died about age 40. For example (Rowe 1992):

> Mothers have always provoked rage and resentment in their adult daughters, while the adult daughters have always provoked anguish and guilt in their mothers. In past centuries, however, such matched misery did not last for long. Daughters could bury their rage and resentment under a concern for duty while they cared for their mothers who, turning 40, rapidly aged, grew frail and died. Now mothers turning 40 are strong and healthy, and only half way through their lives.

This is absurd. As Table 16.7 shows, since life expectancy was first estimated women turning 40 have had average remaining lives of more than 20 years. They did not rapidly age, grow frail, and die.

Life tables have a number of uses, both medical and non-medical. Expectation of life provides a useful summary of mortality without the need for a standard population. The table enables us to predict the future size of and age structure of a population given its present state, called a **population projection**. This can be very useful in predicting such things as the future requirement for geriatric beds in a health district. Life tables are also invaluable in non-medical applications, such as the calculation of insurance premiums, pensions and annuities.

The main difficulty with prediction from a life table is finding a table which applies to the populations under consideration. For the general population of, say, a health district, the national life table will usually be adequate, but for special populations this may not be the case. If we want to predict the future need for care of an institutionalized population, such as in a long stay psychiatric hospital or old peoples' home, the mortality may be considerably greater than that in the general population. Predictions based on the national life table can only be taken as a very rough guide. If possible life tables calculated on that type of population should be used.

16.5 Vital statistics

We have seen a number of occasions where ordinary words have been given quite different meanings in statistics from those they have in common speech; 'Normal' and 'significant' are good examples. 'Vital statistics' is the opposite, a technical term which has acquired a completely unrelated popular meaning. As far as the medical statistician is concerned, vital statistics have nothing to do with the dimensions of female bodies. They are the statistics relating to life and death: birth rates, fertility rates, marriage rates and death rates. We have already dealt with the crude mortality rate, age specific mortality rates, age standardized mortality rate, standardized mortality ratio and expectation of life. In this section we shall define a number of other statistics which are often quoted in the medical literature.

The **infant mortality rate** is the number of deaths under one year of age divided by the number of live births, usually expressed as deaths per 1 000 live births. The **neonatal mortality rate** is the same thing for deaths in the first 4 weeks of life. The **stillbirth rate** is the number of stillbirths divided by the total number of births, live and still. A stillbirth is a child born dead after 28 weeks gestation. The **perinatal mortality rate** is the number of stillbirths and deaths in the first week of life divided by the total births, again usually presented per 1 000 births. Infant and perinatal mortality rates are regarded as particularly sensitive indicators of the health status of the population. The **maternal mortality rate** is the number of deaths of mothers ascribed to problems of pregnancy and birth, divided by the total number of births. The **birth rate** is the number of live births per year divided by the total population. The **fertility rate** is the number of live births per year divided by the number of women of childbearing age, taken as 15–44 years.

The **attack rate** for a disease is the proportion of people exposed to infection who develop the disease. The **case fatality rate** is the proportion of cases who die. The **prevalence** of a disease is the proportion of people who have it at one point in time. The **incidence** is the number of new cases in one year divided by the number at risk.

16.6 The population pyramid

The age distribution of a population can be presented as histogram, using the methods of §4.3. However, because the mortality of males and females is so different the age distributions for males and females are also different. It is usual to present the age distributions for the two sexes separately. Figure 16.1 shows the age distributions for the male and female populations of England and Wales in 1901. Now, these histograms have the same horizontal scale. The conventional way to display them is with the age

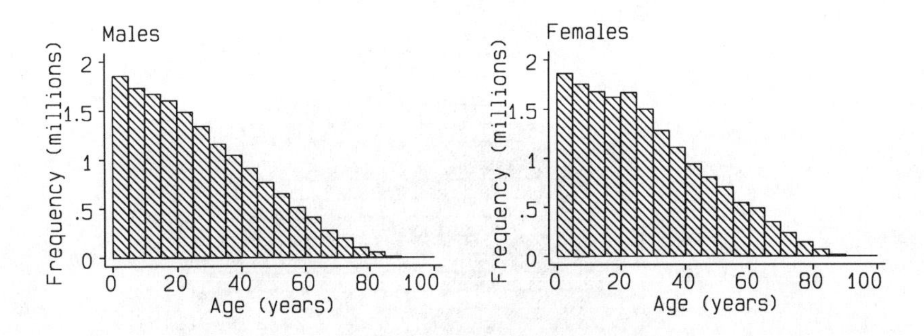

Fig. 16.1. Age distributions for the population of England and Wales, by sex, 1901

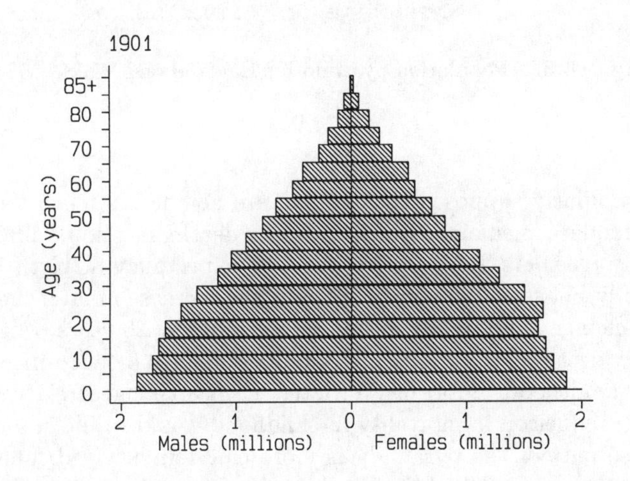

Fig. 16.2. Population pyramid for England and Wales, 1901

scale vertically and the frequency scale horizontally as in Figure 16.2. The frequency scale has zero in the middle and increases to the right for females and to the left for males. This is called a **population pyramid**, because of its shape.

Figure 16.3 shows the population pyramid for England and Wales in 1981. The shape is quite different. Instead of a triangle we have an irregular figure with almost vertical sides which begin to bend very sharply inwards at about age 65. The post-war and 1960s baby booms can be seen as bulges at ages 20 and 35. As major change in population structure has taken place, with a vast increase in the proportion of elderly. This has major implications for medicine, as the care of the elderly has become a large proportion of the work of doctors, nurses and their colleagues. It is interesting to see how this has come about.

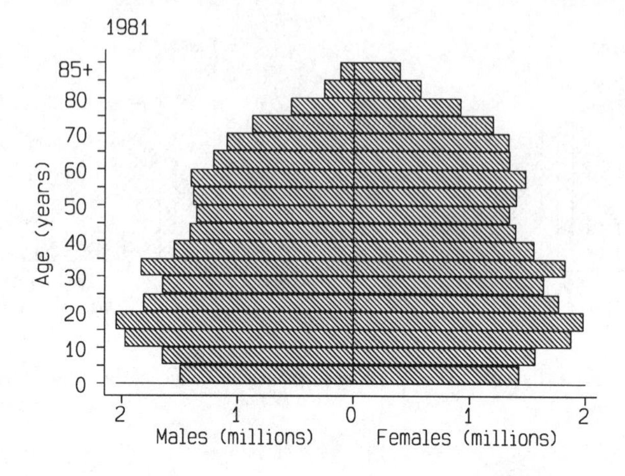

Fig. 16.3. Population pyramid for England and Wales, 1981

It is popularly supposed that people are now living much longer as a result of modern medicine, which prevents deaths in middle life. This is only partly true. As Table 16.7 shows, life expectancy at birth increased dramatically between 1901 and 1981, but the increase in later life is much less. The change is not an extension of every life by 20 years, which would be seen at every age, but mainly a reduction in mortality in childhood and early adulthood. Mortality in later life has changed relatively little. Now, a big reduction in mortality in childhood would result in an increase in the base part of the pyramid, as more children survived, unless there was a corresponding fall in the number of babies being born. In the 19th century, women were having many children and despite the high mortality in childhood the number who survived into adulthood to have children of their own exceeded that of their own parents. The population expanded and this history is embodied in the 1901 population pyramid. In the 20th century, infant mortality fell and people responded to this by having fewer children. In 1841–5, the infant mortality rates were 148 per 1 000 live births, 138 in 1901–5 and only 10 in 1981–5 (OPCS 1992b). The birth rate was 32.2 per 1 000 women in 1841–5, in 1901–5 it was 28.2, and in 1981–5 it was 12.8. The base of the pyramid ceased to expand. As those who were in the base of the 1901 pyramid grew older, the population in the top half of the pyramid increased. The 0–4 age group in the 1901 pyramid are the 80–84 age group in the 1981 pyramid. Had the birth rate not fallen, the population would have continued to expand and we would have as great or greater a proportion of young people in 1981 as we did in 1901, and a vastly larger population. Thus the increase in the proportion of the elderly is not because adult lives have been extended, but because fertility has declined. Life expectancy for the elderly has changed relatively little. Most developed countries have stable population pyramids like Figure 16.3 and those of most developing countries have expanding pyramids like Figure 16.2.

16M Multiple choice questions 87 to 92

(Each branch is either true or false)

87. Age specific mortality rate:

(a) is a ratio of observed to expected deaths;

(b) can be used to compare mortality between different age groups;

(c) is an age adjusted mortality rate;

(d) measures the number of deaths in a year;

(e) measures the age structure of the population.

88. Expectation of life:

(a) is the number of years most people live;

(b) is a way of summarizing age specific death rates;

(c) is the expected value of a particular probability distribution;

(d) varies with age;

(e) is derived from life tables.

89. In a study of post-natal suicide (Appleby 1991), the SMR for suicide among women who had just had a baby was 17 with a 95% confidence interval 14 to 21 (all women = 100). For women who had had a still birth, the SMR was 105 (95% confidence interval 31 to 277). We can conclude that:

(a) women who had just had a baby were less likely to commit suicide than other women of the same age;

(b) women who had just had a still birth were less likely to commit suicide than other women of the same age;

(c) women who had just had a live baby were less likely to commit suicide than women of the same age who had had a still birth;

(d) it is possible that having a still birth increases the risk of suicide;

(e) suicidal women should have babies.

90. In 1971, the SMR for cirrhosis of the liver for men was 773 for publicans and innkeepers and 25 for window cleaners, both being significantly different from 100 (Donnan and Haskey 1978). We can conclude that:

(a) publicans are more than 7 times as likely as the average person to die from cirrhosis of the liver;

(b) the high SMR for publicans may be because they tend to be found in the older age groups;

(c) being a publican causes cirrhosis of the liver;

(d) window cleaning protects men from cirrhosis of the liver;

(e) window cleaners are at high risk of cirrhosis of the liver.

91. The age and sex structure of a population may be described by:

(a) a life table;

(b) a correlation coefficient;

(c) a standardized mortality ratio;

(d) a population pyramid;

(e) a bar chart.

92. The following statistics are adjusted to allow for the age distribution of the population:

(a) age standardized mortality rate;

(b) fertility rate;

(c) perinatal mortality rate;

(d) crude mortality rate;

(e) expectation of life at birth.

16E Exercise: Deaths from volatile substance abuse

Anderson *et al.* (1985) studied mortality associated with volatile substance abuse (VSA), often called glue sniffing. In this study all known deaths associated with VSA from 1971 to 1983 inclusive were collected, using sources including three press cuttings agencies and a six-monthly systematic survey of all coroners. Cases were also notified by the Office of Population Censuses and Surveys for England and Wales and by the Crown Office and procurators fiscal in Scotland.

Table 16.8 shows the age distribution of these deaths for Great Britain and for Scotland alone, with the corresponding age distributions at the 1981 decennial census.

1. Calculate age specific mortality rates for VSA per year and for the whole period. What is unusual about these age specific mortality rates?

2. Calculate the SMR for VSA deaths for Scotland.

3. Calculate the 95% confidence interval for this SMR.

Table 16.8. Volatile substance abuse mortality and population size, Great Britain and Scotland, 1971–83 (Anderson *et al.* 1985)

Age group (years)	Great Britain		Scotland	
	VSA deaths	Population (thousands)	VSA deaths	Population (thousands)
0–9	0	6 770	0	653
10–14	44	4 271	13	425
15–19	150	4 467	29	447
20–24	45	3 959	9	394
25–29	15	3 616	0	342
30–39	8	7 408	0	659
40–49	2	6 055	0	574
50–59	7	6 242	0	579
60+	4	10 769	0	962

4. Does the number of deaths in Scotland appear particularly high? Apart from a lot of glue sniffing, are there any other factors which should be considered as possible explanations for this finding?

17

Multifactorial methods

17.1 Multiple regression

In Chapters 10 and 11 we looked at methods of analysing the relationship between an outcome variable and a predictor. The predictor could be quantitative, as in regression, or qualitative, as in one way analysis of variance. In this chapter we shall look at the extension of these methods to more than one predictor variable, and describe related methods for use when the outcome is dichotomous or censored survival data. These methods are very difficult to do by hand and computer programs are always used. I shall omit the formulae.

Table 17.1 shows the ages, heights and maximum voluntary contraction of the quadriceps muscle (MVC) in a group of male alcoholics. The outcome variable is MVC. Figure 17.1 shows the relationship between MVC and height. We can fit a regression line of the form MVC $= a + b \times$ height (§11.2–3). This enables us to predict what the mean MVC would be for men of any given height. But MVC varies with other things beside height. Figure 17.2 shows the relationship between MVC and age.

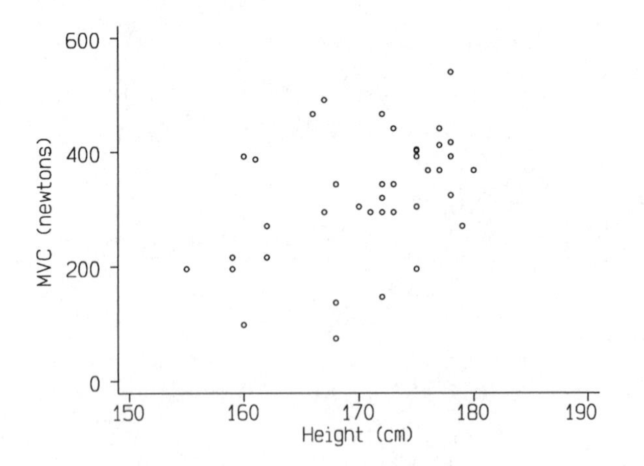

Fig. 17.1. Muscle strength (MVC) against height

Table 17.1. Maximum voluntary contraction (MVC) of quadriceps muscle, age and height, of 41 male alcoholics (Hickish *et al.* 1989)

Age (years)	Height (cm)	MVC (newtons)	Age (years)	Height (cm)	MVC (newtons)
24	166	466	42	178	417
27	175	304	47	171	294
28	173	343	47	162	270
28	175	404	48	177	368
31	172	147	49	177	441
31	172	294	49	178	392
32	160	392	50	167	294
32	172	147	51	176	368
32	179	270	53	159	216
32	177	412	53	173	294
34	175	402	53	175	392
34	180	368	53	172	466
35	167	491	55	170	304
37	175	196	55	178	324
38	172	343	55	155	196
39	172	319	58	160	98
39	161	387	61	162	216
39	173	441	62	159	196
40	173	441	65	168	137
41	168	343	65	168	74
41	178	540			

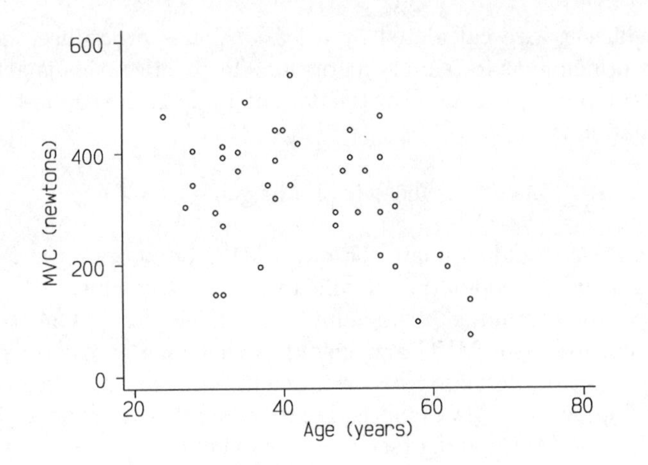

Fig. 17.2. Muscle strength (MVC) against age

Table 17.2. Correlation matrix for the data of Table 17.1

	age	height	MVC
age	1.000	−0.338	−0.417
height	−0.338	1.000	0.419
MVC	−0.417	0.419	1.000

We can show the strengths of the linear relationships between all three variables by their **correlation matrix.** This is a tabular display of the correlation coefficients between each pair of variables, **matrix** being used in its mathematical sense as a rectangular array of numbers. The correlation matrix for the data of Table 17.1 is shown in Table 17.2. The coefficients of the main diagonal are all 1.0, because they show the correlation of the variable with itself, and the correlation matrix is symmetrical about this diagonal. Because of this symmetry many computer programs print only the part of the matrix below the diagonal. Inspection of Table 17.2 shows that older men were shorter and weaker than younger men, that taller men were stronger than shorter men, and that the magnitudes of all three relationships was similar. Reference to Table 11.2 with $41 - 2 = 39$ degrees of freedom shows that all three correlations are significant.

We could fit a regression line of the form MVC $= a + b \times$ age, from which we could predict the mean MVC for any given age. However, MVC would still vary with height. To investigate the effect of both age and height, we can use multiple regression to fit a regression equation of the form

$$\text{MVC} = b_0 + b_1 \times \text{age} + b_2 \times \text{height}$$

The coefficients are calculated by a least squares procedure, exactly the same in principle as for simple regression. In practice, this is always done using a computer program. For the data of Table 17.1, the multiple regression equation is

$$\text{MVC} = -466 + 5.40 \times \text{height} - 3.08 \times \text{age}$$

From this, we would estimate the mean MVC of men with any given age and height, in the population of which these are a sample.

There are a number of assumptions implicit here. One is that the relationship between MVC and height is the same at each age, that is, that there is no interaction between height and age. Another is that the relationship between MVC and height is linear, that is of the form MVC $= a + b \times$ height. Multiple regression analysis enables us to test both of these assumptions.

Multiple regression is not limited to two predictor variables. We can have any number, although the more variables we have the more difficult it

Table 17.3. Analysis of variance for the regression of FEV1 on height

Source of variation	Degrees of freedom	Sum of squares	Mean square	Variance ratio (F)	Probability
Total	19	9.438 68			
Due to regression	1	3.189 37	3.189 37	9.19	0.007
Residual (about regression)	18	6.249 31	0.347 18		

becomes to interpret the regression. We must, however, have more points than variables, and as the degrees of freedom for the residual variance are $n - 1 - q$ if q variables are fitted, and this should be large enough for satisfactory estimation of confidence intervals and tests of significance. This will become clear after the next section.

17.2 Significance tests in multiple regression

As we saw in §11.5, the significance of a simple linear regression line can be tested using the t distribution. We can carry out the same test using analysis of variance. For the FEV1 and height data of Table 11.1 the sums of squares and products were calculated in §11.3. The total sum of squares for FEV1 is $S_{yy} = 9.438\,68$, with $n - 1 = 19$ degrees of freedom. The sum of squares due to regression was calculated in §11.5 to be 3.189 37. The residual sum of squares, i.e. the sum of squares about the regression line, is found by subtraction as $9.438\,68 - 3.189\,37 = 6.249\,31$, and this has $n - 2 = 18$ degrees of freedom. We can now set up an analysis of variance table as described in §10.9, shown in Table 17.3.

Note that the square root of the variance ratio is 3.03, the value of t found in §11.5. The two tests are equivalent. Note also that the regression sum of squares divided by the total sum of squares $= 3.189\,37/9.438\,68 = 0.337\,9$ is the square of the correlation coefficient, $r = 0.58$ (§11.5, §11.10). This ratio, sum of squares due to regression over total sum of squares, is the proportion of the variability accounted for by the regression. The percentage variability accounted for or explained by the regression is 100 times this, i.e. 34%.

Returning to the MVC data, we can test the significance of the regression of MVC on height and age together by analysis of variance. If we fit the regression model in §17.1, the regression sum of squares has two degrees of freedom, because we have fitted two regression coefficients. The analysis of variance for the MVC regression is shown in Table 17.4.

The regression is significant; it is unlikely that this association could have arisen by chance if the null hypothesis were true. The proportion of variability accounted for, denoted by R^2, is $131\,495/503\,344 = 0.26$. The square root of this is called the multiple correlation coefficient, R. R^2 must

Table 17.4. Analysis of variance for the regression of MVC on height and age

Source of variation	Degrees of freedom	Sum of squares	Mean square	Variance ratio (F)	Probability
Total	40	503 344			
Regression	2	131 495	65 748	6.72	0.003
Residual	38	371 849	9 785		

lie between 0 and 1, and as no meaning can be given to the direction of correlation in the multivariate case, R is also taken as positive. The larger R is, the more closely correlated with the outcome variable the the set of variables are. When $R = 1$ the variables are perfectly correlated in the sense that the outcome variable is a linear combination of the others. When the outcome variable is not linearly related to any of the predictor variables, R will be small, but not zero.

We may wish to know whether both or only one of our variables leads to the association. To do this, we can calculate a standard error for each regression coefficient. In the example, we have $\text{SE}(b_1) = 2.55$, $\text{SE}(b_2) = 1.47$, giving for b_1, $t = 5.40/2.55 = 2.12$, 38 df, P = 0.04, and for b_2, $t = -3.08/1.47 = -2.10$, 38 df, P = 0.04. We can conclude that both age and height are significantly associated with MVC.

A difficulty arises when the predictor variables are correlated with one another. This increases the standard error of the estimates, and variables may have a multiple regression coefficient which is not significant despite being related to the outcome variable. We can see that this will be so most clearly by taking an extreme case. Suppose we try to fit

$$\text{MVC} = b_0 + b_1 \times \text{height} + b_2 \times \text{height}$$

For the MVC data

$$\text{MVC} = -908 + 6.20 \times \text{height} + 1.00 \times \text{height}$$

is a regression equation which minimizes the residual sum of squares. However, it is not unique, because

$$\text{MVC} = -908 + 5.20 \times \text{height} + 2.00 \times \text{height}$$

will do so too. The two equations give they same predicted MVC. There is no unique solution, and so no regression equation can be fitted, even though there is a clear relationship between PEFR and height. When the predictor variables are correlated the individual coefficients will be poorly estimated and have large standard errors. Correlated predictor variables may obscure the relationship of each with the outcome variable.

A different (and equivalent) way of testing the effects of two correlated predictor variables separately is to proceed as follows. We fit three models:

Table 17.5. Analysis of variance for the regression of MVC on height and age, showing adjusted sums of squares

Source of variation	Degrees of freedom	Sum of squares	Mean square	Variance ratio (F)	Probability
Total	40	503 344			
Regression	2	131 495	65 748	6.72	0.003
age alone	1	87 471	87 471	8.94	0.005
height given age	1	44 024	44 024	4.50	0.04
height alone	1	88 511	88 511	9.05	0.005
age given height	1	42 984	42 984	4.39	0.04
Residual	38	371 849	9 785		

1. MVC on height and age, regression sum of squares = 131 495, d.f. = 2
2. MVC on height, regression sum of squares = 88 511, d.f. = 1
3. MVC on age, regression sum of squares = 87 471, d.f. = 1

Note that 88 511 + 87 471 = 175 982 is greater than 131 495. This is because age and height are correlated. We then test the effect of height if age is taken into account, referred to as the effect of height given age. The regression sum of squares for height given age is the regression sum of squares (age and height) minus regression sum of squares (age only), which is 131 495 − 87 471 = 44 024. This has degrees of freedom= 2 − 1 = 1. Similarly, the effect of age allowing for height, i.e. age given height, is tested by regression sum of squares (age and height) minus regression sum of squares (height only)= 131 495 − 88 511 = 42 984, with degrees of freedom= 2 − 1 = 1. We can set all this out in an analysis of variance table (Table 17.5). The third to sixth rows of the table are indented for the source of variation, degrees of freedom and sum of squares columns, to indicate that they are different ways of looking at variation already accounted for in the second row. The indented rows are not included when the degrees of freedom and sums of squares are added to give the total. After adjustment for age there is still evidence of a relationship between MVC and height, and after adjustment for height there is still evidence of a relationship between MVC and age. Note that the P values as the same as those found by t test. This approach is essential for qualitative predictor variables (§17.6), when the t tests may be invalid.

17.3 Interaction in multiple regression

An interaction between two predictor variables arises when the effect of one on the outcome depends on the value of the other. For example, tall men may be stronger than short men when they are young, but the difference may disappear as they age.

Table 17.6. Analysis of variance for the interaction of height and age

Source of variation	Degrees of freedom	Sum of squares	Mean square	Variance ratio (F)	Probability
Total	40	503 344			
Regression	3	202 719	67 573	8.32	0.000 2
Height and age	2	131 495	65 748	8.09	0.001
Height × age	1	71 224	71 224	8.77	0.005
Residual	37	300 625	8 125		

We can test for interaction as follows. We have fitted

$$\text{MVC} = b_0 + b_1 \times \text{height} + b_2 \times \text{age}$$

An interaction may take two simple forms. As height increases, the effect of age may increase so that the difference in MVC between young and old tall men is greater than the difference between young and old short men. Alternatively, as height increases, the effect of age may decrease. More complex interactions are beyond the scope of this discussion. Now, if we fit

$$\text{MVC} = b_0 + b_1 \times \text{height} + b_2 \times \text{age} + b_3 \times \text{height} \times \text{age}$$

for fixed height the effect of age is $b_2 + b_3 \times \text{height}$. If there is no interaction, the effect of age is the same at all heights, and b_3 will be zero. Of course, b_3 will not be exactly zero, but only within the limits of random variation. We can fit such a model just as we fitted the first one. We get

$$\text{MVC} = 4\,661 - 24.7 \times \text{height} - 112.8 \times \text{age} + 0.650 \times \text{height} \times \text{age}$$

The regression is still significant, as we would expect. However, the coefficients of height and age have changed; they have even changed sign. The coefficient of height depends on age. The regression equation can be written

$$\text{MVC} = 4\,661 + (-24.7 + 0.650 \times \text{age}) \times \text{height} - 112.8 \times \text{age}$$

The coefficient of height depends on age, the difference in strength between short and tall subjects being greater for older subjects than for younger.

The analysis of variance for this regression equation is shown in Table 17.6. The regression sum of squares is divided into two parts: that due to age and height, and that due to the interaction term after the main effects of age and height have been accounted for. The interaction row is the difference between the regression row in Table 17.6, which has 3 degrees of freedom, and the regression row in Table 17.4, which has 2. From this we see that the interaction is highly significant. The effects of height and age on MVC are not additive.

17.4 Polynomial regression

So far, we have assumed that all the regression relationships have been linear, i.e. that we are dealing with straight lines. This is not necessarily so. We may have data where the underlying relationship is a curve rather than a straight line. Unless there is a theoretical reason for supposing that a particular form of the equation, such as logarithmic or exponential, is needed, we test for non-linearity by using a polynomial. Clearly, if we can fit a relationship of the form

$$\text{MVC} = b_0 + b_1 \times \text{height} + b_2 \times \text{age}$$

we can also fit one of the form

$$\text{MVC} = b_0 + b_1 \times \text{height} + b_2 \times \text{height}^2$$

to give a quadratic equation, and continue adding powers of height to give equations which are cubic, quartic, etc.

Height and height squared are highly correlated, which can lead to problems in estimation. To reduce the correlation, we can subtract a number close to mean height from height before squaring. For the data of Table 17.1, the correlation between height and height squared is 0.999 8. Mean height is 170.7 cm, so 170 is a convenient number to subtract. The correlation between height and height minus 170 squared is -0.44, so the correlation has been reduced, though not eliminated. The regression equation is

$$\text{MVC} = -961 + 7.49 \times \text{height} + 0.092 \times (\text{height} - 170)^2$$

To test for non-linearity, we proceed as in §17.2. We fit two regression equations, a linear and a quadratic. The non-linearity is then tested by the difference between the sum of squares due to the quadratic equation and the sum of squares due to the linear. The analysis of variance is shown in Table 17.7. In this case the quadratic term is not significant, so there is no evidence of of non-linearity. Were the quadratic term significant, we could fit a cubic equation and test the effect of the cubic term in the same way. Polynomial regression of one variable can be combined with ordinary linear regression of others to give regression equations of the form

$$\text{MVC} = b_0 + b_1 \times \text{height} + b_2 \times \text{height}^2 + b_3 \times \text{age}$$

and so on. Royston and Altman (1994) have shown that quite complex curves can be fitted with a small number of coefficients if we use $\log(x)$ and powers -1, 0.5, 0.5, 1 and 2 in the regression equation.

Table 17.7. Analysis of variance for polynomial regression of MVC on height

Source of variation	Degrees of freedom		Sum of squares		Mean square	Variance ratio (F)	Prob-ability
Total	40		503 344				
Regression	2		89 103		44 552	4.09	0.02
linear		1		88 522	88 522	7.03	0.01
quadratic		1		581	581	0.05	0.8
Residual	38		414 241		12 584		

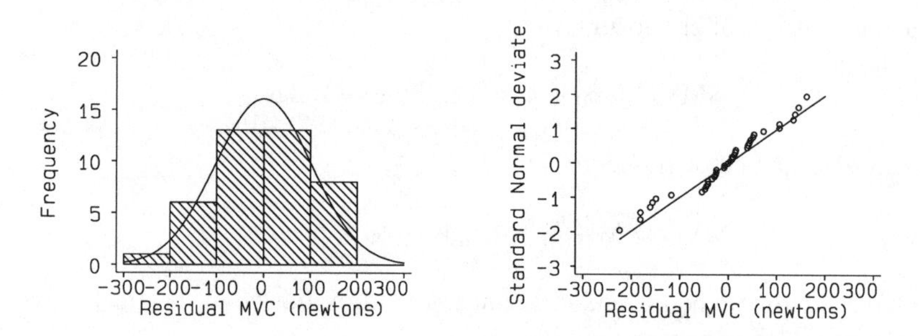

Fig. 17.3. Histogram and Normal plot of residuals of MVC about height and age

17.5 Assumptions of multiple regression

For the regression estimates to be optimal and the F tests valid, the residuals (the difference between observed values of the dependent variable and those predicted by the regression equation) should be Normally distributed and have the same variance throughout the range. We also assume that the relationships which we are modelling are linear. These assumptions are the same as for simple linear regression (§11.8) and can be checked graphically in the same way, using histograms, Normal plots and scatter diagrams. If the assumptions of Normal distribution and uniform variance are not met, we can use a transformation as described in §10.4 and §11.8. Non-linearity can be dealt with using polynomial regression.

The regression equation of strength on height and age is $\text{MVC} = -466 + 5.40 \times \text{height} - 3.08 \times \text{age}$ and the residuals are given by

$$\textit{residual} = \text{MVC} - (-466 + 5.40 \times \text{height} - 3.08 \times \text{age})$$

Figure 17.3 shows a histogram and a Normal plot of the residuals for the MVC data. The distribution looks quite good. Figure 17.4 shows a plot of residuals against MVC. The variability looks uniform. We can also check the linearity by plotting residuals against the predictor variables. Figure 17.4 also shows the residual against age. There is an indication that

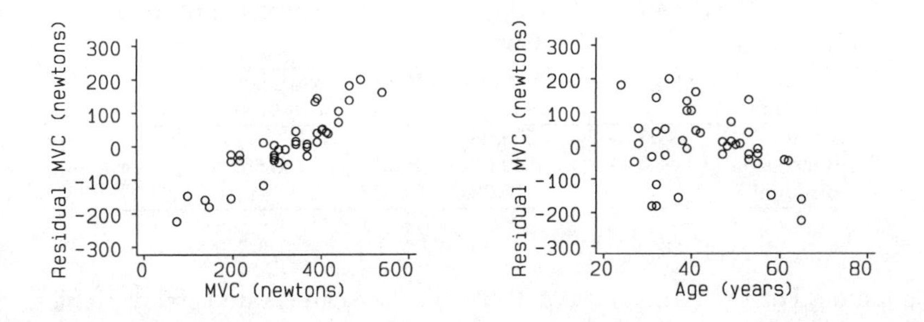

Fig. 17.4. Residuals against observed MVC, to check uniformity of variance, and age, to check linearity

residual may be related to age. The possibility of a nonlinear relationship can be checked by polynomial regression, which, in this case, does not produce a quadratic term which approaches significance.

17.6 Qualitative predictor variables

In §17.1 the predictor variables, height and age, were quantitative. In the study from which these data come, we also recorded whether or not subjects had cirrhosis of the liver. Cirrhosis was recorded as 'present' or 'absent', so the variable was dichotomous. It is easy to include such variables as predictors in multiple regression. We create a variable which is 0 if the characteristic is absent, 1 if present, and use this in the regression equation just as we did height. The regression coefficient of this dichotomous variable is the difference in the mean of the outcome variable between subjects with the characteristic and subjects without. If the coefficient in this example were negative, it would mean that subjects with cirrhosis were not as strong as subjects without cirrhosis. In the same way, we can use sex as a predictor variable by creating a variable which is 0 for females and 1 for males. The coefficient then represents the difference in mean between male and female. If we use only one, dichotomous predictor variable in the equation, the regression is exactly equivalent to a two sample t test between the groups defined by the variable (§10.3).

A predictor variable with more than two categories or classes is called a **class variable** or a **factor**. We cannot simply use a class variable in the regression equation, unless we can assume that the classes are ordered in the same way as their codes, and that adjoining classes are in some sense the same distance apart. For some variables, such as the diagnosis data of Table 4.1 and the housing data of Table 13.1, this is absurd. For others, such as the AIDS categories of Table 10.7, it is a very strong assumption. What we do instead is to create a set of dichotomous variables to represent

Table 17.8. Analysis of variance for the regression of mannitol excretion on HIV status

Source of variation	Degrees of freedom	Sum of squares	Mean square	Variance ratio (F)	Probability
Total	58	1 559.035			
Regression	3	49.011	16.337	0.60	0.6
Residual	55	1 510.024	27.455		

the factor. For the AIDS data of Table 10.7, we can create three variables:

$hiv_1 = 1$ if subject has AIDS, 0 otherwise
$hiv_2 = 1$ if subject has ARC, 0 otherwise
$hiv_3 = 1$ if subject is HIV positive but has no symptoms, 0 otherwise

If the subject is HIV negative, all three variables are zero. hiv_1, hiv_2, and hiv_3 are called **dummy variables**. Some computer programs will calculate the dummy variables automatically if the variable is declared to be a factor, for others the user must define them. We put the three dummy variables into the regression equation. This gives the equation:

$$mannitol = 11.4 - 0.066 \times hiv_1 - 2.56 \times hiv_2 - 1.69 \times hiv_3$$

Each coefficient is the difference in mannitol absorption between the class represented by that variable and the class represented by all dummy variables being zero, HIV negative, called the **reference class**. The analysis of variance for this regression equation is shown in Table 17.8, and the F test shows that there is no significant relationship between mannitol absorption and HIV status. The regression program prints out standard errors and t tests for each dummy variable, but these t tests should be ignored, because we cannot interpret one dummy variable in isolation from the others.

17.7 Multi-way analysis of variance

A different approach to the analysis of multi-factorial data is provided by the direct calculation of analysis of variance. Table 17.8 is identical to the one way analysis of variance for the same data in Table 10.8. We can also produce analyses of variance for several factors at once. Table 17.9 shows the two-way analysis of variance for the mannitol data, the factors being HIV status and presence or absence of diarrhoea. This could be produced equally well by multiple regression with two categorical predictor variables. If there were the same number of patients with and without diarrhoea in each HIV group the factors would be **balanced**. The model sum of squares would then be the sum of the sums of squares for HIV and for diarrhoea, and these could be calculated very simply from the total of the HIV groups and the diarrhoea groups. For balanced data we can assess many categorical factors and there interactions quite easily by manual calculation. See

Table 17.9. Two-way analysis of variance for mannitol excretion, with HIV status and diarrhoea as factors

Source of variation	Degrees of freedom	Sum of squares	Mean square	Variance ratio (F)	Probability
Total	58	1 559.035			
Model	4	134.880	33.720	1.28	0.3
HIV	3	58.298	19.432	0.74	0.5
Diarrhoea	1	85.869	85.869	3.26	0.08
Residual	54	1 424.155	26.373		

Armitage and Berry (1987) for details. Complex multifactorial balanced experiments are rare in medical research, and they can be analysed by regression anyway to get identical results. Most computer programs in fact use the regression method to calculate analyses of variance.

Multiple regression in which qualitative and quantitative predictor variables are both used is also known as **analysis of covariance**. For ordinal data, there is a two-way analysis of variance using ranks, the Friedman test (see Conover 1980, Altman 1991)

17.8 Logistic regression

Logistic regression is used when the outcome variable is dichotomous, a 'yes or no', whether or not the subject has a particular characteristic such as a symptom. We want a regression equation which will predict the proportion of individuals who have the characteristic, or estimate the probability that an individual will have the symptom. We cannot use an ordinary linear regression equation, because this might predict proportions less than zero or greater than one, which would be meaningless. Instead we use the logit of the proportion as the outcome variable. The **logit** of a proportion p is the log odds (§13.7):

$$\text{logit}(p) = \log_e\left(\frac{p}{1-p}\right)$$

The logit can take any value from minus infinity, when $p = 0$, to plus infinity, when $p = 1$. We can fit regression models to the logit which are very similar to the ordinary multiple regression and analysis of variance models found for data from a Normal distribution. We assume that relationships are linear on the logistic scale:

$$\log_e\left(\frac{p}{1-p}\right) = b_0 + b_1x_1 + b_2x_2 + \cdots + b_mx_m$$

where $x_1, \ldots, x_m$ are the predictor variables and p is the proportion to be predicted. The method is call **logistic regression**, and the calculation is computer intensive. We will look at the interpretation in an example.

Table 17.10. Smoking and cough first thing in the morning reported by schoolboys (Bland *et al.* 1978)

Boys' smoking	coughs		does not cough		Total	
never smoked	41	(3.2%)	1 260	(96.8%)	1301	(100.0%)
once only	28	(2.9%)	947	(97.1%)	975	(100.0%)
sometimes	16	(4.0%)	380	(96.0%)	396	(100.0%)
at least 1 per week	33	(19.2%)	139	(80.8%)	172	(100.0%)

$\chi^2 = 105.1$, 3 d.f., $P < 0.0001$

Table 17.11. Parents' smoking and boy's cough first thing in the morning, reported by schoolboys (Bland *et al.* 1978)

Boys'	Parents' smoking					
cough	neither		one		both	
coughs	24	(2.8%)	46	(4.5%)	48	(5.0%)
does not cough	823	(97.2%)	985	(95.5%)	918	(95.0%)
Total	1847	(100.0%)	1031	(100.0%)	966	(100.0%)

$\chi^2 = 5.6$, 2 d.f., $P = 0.06$, trend $\chi^2 = 5.1$, 1 d.f., $P = 0.02$

Table 17.12. Smoking by parents and by themselves reported by schoolboys (Bland *et al.* 1978)

Boys'	Parents' smoking					
smoking	neither		one		both	
never smoked	479	(56.6%)	431	(41.8%)	391	(40.5%)
once only	256	(30.2%)	394	(38.2%)	325	(33.6%)
sometimes	90	(10.6%)	147	(14.3%)	159	(16.5%)
at least 1 per week	22	(2.6%)	59	(5.7%)	91	(9.4%)
Total	847	(100.0%)	1031	(100.0%)	966	(100.0%)

$\chi^2 = 86.0$, 6 d.f., $P < 0.0001$

Consider the following problem. Schoolboys who smoke cigarettes are more likely than others to say that they cough first thing in the morning (Table 17.10). Children whose parents smoke are more likely than others to say that they cough first thing in the morning (Table 17.11). Children whose parents smoke are more likely than others themselves to smoke (Table 17.12). Could it be that the relationship between morning cough and child's smoking arises because both are caused by parental smoking?

Table 17.13 shows the relationship of cough to both smoking by the child and by the parents. We have a two factor layout, with a binomial response, i.e. the dependent variable becomes a proportion. In this case, inspection of the proportions suggests that only the child's smoking is important, as does inspection of the logits (Table 17.14). We can fit models containing child's smoking only, parents' smoking only, or child's smoking and parents' smoking together. We use dummy variables (§17.6) for the factors. For each model the program finds the values for the coefficients which give predicted values closest to the observed values. It also gives us a measure of how far from the observed frequencies the predicted frequencies are. This measure,

Table 17.13. Proportion reporting morning cough, by parents' smoking and by boys' own smoking

Boys' smoking	Parents' smoking		
	neither	one	both
never	11/479 = 0.023	16/431 = 0.037	14/391 = 0.036
once	6/256 = 0.023	13/394 = 0.033	9/325 = 0.028
sometimes	3/90 = 0.033	6/147 = 0.041	7/159 = 0.044
at least 1 per week	4/22 = 0.181	11/59 = 0.186	18/91 = 0.198

called the **deviance**, corresponds to the sum of squares about regression in a Normal theory multiple regression. The deviance follows a Chi-squared distribution if the model is correct, i.e. if the deviations from what the model predicts are due to chance. For the example we have:

Model:	d.f.	χ^2
child's smoking	8	2.7
parents' smoking	9	58.4
child's smoking + parents' smoking	6	0.6

The expected value of χ^2 is equal to its degrees of freedom (§7A). Any value of χ^2 which exceeds the degrees of freedom greatly indicates that the model does not fit the data. Thus either the child's smoking or the child's smoking + the parents' smoking would be good fits here. We can subtract these χ^2 statistics and their associated degrees of freedom, just like sums of squares. This is because the sum of two χ^2 variables is another χ^2 variable, with degrees of freedom given by the sum of the degrees of freedom (§7A). To see whether adding the parents' smoking to the model with the child's smoking improves the prediction, we can subtract the χ^2 for child's smoking from that for child + parent, giving a χ^2 for parents' smoking given the child's own smoking of $2.7 - 0.6 = 2.1$, with $8 - 6 = 2$ degrees of freedom. As this is so close to what would be expected by chance, there no evidence of any effect of parents' smoking other than its effect via the child's own smoking. Thus there is no evidence for a passive smoking effect here.

The coefficients of the logistic regression equation are shown in Table 17.15. 'Never smoked' was chosen as the reference class, so its coefficient is set to zero and the other coefficients then measure the difference between the smoking groups and never-smokers (§17.6). For a boy who has never smoked the odds on reporting a cough are given by log odds $= -3.425$, so odds $e^{-3.425} = 0.033$. The probability is given by $p/(1-p) = 0.033$

Table 17.14. Logit of proportion reporting morning cough, by parents' smoking and by boys' own smoking

Boys' smoking	Parents' smoking		
	neither	one	both
never smoked	−3.75	−3.26	−3.29
once only	−3.73	−3.38	−3.56
sometimes	−3.37	−3.16	−3.08
at least 1 per week	−1.50	−1.47	−1.40

Table 17.15. Coefficients in the logistic model for morning cough and boys' smoking

Parameter	Coefficient	Standard error	
constant	−3.425	0.159	
never smoked	0.000	0.000	(arbitrary)
smoked once	−0.096	0.249	
smoked sometimes	0.258	0.301	
smoked 1/week	1.987	0.250	

so $p = (0.033)/(1 + 0.033) = 0.032$. For a boy smoking at least one cigarette per week, the log odds$= -3.425 + 1.987 = -1.438$, so odds $= 0.237$ and the probability of reporting a cough is given by $p/(1-p) = 0.237$ so $p = 0.237/(1+0.237) = 0.192$. Of more interest is the odds ratio comparing the effect of smoking at least once per week with never having smoked. The log odds ratio is the coefficient of 'smoked 1/week' $= 1.987$, the antilog is $e^{1.987} = 7.29$. The odds ratio is thus 7.29.

If we have a continuous predictor variable, the coefficient is the change in log odds for an increase of one unit in the predictor variable, and the antilog of the coefficient is the factor by which the odds must be multiplied for a unit increase in the predictor. When we have a case control study, we can analyse the data by using the case or control status as the outcome variable in a logistic regression. The coefficients are then the approximate log relative risks due to the factors.

17.9 Survival data using Cox regression

One problem of survival data, the censoring of individuals who have not died at the time of analysis, has been discussed in §15.6. There is another which is important for multifactorial analysis. We often have no suitable mathematical model of the way survival is related to time, i.e. the survival curve. The solution now widely adopted to this problem was proposed by Cox (1972), and is known as **Cox regression** or the **proportional hazards model**. In this approach, we say that for subjects who have lived to time t, the probability of an endpoint (e.g. dying) instantaneously at time t is $h(t)$, which is an unknown function of time. We call the probability of an endpoint the **hazard**, and $h(t)$ is the **hazard function**. We then assume that anything which affects the hazard does so by the same ratio at all times. Thus, something which doubles the risk of an endpoint on day one will also double the risk of an endpoint on day two, day three and so on. Thus, if $h_0(t)$ is the hazard function for subjects with all the predictor variables equal to zero, and $h(t)$ is the hazard function for a subject with some other values for the predictor variables, $h(t)/h_0(t)$ depends only on the predictor variables, not on time t. We call $h(t)/h_0(t)$ the **hazard ratio**. It is the relative risk of an endpoint occurring at any given time.

In statistics, it is convenient to work with differences rather than ratios, so we take the logarithm of the ratio (see §5A) and have a regression-like equation:

$$\log_e\left(\frac{h(t)}{h_0(t)}\right) = b_1x_1 + b_2x_2 + \cdots + b_px_p$$

where $x_1, \ldots, x_p$ are the predictor variables and $b_1, \ldots, b_p$ are the coefficients which we estimate from the data. This is Cox's proportional hazards model. Cox regression enables us to estimate the values of $b_1, \ldots, b_p$ which best predict the observed survival. There is no constant term b_0, its place being taken by the baseline hazard function $h_0(t)$.

Table 15.7 shows the time to recurrence of gallstones, or the time for which patients are known to have been gallstone-free, following dissolution by bile acid treatment or lithotrypsy, with the number of previous gallstones, their maximum diameter, and the time required for their dissolution. The difference between patients with a single and with multiple previous gallstones was tested using the logrank test (§15.6). Cox regression enables us to look at continuous predictor variables, such as diameter of gallstone, and to examine several predictor variables at once. Table 17.16 shows the result of the Cox regression. We can carry out an approximate test of significance dividing the coefficient by its standard error, and if the null hypothesis that the coefficient would be zero in the population is true, this follows a Standard Normal distribution. The chi-squared statistic tests the relationship between the time to recurrence and the three variables together. The maximum diameter has no significant relationship to time to recurrence, so we can try a model without it (Table 17.17). As the change in overall chi-squared shows, removing diameter has had very little effect.

The coefficients in Table 17.17 are the log hazard ratios. The coefficient for multiple gallstones is 0.963. If we antilog this, we get exp(0.963) = 2.62. As multiple gallstones is a 0 or 1 variable, the coefficient measures the difference between those with single and multiple stones. A patient with multiple gallstones is 2.62 time as likely to have a recurrence at any time than a patient with a single stone. The 95% confidence interval for this estimate is found from the antilogs of the confidence interval in Table 17.17,

Table 17.16. Cox regression of time to recurrence of gallstones on presence of multiple stones, maximum diameter of stone and months to dissolution

Variable	Coef.	Std. Err.	z	P	95% Conf. Interval
Mult. gall st.	0.838	0.401	2.09	0.038	0.046 to 1.631
Max. diam.	−0.023	0.036	−0.63	0.532	−0.094 to 0.049
Months. to dissol.	0.044	0.017	2.64	0.009	0.011 to 0.078

$\chi^2 = 12.57$, 3 d.f., P = 0.006

Table 17.17. Cox regression of time to recurrence of gallstones on presence of multiple stones and months to dissolution

Variable	Coef.	Std. Err.	z	P	95% Conf. Interval
Mult. gall st.	0.963	0.353	2.73	0.007	0.266 to 1.661
Months. to dissol.	0.043	0.017	2.59	0.011	0.010 to 0.076

$\chi^2 = 12.16$, 2 d.f., P = 0.002

1.30 to 5.26. Note that a positive coefficient means an increased risk of the event, in this case recurrence. The coefficient for months to dissolution is 0.043, which has antilog = 1.04. This is a quantitative variable, and for each month to dissolve the hazard ratio increases by a factor of 1.04. Thus a patient whose stone took two months to dissolve has a risk of recurrence 1.04 times that for a patient whose stone took one month, a patient whose stone took three months has a risk 1.04^2 times that for a one month patient, and so on.

If we have only the dichotomous variable multiple gallstones in the Cox model, we get for the overall test statistic $\chi^2 = 6.11$, 1 d.f. In §15.6 we analysed these data by comparison of two groups using the logrank test which gave $\chi^2 = 6.62$, 1 d.f. The two methods give similar, but not identical results. The logrank test is non-parametric, making no assumption about the distribution of survival time. The Cox method is said to **semi-parametric**, because although it makes no assumption about the shape of the distribution of survival time, it does require assumptions about the hazard ratio. Fuller accounts of Cox regression are given by Altman (1991) and Matthews and Farewell (ref).

17.10 Stepwise regression

Stepwise regression is a technique for choosing predictor variables from a large set. The stepwise approach can be used with multiple linear, logistic and Cox regression.

There are two basic strategies: step-up and step-down, also called forward and backward. In **step-up** or **forward** regression, we fit all possible one-way regression equations. Having found the one which accounts for the greatest variance, all two-way regressions including this variable are fitted.

The equation accounting for the most variation is chosen, and all three-way regressions including these are fitted, and so on. This continues until no significant increase in variation accounted for is found. In the **step-down** or **backward** method, we first fit the regression with all the predictor variables, and then the variable is removed which reduces the amount of variation accounted for by the least amount, and so on. There are also more complex methods, in which variables can both enter and leave the regression equation.

These methods must be treated with care. Different stepwise techniques may produce different sets of predictor variables in the regression equation. This is especially likely when the predictor variables are correlated with one another. The technique is very useful for selecting a small set of predictor variables for purposes of standardization and prediction. For trying to get an understanding of the underlying system, stepwise methods can be very misleading. When predictor variables are highly correlated, once one has entered the equation in a step-up analysis, the other will not enter, even though it is related to the outcome. Thus it will not appear in the final equation.

17.11 Meta-analysis: data from several studies

Meta-analysis is the combination of data from several studies to produce a single estimate. From the statistical point of view, meta-analysis is a straightforward application of multifactorial methods. We have several studies of the same thing, which might be clinical trials or epidemiological studies, perhaps carried out in different countries. Each trial gives us an estimate of an effect. We assume that these are estimates of the same global population value. We check the assumptions of the analysis, and, if these assumptions are satisfied, we combine the separate study estimates to make a common estimate. This is a multifactorial analysis, where the treatment or risk factor is one predictor variable and the study is another, categorical variable.

The main problems of meta-analysis arise before we begin the analysis of the data. First, we must have a clear definition of the question so that we only include studies which address this. For example, if we want to know whether lowering serum cholesterol reduces mortality from coronary artery disease, we would not want to include a study where the attempt to lower cholesterol failed. On the other hand, if we ask whether dietary advice lowers mortality, we would include such a study. Which studies we include may have a profound influence on the conclusions (Thompson 1993). Second, we must have all the relevant studies. A simple literature search is not enough. Not all studies which have been started are published; studies which produce significant differences are more likely to be published than those which do not (e.g. Pocock and Hughes 1990; Easterbrook *et al.* 1991). Within a study, results which are significant may be emphasized and parts of the data which produce no differences may be ignored by the investigators as uninteresting. Publication of unfavourable results may be discouraged by the sponsors of research. Researchers who are not native English speakers may feel that publication in the English language literature is more prestigious as it will reach a wider audience, and so try there first, only publishing in their own language if they cannot publish in English. The English language literature may contain more positive results than do other literatures. The phenomenon by which significant and positive results are more likely to be reported, and reported more prominently, than non-significant and negative ones is called **publication bias**. Thus we must not only trawl the published literature for studies, but use personal knowledge of ourselves and others to locate all the unpublished studies. Only then than we carry out the meta-analysis.

When we have all the studies which meet the definition, we combine them to get a common estimate of the effect of the treatment or risk factor. We regard the studies as providing several observations of the the same population value. There are two stages in meta-analysis. First we check that the studies do provide estimates of the same thing. Second, we calculate the common estimate and its confidence interval. To do this we may have the original data from all the studies, which can combine into one large data file with study as one of the variables, or we may only have summary statistics obtained from publications.

If the outcome measure is continuous, such as mean fall in blood pressure, we can check that subjects are from the same population by analysis of variance, with treatment or risk factor, study, and interaction between them in the model. Multiple regression can also be used, remembering that study is a categorical variable and dummy variables are required. We test the treatment times study interaction in the usual way. If the interaction is significant this indicates that the treatment effect is not the same in all studies, and so we cannot combine the studies. It is the interaction which is important. It does not matter much if the mean blood pressure varies from study to study. What matters is whether the effect of the treatment on blood pressure varies more than we would expect. We may want to examine the studies to see whether any characteristic of the studies explains this variation. This might be a feature of the subjects, the treatment or the data collection. If there is no interaction, we can drop the interaction term from the model and the treatment or risk factor effect is then the estimate we want. Its standard error and confidence interval are found as described in §17.2.

If the outcome measure is dichotomous, such as survived or died, the estimate of the treatment or risk factor effect will be in the form of an

Table 17.18. Odds ratios and confidence intervals in five studies of vitamin A supplementation in infectious disease (Glasziou and Mackerras 1993)

Study	Dose regime	Vitamin A		Controls	
		deaths	number	deaths	number
1	200 000 IU six-monthly	101	12 991	130	12 209
2	200 000 IU six-monthly	39	7 076	41	7 006
3	8 333 IU weekly	37	7 764	80	7 755
4	200 000 IU four-monthly	152	12 541	210	12 264
5	200 000 IU once	138	3 786	167	3 411

Table 17.19. Odds ratios and confidence intervals in five studies of vitamin A supplementation in infectious disease

Study	Odds ratio	95% confidence interval
1	0.73	0.56 to 0.95
2	0.94	0.61 to 1.46
3	0.46	0.31 to 0.68
4	0.70	0.57 to 0.87
5	0.73	0.58 to 0.93

odds ratio (§13.7). We can proceed in the same way as for a continuous outcome, using logistic regression (§17.8). Several other methods exist for checking the homogeneity of the odds ratios across studies, such as Woolf's test (see Armitage and Berry 1987) or that of Breslow and Day (1980). They all give similar answers, and, since they are based on different large-sample axproximations, the larger the study samples the more similar the results will be. Provided the odds ratios are homogeneous across studies, we can then estimate the common odds ratio. This can be done using the Mantel–Haenszel method (see Armitage and Berry 1987) or by logistic regression.

For example, Glasziou and Mackerras (1993) carried out a meta-analysis of vitamin A supplementation in infectious disease. Their data for five community studies are shown in Table 17.18. We can obtain odds ratios and confidence intervals as described in §13.7, shown in Table 17.19.

The common odds ratio can be found in several ways. To use logistic regression, we regress the event of death on vitamin A treatment and study. I shall treat the treatment as a dichotomous variable, set to 1 if treated with vitamin A, 0 if control. Study is a categorical variable, so we create dummy variables *study*1 to *study*4, which are set to one for studies 1 to 4 respectively, and to zero otherwise. We test the interaction by creating another set of variables, the products of *study*1 to *study*4 and vitamin A. Logistic regression of death on vitamin A, study and interaction gives a chi-squared statistic for the model of 496.99 with 9 d.f., which is highly significant. Logistic regression without the interaction terms gives 490.33

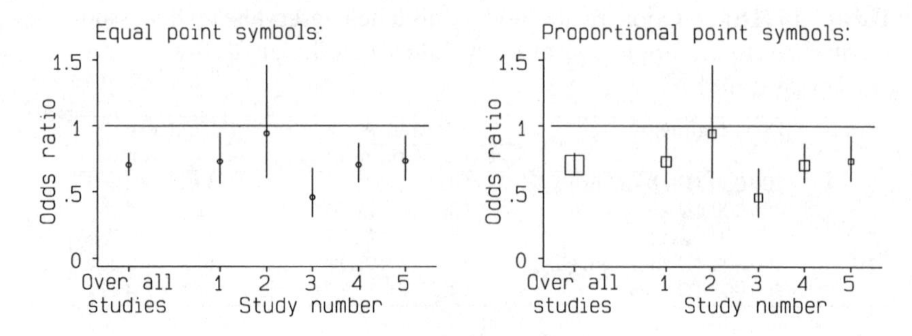

Fig. 17.5. Meta-analysis of five Vitamin A trials (data of Glasziou and Mackerras 1993). The vertical lines are the confidence intervals.

with 5 d.f. The difference is 496.99 − 490.33 = 6.66 with 9 − 5 = 4 d.f., which has P = 0.15, so we can drop the interaction from the model. The adjusted odds ratio for vitamin A is 0.70, 95% confidence interval 0.62 to 0.79, P < 0.0001.

The odds ratios and their confidence intervals are shown in Figure 17.5. The confidence interval is indicated by a line, the point estimate of the odds ratio by a circle. In this picture the most important trial appears to be Study 2, with the widest confidence interval. In fact, it is the study with the least effect on the whole estimate, because it is the study where the odds ratio is least well estimated. In the second picture, the odds ratio is indicated by the middle of a square. The area of the square is proportional to the number of subjects in the study. This now makes Study 2 appear relatively unimportant, and makes the overall estimate stand out.

17.M Multiple choice questions 93 to 97

(Each answer is true or false)

93. In multiple regression, R^2:

(a) is the square of the multiple correlation coefficient;

(b) would be unchanged if we exchanged the outcome (dependent) variable and one of the predictor (independent) variables;

(c) is called the proportion of variability explained by the regression;

(d) is the ratio of the error sum of squares to the total sum of squares;

(e) would increase if more predictor variables were added to the model.

Table 17.20. Analysis of variance for the effects of age, sex and ethnic group (Afro-Caribbean vs White) on inter-pupil distance (Imafedon, personal communication)

Source of variation	Degrees of freedom	Sum of squares	Mean square	Variance ratio (F)	Probability
Total	37	603.586			
Age group	2	124.587	62.293	6.81	0.003
Sex	1	1.072	1.072	0.12	0.7
Ethnic group	1	134.783	134.783	14.74	0.000 5
Residual	33	301.782	9.145		

94. The analysis of variance table for a study of the distance between the pupils of the eyes is shown in Table 17.20:

(a) There were 34 observations;

(b) There is good evidence of an ethnic group difference in the population;

(c) We can conclude that there is no difference in inter-pupil distance between men and women;

(d) There were two age groups;

(e) The difference between ethnic groups is likely to be due to a relationship between ethnicity and age in the sample.

95. Table 17.21 shows the logistic regression of vein graft failure on some potential explanatory variables. From this analysis:

(a) patients with high white cell counts were more likely to have graft failure;

(b) the log odds of graft failure for a diabetic is between 0.389 less and 2.435 greater than that for a non-diabetic;

(c) grafts were more likely to fail in female subjects, though this is not significant;

(d) there were four types of graft;

(e) any relationship between white cell count and graft failure may be due to smokers having higher white cell counts.

Table 17.21. Logistic regression of graft failure after 6 months (Thomas *et al.* 1993)

Variable	Coef.	Std. Err.	z=coef/se	P	95% Conf.	Interval
white cell count	1.238	0.273	4.539	< 0.001	0.695	1.781
graft type 1	0.175	0.876	0.200	0.842	−1.570	1.920
graft type 2	0.973	1.030	0.944	0.348	−1.080	3.025
graft type 3	0.038	1.518	0.025	0.980	−2.986	3.061
female	−0.289	0.767	−0.377	0.708	−1.816	1.239
age	0.022	0.035	0.633	0.528	−0.048	0.092
smoker	0.998	0.754	1.323	0.190	−0.504	2.501
diabetic	1.023	0.709	1.443	0.153	−0.389	2.435
constant	−13.726	3.836	−3.578	0.001	−21.369	−6.083

Number of observations = 84, chi squared = 38.05, d.f.= 8, P < 0.0001

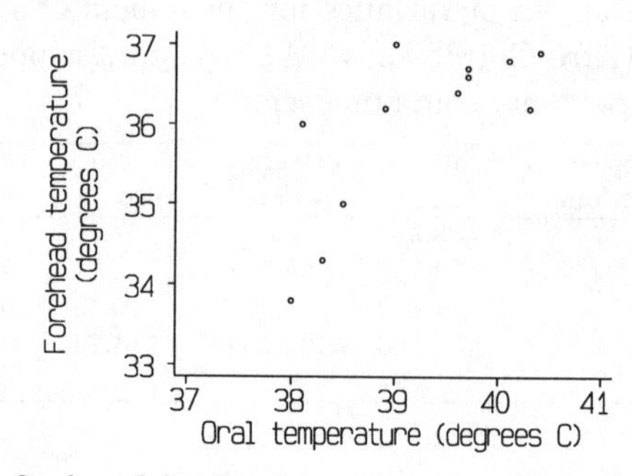

Fig. 17.6. Oral and forehead temperature measurements made in a group of pyrexic patients

96. For the data in Figure 17.6:

(a) the relationship could be investigated by linear regression;

(b) an 'oral squared' term could be used to test whether there is any evidence that the relationship is not a straight line;

(c) if an 'oral squared' term were included there would be 2 degrees of freedom for the model;

(d) the coefficients of an 'oral' and an 'oral squared' term would be uncorrelated;

(e) the estimation of the coefficient of a quadratic term would be improved by subtracting the mean from the oral temperature before squaring.

97. Table 17.22 shows the results of an observational study following up asthmatic children discharged from hospital. From this table:

(a) the analysis could only have been done if all children had been readmitted to hospital;

(b) the proportional hazards model would have been better than Cox regression;

(c) Boys have a shorter average time before readmission than do girls;

(d) the use of theophyline prevents readmission to hospital;

(e) children with several previous admissions have an increased risk of readmission.

Table 17.22. Cox regression of time to readmission for asthmatic children following discharge from hospital (Mitchell *et al.* 1994)

Variable	Coef.	Std. Err.	coef/se	P
boy	−0.197	0.088	−2.234	0.026
age	−0.126	0.017	−7.229	< 0.001
previous admissions (square root)	0.395	0.034	11.695	< 0.001
inpatient i.v. therapy	0.267	0.093	2.876	0.004
inpatient theophyline	−0.728	0.295	−2.467	0.014

Number of observations = 1 024, $\chi^2 = 167.15$, 5 d.f., $P < 0.0001$

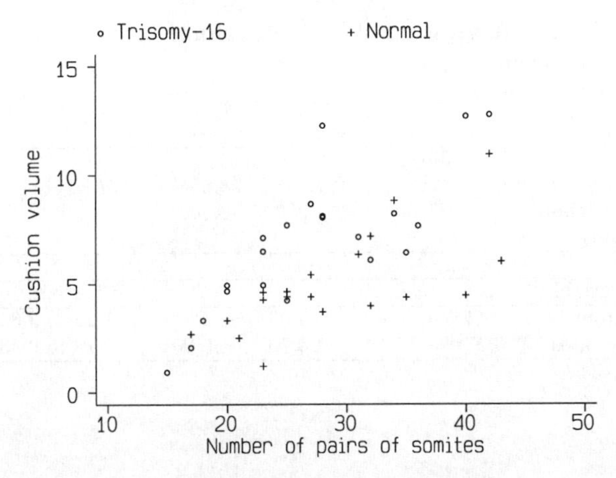

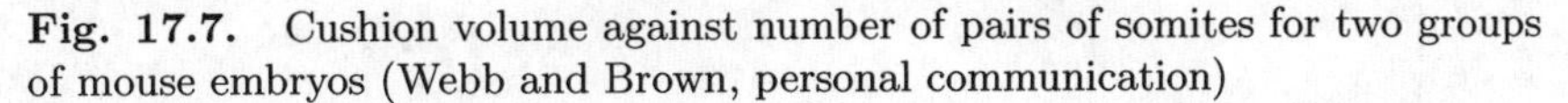

Fig. 17.7. Cushion volume against number of pairs of somites for two groups of mouse embryos (Webb and Brown, personal communication)

Table 17.23. Number of somites and cushion volume in mouse embryos

Normal				Trisomy-16			
som.	c.vol.	som.	c.vol.	som.	c.vol.	som.	c.vol.
17	2.674	28	3.704	15	0.919	28	8.033
20	3.299	31	6.358	17	2.047	28	12.265
21	2.486	32	3.966	18	3.302	28	8.097
23	1.202	32	7.184	20	4.667	31	7.145
23	4.263	34	8.803	20	4.930	32	6.104
23	4.620	35	4.373	23	4.942	34	8.211
25	4.644	40	4.465	23	6.500	35	6.429
25	4.403	42	10.940	23	7.122	36	7.661
27	5.417	43	6.035	25	7.688	40	12.706
27	4.395			25	4.230	42	12.767
				27	8.647		

17.E Exercise: A multiple regression analysis

Trisomy-16 mice can be used as an animal model for Down's syndrome. This analysis looks at the volume of a region of the heart, the atrioventricular cushion, of a mouse embryo, compared between trisomic and normal embryos. The embryos were at varying stages of development, indicated by the number of pairs of somites (precursors of vertebrae). Figure 17.7 and Table 17.23 show the data. The group was coded 1=normal, 2=trisomy-16. Table 17.24 shows the results of a regression analysis and Figure 17.8 shows residual plots.

1. Is there any evidence of a difference in volume between groups for given stage of development?
2. Figure 17.8 shows residual plots for the analysis of Table 17.24. Are

Table 17.24. Regression of cushion volume on number of pairs of somites and group in mouse embryos

Source of variation	Degrees of freedom	Sum of squares	Mean square	Variance ratio (F)	Probability
Total	39	328.976			
Due to regression	2	197.708	98.854	27.86	P < 0.0001
Residual (about regression)	37	131.268	3.548		

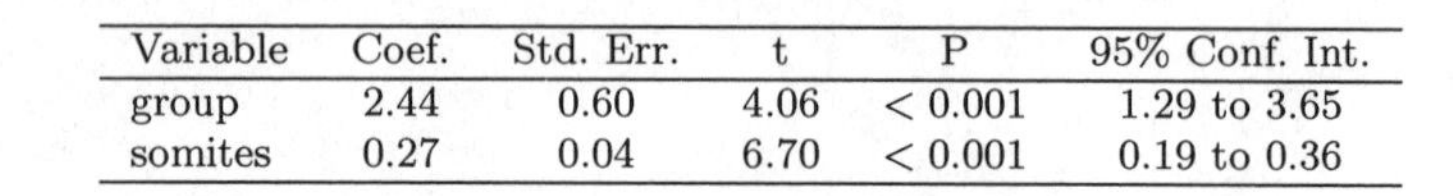

Variable	Coef.	Std. Err.	t	P	95% Conf. Int.
group	2.44	0.60	4.06	< 0.001	1.29 to 3.65
somites	0.27	0.04	6.70	< 0.001	0.19 to 0.36

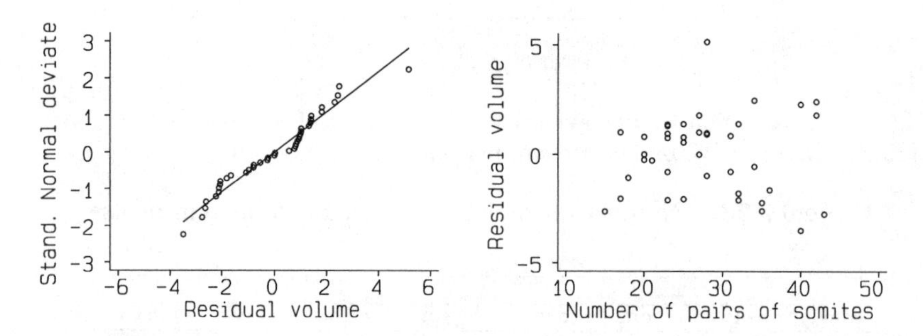

Fig. 17.8. Residual against number of pairs of somites and Normal plot of residuals for the analysis of Table 17.24

Table 17.25. Analysis of variance for regression with number of pairs of somites × group interaction

Source of variation	Degrees of freedom	Sum of squares	Mean square	Variance ratio (F)	Probability
Total	39	328.976			
Due to regression	3	207.139	69.046	20.40	P < 0.0001
Residual (about regression)	36	121.837	3.384		

there any features of the data which might make the analysis invalid?
3. It appears from Figure 17.7 that the relationship between volume and number of pairs of somites may not be the same in the two groups. Table 17.25 shows the analysis of variance for regression analysis including an interaction term. Calculate the F-ratio to test the evidence that the relationship is different in normal and trisomy-16 embryos. You can find the probability from Table 10.1, using the fact that the square root of F with 1 and n degrees of freedom is t with n degrees of freedom.

18

Determination of sample size

18.1 Estimation of a population mean

One of the questions most frequently asked of a medical statistician is 'How large a sample should I take?' In this chapter we shall see how statistical methods for deciding sample sizes can be used in practice as an aid in designing investigations. The methods we shall use are large sample methods, that is, they assume that large sample methods will be used in the analysis and so take no account of degrees of freedom.

We can use the concepts of standard error and confidence interval to help decide how many subjects should be included in a sample. If we want to estimate some population quantity, such as the mean, and we know how the standard error is related to the sample size then we can calculate the sample size required to give a confidence interval with the desired width. The difficulty is that the standard error may also depend either on the quantity we wish to estimate, or on some other property of the population, such as the standard deviation. We must estimate these quantities from data already available, or carry out a pilot study to obtain a rough estimate. The calculation of sample size can only be approximate anyway, so the estimates used to do it need not be precise.

If we want to estimate the mean of a population, we can use the formula for the standard error of a mean, $s/\sqrt{n}$, to estimate the sample size required. For example, suppose we wish to estimate the mean FEV1 in a population of young men. We know that in another study FEV1 had standard deviation $s = 0.67$ litre (§4.8). We therefore expect the standard error of the mean to be $0.67/\sqrt{n}$. We can set the size of standard error we want and choose the sample size to achieve this. We might decide that a standard error of 0.1 litre is what we want, so that we would estimate the mean to within $1.96 \times 0.1 = 0.2$ litre. Then: $\text{SE} = 0.67/\sqrt{n}$, $n = 0.67^2/\text{SE}^2 = 0.67^2/0.1^2 = 45$. We can also see what the standard error would be for different values of n :

n	10	20	50	100	200	500
standard error	0.21	0.15	0.09	0.07	0.05	0.03

So that if we had a sample size of 200, we would expect the 95% confidence interval to be 0.14 litre on either sided of the sample mean (1.96 standard errors) whereas with a sample of 50 the 95% confidence interval would be 0.30 litre on either side of the mean.

18.2 Estimation of a population proportion

When we wish to estimate a proportion we have a further problem. The standard error depends on the very quantity which we wish to estimate. We must guess the proportion first. For example, suppose we wish to estimate the prevalence of a disease, which we suspect to be about 2%, to the nearest 1 per 1 000. The unknown proportion, p, is guessed to be 0.02 and we want the 95% confidence interval to be 0.001 on either side, so the standard error must be half this, 0.000 5.

$$0.0005 = \sqrt{\frac{p(1-p)}{n}} = \sqrt{\frac{0.02(1-0.02)}{n}}$$

$$n = \frac{0.02(1-0.02)}{0.0005^2} = 78\,400$$

The accurate estimation of very small proportions requires very large samples. This is a rather extreme example and we do not usually need to estimate proportions with such accuracy. A wider confidence interval, obtainable with a smaller sample is usually acceptable. We can also ask 'If we can only afford a sample size of 1 000, what will be the standard error?'

$$\sqrt{\frac{0.02(1-0.02)}{1000}} = 0.0044$$

The 95% confidence limits would be, roughly, $p \pm 0.009$. For example, if the estimate were 0.02, the 95% confidence limits would be 0.011 to 0.029. If this accuracy were sufficient we could proceed.

These estimates of sample size are based on the assumption that the sample is large enough to use the Normal distribution. If a very small sample is indicated it will be inadequate and other methods must be used which are beyond the scope of this book.

18.3 Sample size for significance tests

We often want to demonstrate the existence of a difference or relationship as well as wanting to estimate its magnitude, as in a clinical trial, for example. We base these sample size calculations on significance tests, using the power of a test (§9.9) to help choose the sample size required to detect a difference if it exists. The power of a test is related to the postulated difference in the population, the standard error of the sample difference (which in turn depends on the sample size), and the significance level, which we usually take to be $\alpha = 0.05$. These quantities are linked by an equation which enables us to determine any one of them given the others. We can then say what sample size would be required to detect any given difference. We then decide what difference we need to be able to detect. This might be a difference which would have clincial importance, or a difference which we think the treatment may produce.

Suppose we have a sample which gives an estimate d of the population difference μ_d. We assume d comes from a Normal distribution with mean μ_d and has standard error SE(d). Here d might be the difference between two means, two proportions, or anything else we can calculate from data. We are interested in testing the null hypothesis that there is no difference in the population, i.e. $\mu_d = 0$. We are going to use a significance test at the α level, and want the power, the probability of detecting a significant difference, to be P.

I shall define u_α to be the value such that the Standard Normal distribution (mean 0 and variance 1) is less than $-u_\alpha$ or greater than u_α with probability α. Thus, for example, $u_{0.05} = 1.96$. The probability of lying between $-u_\alpha$ and u_α is $1 - \alpha$.

If the null hypothesis were true, the test statistic $d/\text{SE}(d)$ would be from a Standard Normal distribution. We reject the null hypothesis at the α level if the test statistic is greater than u_α or less than $-u_\alpha$, 1.96 for the usual 5% significance level. For significance we must have:

$$\frac{d}{\text{SE}(d)} < -u_\alpha \text{ or } \frac{d}{\text{SE}(d)} > u_\alpha$$

Let us assume that we are trying to detect a difference such that d will be greater than 0. The first alternative is then extremely unlikely and can be ignored. Thus we must have, for a significant difference: $d/\text{SE}(d) > u_\alpha$ so $d > u_\alpha\text{SE}(d)$. The critical value which d must exceed is $u_\alpha\text{SE}(d)$.

Now, d is a random variable, and for some samples it will be greater than its mean, μ_d, for some it will be less than its mean. d is an observation from a Normal distribution with mean μ_d and variance $\text{SE}(d)^2$. We want d to exceed the critical value with probability P, the chosen power of the test. The value of the Standard Normal distribution which is exceeded with probability P is $-u_{2(1-P)}$ (see Figure 18.1). $(1-P)$ is often represented as β (beta). This is the probability of failing to obtain a significant difference when the null hypothesis is false and the population difference is μ_d. It is the probability of a Type II error (§9.4). The value which d exceeds with probability P is the mean minus $-u_{2(1-P)}$ standard deviations: $\mu_d - u_{2(1-P)}\text{SE}(d)$. Hence for significance this must exceed the critical value, $u_\alpha\text{SE}(d)$. This gives

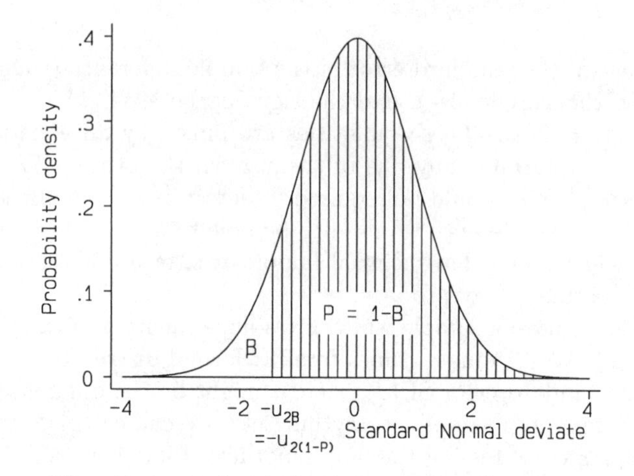

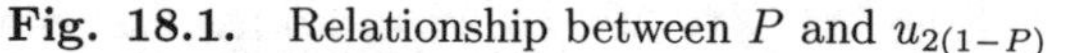
Fig. 18.1. Relationship between P and $u_{2(1-P)}$

Table 18.1. Values of $f(\alpha, P) = (u_\alpha + u_{2(1-P)})^2$ for different P and α

Power, P	Significance level, α	
	0.05	0.01
0.50	3.8	6.6
0.70	6.2	9.6
0.80	7.9	11.7
0.90	10.5	14.9
0.95	15.2	20.4
0.99	18.4	24.0

$$\mu_d - u_{2(1-P)}\text{SE}(d) = u_\alpha \text{SE}(d)$$

Putting the correct standard error formula into this will yield the required sample size. We can rearrange it as

$$\mu_d^2 = (u_\alpha + u_{2(1-P)})^2 \text{SE}(d)^2$$

This is the condition which must be met if we are to have a probability P of detecting a significant difference at the α level. We shall use the expression $(u_\alpha + u_{2(1-P)})^2$ a lot, so for convenience I shall denote it by $f(\alpha, P)$. Table 18.1 shows the values of the factor $f(\alpha, P)$ for different values of α and P. The usual value used for α is 0.05, and P is usually 0.80, 0.90, or 0.95.

18.4 Comparison of two means

When we are comparing the means of two samples, sample sizes n_1 and n_2, from populations with means μ_1 and μ_2, with the variance of the measurements being σ^2, we have $\mu_d = \mu_1 - \mu_2$ and

$$\mathrm{SE}(d) = \sqrt{\frac{\sigma^2}{n_1} + \frac{\sigma^2}{n_2}}$$

so the equation becomes:

$$(\mu_1 - \mu_2)^2 = f(\alpha, P)\sigma^2 \left(\frac{1}{n_1} + \frac{1}{n_2}\right)$$

For example, suppose we want to compare biceps skinfold in patients with Crohn's disease and coeliac disease, following up the inconclusive comparison of biceps skinfold in Table 10.4 with a larger study. We shall need an estimate of the variability of biceps skinfold in the population we are considering. We can usually get this from the medical literature, or as here from our own data. If not we must do a pilot study, a small preliminary investigation to collect some data and calculate the standard deviation. For the data of Table 10.4, the within-groups standard deviation is 2.3 mm. We must decide what difference we want to detect. In practice this may be difficult. In my small study the mean skinfold thickness in the Crohn's pateints was 1 mm greater than in my coeliac patients. I will design my larger study to detect a difference of 0.5 mm. I shall take the usual significance level of 0.05. I want a fairly high power, so that there is a high probability of detecting a difference of the chosen size should it exist. Usual values for the power required are 0.90 or 0.95. I shall take 0.90, which gives $f(\alpha, P) = 10.5$ from Table 18.1. The equation becomes:

$$0.5^2 = 10.5 \times 2.3^2 \times \left(\frac{1}{n_1} + \frac{1}{n_2}\right)$$

We have one equation with two unknowns, so we must decide on the relationship between n_1 and n_2. I shall try to recruit equal numbers in the two groups:

$$0.5^2 = 10.5 \times 2.3^2 \times \left(\frac{1}{n} + \frac{1}{n}\right)$$

$$n = \frac{10.5 \times 2.3^2 \times 2}{0.5^2} = 444.36$$

and I need 444 subjects in each group.

It may be that we do not know exactly what size of difference we are interested in. A useful approach is to look at the size of the difference we could detect using different sample sizes, as in Table 18.2. This done by putting different values of n in the sample size equation.

Table 18.2. Difference in mean biceps skinfold thickness (mm) detected at the 5% significance level with power 90% for different sample sizes, equal groups

Size of each group, n	Difference detected with probability 0.90
10	3.33
20	2.36
50	1.49
100	1.05
200	0.75
500	0.47
1 000	0.33

Table 18.3. Sample size required in each group to detect a difference between two means at the 5% significance level with power 90%, using equally sized samples

Difference in standard deviations	n	Difference in standard deviations	n	Difference in standard deviations	n
0.01	210 000	0.1	2 100	0.6	58
0.02	52 500	0.2	525	0.7	43
0.03	23 333	0.3	233	0.8	33
0.04	13 125	0.4	131	0.9	26
0.05	8 400	0.5	84	1.0	21

If we measure the difference in terms of standard deviations, we can make a general table. Table 18.3 gives the sample size required to detect differences between two equally sized groups. Altman (1983) gives a neat graphical method of calculation.

We do not need to have $n_1 = n_2 = n$. We can calculate $\mu_1 - \mu_2$ for different combinations of n_1 and n_2. The size of difference, in terms of standard deviations, which would be detected is given in Table 18.4. We can see from this that what matters is the size of the smaller sample. For example, if we have 10 in group 1 and 20 in group 2, we do not gain very much by increasing the size of group 2; increasing group 2 from 20 to 100 produces less advantage than increasing group 1 from 10 to 20. In this case the optimum is clearly to have samples of equal size.

The above enables us to compare independent samples. When we have paired observations, as in a cross-over trial, we need to take into account the pairing. If we have data on the distribution of differences, and hence their variance, s_d^2, the standard error of the mean difference is $\text{SE}(d) = \sqrt{s_d^2/n}$. If we do not, but have an estimate of the correlation, r, between repeated measurements of the quantity over a time similar to that proposed, then

Table 18.4. Difference (in standard deviations) detectable at the 5% significance level with power 90% for different sample sizes, unequal groups

n_2	n_1						
	10	20	50	100	200	500	1 000
10	1.45	1.25	1.13	1.08	1.05	1.03	1.03
20	1.25	1.03	0.85	0.80	0.75	0.75	0.73
50	1.13	0.85	0.65	0.55	0.50	0.48	0.48
100	1.08	0.80	0.55	0.45	0.40	0.35	0.35
200	1.05	0.75	0.50	0.40	0.33	0.28	0.25
500	1.03	0.75	0.48	0.35	0.28	0.20	0.18
1 000	1.03	0.73	0.48	0.35	0.25	0.18	0.15

$\text{SE}(d) = \sqrt{2s^2(1-r^2)/n}$, where s is the usual standard deviation between subjects. If we have neither of these, which happens often, we need a pilot study. As we need about 20 subjects for such a study, and many cross-over trials are of this order (§2.6), we could carry out a small trial. The difference will either be so large that we have the answer, or, if not, we will have sufficient data with which to design a much larger study.

18.5 Comparison of two proportions

Using the same approach, we can also calculate the sample sizes for comparing two proportions. If we have two sample with sizes n_1 and n_2 from Binomial populations with proportions p_1 and p_2, the difference is $\mu_d = p_1 - p_2$, the standard error of the difference between the sample proportions (§8.6) is:

$$\text{SE}(d) = \sqrt{\frac{p_1(1-p_1)}{n_1} + \frac{p_2(1-p_2)}{n_2}}$$

If we put these into the previous formula we have:

$$(p_1 - p_2)^2 = f(\alpha, P)\left(\frac{p_1(1-p_1)}{n_1} + \frac{p_2(1-p_2)}{n_2}\right)$$

The size of the proportions, p_1 and p_2, is important, as well as their difference. (The significance test implied here is similar to the chi-squared test for a 2 by 2 table). When the sample sizes are equal, i.e. $n_1 = n_2 = n$, we have

$$n = \frac{f(\alpha, P)\,(p_1(1-p_1) + p_2(1-p_2))}{(p_1 - p_2)^2}$$

Suppose we wish to compare the survival rate with a new treatment with that with an old treatment, where it is about 60%. What values of n_1 and n_2 will have 90% chance of giving significant difference at the 5% level for different values of p_2? For $P = 0.90$ and $\alpha = 0.05$, $f(\alpha, P) = 10.5$.

Table 18.5. Sample size in each group required to detect different proportions p_2 when $p_1 = 0.6$ at the 5% significance level with power 90%, equal groups

p_2	n
0.90	39
0.80	105
0.70	473
0.65	1 964

Suppose we wish to detect an increase in the survival rate on the new treatment to 80%, so $p_2 = 0.80$, and $p_1 = 0.60$.

$$\begin{aligned} n &= \frac{10.5 \times (0.8(1-0.8) + 0.6(1-0.6))}{(0.8-0.6)^2} \\ &= \frac{10.5 \times (0.16 + 0.24)}{0.2^2} \\ &= 105 \end{aligned}$$

We would require 105 in each group to have a 90% chance of showing a significant difference if the population proportions were 0.6 and 0.8.

When we don't have a clear idea of the value of p_2 in which we are interested, we can calculate the sample size required for several proportions, as in Table 18.5. It is immediately apparent that to detect small differences between proportions we need very large samples.

The case where samples are of equal size is usual in experimental studies, but not in observational studies. Suppose we wish to compare the prevalence of a certain condition in two populations. We expect that in one population it will be 5% and that it may be more common the second. We can rearrange the equation:

$$n_2 = \frac{f(\alpha, P)p_2(1-p_2)}{(p_1 - p_2)^2 + f(\alpha, P)\frac{p_1(1-p_1)}{n_1}}$$

Table 18.6 shows n_2 for different n_1 and p_2. For some values of n_1 we get a negative value of n_2. This means that no value of n_2 is large enough. It is clear that when the proportions themselves are small, the detection of small differences requires very large samples indeed.

18.6 Detecting a correlation

Investigations are often set up to look for a relationship between two continuous variables. It is convenient to treat this as an estimation of or test

Table 18.6. n_2 for different n_1 and p_2 when $p_1 = 0.05$ at the 5% significance level with power 90%

p_2	n_1								
	50	100	200	500	1 000	2 000	5 000	10 000	100 000
0.06	.	.	.	.	.	.	237 000	11 800	7 900
0.07	.	.	.	.	.	4 500	2 300	2 000	1 800
0.08	.	.	.	.	1 900	1 200	970	900	880
0.10	.	.	1 500	630	472	420	390	390	380
0.15	5 400	270	180	150	140	140	140	140	130
0.20	134	96	84	78	76	76	75	75	75

of a correlation coefficient. The correlation coefficient has an awkward distribution, which tends only very slowly to the Normal, even when both variables themselves follow a Normal distribution. We can use Fisher's z transformation:

$$z = \frac{1}{2}\log_e\left(\frac{1+r}{1-r}\right)$$

which follows a Normal distribution with mean

$$z_\rho = \frac{1}{2}\log_e\left(\frac{1+\rho}{1-\rho}\right) + \frac{\rho}{2(n-1)}$$

and variance $1/(n-3)$ approximately, where ρ is the population correlation coefficient and n is the sample size (§11.10). For sample size calculations we can approximate z_ρ by

$$z_\rho = \frac{1}{2}\log_e\left(\frac{1+\rho}{1-\rho}\right)$$

The 95% confidence interval for z will be $z_\rho \pm 1.96\sqrt{1/(n-3)}$, approximately. Given a rough idea of ρ we can estimate n required for any accuracy. For example, suppose we want to estimate a correlation coefficient, which we guess to about 0.5, and we want it to within 0.1 either way, i.e. we want a confidence interval like 0.4 to 0.6. The z transformations of these values of r are $z_{0.4} = 0.42365$, $z_{0.5} = 0.54931$, $z_{0.6} = 0.69315$, the differences are $z_{0.5} - z_{0.4} = 0.12566$ and $z_{0.6} - z_{0.5} = 0.14384$, and so to get the sample size we want we need to set 1.96 standard errors to the smaller of these differences. We get $1.96\sqrt{1/(n-3)} = 0.12566$ giving $n = 246$.

We more often want to see whether there is any evidence of a relationship. When $r = 0$, $z_r = 0$, so to test the null hypothesis that $\rho = 0$ we can test the null hypothesis that $z_\rho = 0$. The difference we wish to test is $\mu_d = z_\rho$, which has $\text{SE}(d) = \sqrt{1/(n-3)}$. Putting this into the formula of §18.3 we get

$$z_\rho^2 = f(\alpha, P)\frac{1}{n-3}$$

Table 18.7. Approximate sample size required to detect a correlation at the 5% significance level with power 90%

ρ	n	ρ	n	ρ	n
0.01	100 000	0.1	1 000	0.6	25
0.02	26 000	0.2	260	0.7	17
0.03	12 000	0.3	110	0.8	12
0.04	6 600	0.4	62	0.9	8
0.05	4 200	0.5	38		

Thus we have

$$\left(\frac{1}{2}\log_e\left(\frac{1+\rho}{1-\rho}\right)\right)^2 = f(\alpha, P)\frac{1}{n-3}$$

and we can estimate n, ρ or P given the other two. Table 18.7 shows the sample size required to detect a correlation coefficient with a power of $P = 0.9$ and a significance level $\alpha = 0.05$.

18.7 Accuracy of the estimated sample size

In this section we have assumed that samples are sufficiently large for distributions to be approximately Normal and for estimates of variance to be good estimates. With very small samples this may not be the case. Various more accurate methods exist, but any size calculation is approximate and except for very small samples, say less than 10, the methods described above should be adequate.

These methods depend on assumptions about the size of difference sought and the variability of the observations. It may be that the population to be studied may not have exactly the same characteristics as those from which the standard deviation or porportions were estimated. The likely effects of changes in these can be examined by putting different values of them in the formula. However, there is always an element of venturing into the unknown when embarking on a study and we can never by sure that the sample and population will be as we expect. The determination of sample size as described above is thus only a guide, and it is probably as well always to err on the side of a larger sample when coming to a final decision.

The choice of power is arbitrary, in that there is not optimum choice of power for a study. I usually recommend 90%, but 80% is often quoted. This gives smaller estimated sample sizes, but, of course, a greater chance of failing to detect effects.

For a fuller treatment of sample size estimation and fuller tables see Machin and Campbell (1987) and Lemeshow *et al.* (1990).

18.M Multiple choice questions 98 to 100

(Each answer is true or false)

98. The power of a two-sample t test:

(a) increases if the sample sizes are increased;

(b) depends on the difference between the population means which we wish to detect;

(c) depends on the difference between the sample means;

(d) is the probability that the test will detect a given population difference;

(e) cannot be zero.

99. The sample size required for a study to compare two proportions:

(a) depends on the magnitude of the effect we wish to detect;

(b) depends on the significance level we wish to employ;

(c) depends on the power we wish to have;

(d) depends on the anticipated values of the proportions themselves;

(e) should be decided by adding subjects until the difference is significant.

100. The sample size required for a study to estimate a mean:

(a) depends on the width of the confidence interval which we want;

(b) depends on the variability of the quantity being studied;

(c) depends on the power we wish to have;

(d) depends on the anticipated value of the mean;

(e) depends on the anticipated value of the standard deviation.

18.E Exercise: Estimation of sample sizes

1. What sample size would be required to estimate a 95% reference interval using the Normal distribution method, so that the 95% confidence interval for the reference limits were at most 20% of the reference interval size?
2. How big a sample would be required for an opinion pollster to estimate voter preferences to within two percentage points?
3. Mortality from myocardial infarction after admission to hospital is about 15%. How many patients would be required for a clinical trial to detect a 10% reduction in mortality, i.e. to 13.5%, if the power required was 90%? How many would be needed if the power were only 80%?
4. How many patients would be required in a clinical study to compare an enzyme concentration in patients with a particular disease and controls, if differences of less than one standard deviation would not be clinically important? If there was already a sample of measurements from 100 healthy controls, how many disease cases would be required?

19

Solutions to exercises

Some of the multiple choice questions are quite hard. If you score +1 for a correct answer, −1 for an incorrect answer, and 0 for a part which you omitted, I would regard 40% as the pass level, 50% as good, 60% as very good, and 70% as excellent. These questions are hard to set and some may be ambiguous, so you will not score 100%.

Solution to Exercise 2M: multiple choice questions 1 to 6

1. FFFFF. Controls should be treated in the same place at the same time, under the same conditions other than the treatment under test (§2.1). All must be willing and eligible to receive either treatment (§2.4).
2. FTFTF. Random allocation is done to achieve comparable groups, allocation being unrelated to the subjects' characteristics (§2.2). The use of random numbers helps to prevent bias in recruitment (§2.3).
3. TFFTF. Patients do not know their treatment, but they usually do know that they are in a trial (§2.9). Not the same as a crossover trial (§2.6).
4. FFFFF. Vaccinated and refusing children are self-selected (§2.4). We analyse by intention to treat (§2.5). We can compare effect of a vaccination programme by comparing whole vaccination group, vaccinated and refusers to the controls.
5. TFTTT. §2.6. The order is randomized.
6. FFTTT. §2.8, §2.9. The purpose of placebos is make dissimilar treatments appear similar. Only in randomized trials can we rely on comparability, and then only within the limits of random variation (§2.2).

Solution to Exercise 2E

1. It was hoped that women in the KYM group would be more satisfied with their care. The knowledge that they would receive continuity of care was an important part of the treatment, and so the lack of blindness is essential. More difficult is that KYM women were given a choice and so may have felt more committed to whichever scheme, KYM or standard, they had chosen, than did the control group. We must accept this element of patient control as part of the treatment.

Table 19.1. Method of delivery in the KYM study

Method of delivery	Allocated to KYM %	n	Allocated to control %	n
Normal	79.7	382	74.8	354
Instrumental	12.5	60	17.8	84
Caesarian	7.7	37	7.4	35

2. The study should be (and was) analysed by intention to treat (§2.5). As often happens, the refusers did worse than did the acceptors of KYM, and worse than the control group. When we compare all those allocated to KYM with those allocated to control, there is very little difference (Table 19.1).

3. Women had booked for hospital antenatal care expecting the standard service. Those allocated to this therefore received what they had requested. Those allocated to the KYM scheme were offered a treatment which they could refuse if they wished, refusers getting the care for which they had originally booked. No extra examinations were carried out for research purposes, the only special data being the questionnaires, which could be refused. There was therefore no need to get the women's permission for the randomization. I thought this was a convincing argument.

Solution to Exercise 3M: multiple choice questions 7 to 13

7. FTTTT. A population can be anything (§3.3).

8. TFFFT. A census tells us who is there on that day, and only applies to current in-patients. The hospital could be quite unusual. Some diagnoses are less likely than others to lead to admission or to long stay (§3.2).

9. TFFTF. All members and all samples have equal chances of being chosen (§3.4). We must stick to the sample the random process produces. Errors can be estimated using confidence intervals and significance tests. Choice does not depend on the subject's characteristics at all, except for its being in the population.

10. FTTFT. Some populations are unidentifiable and some cannot be listed easily (§3.4).

11. TTTTF. This is a random cluster sample (§3.4). Each patient had the same chance of their hospital being chosen and then the same chance of being chosen within the hospital. This would not be so if we chose a fixed number from each hospital rather than a fixed proportion, as those in small hospitals would be more likely to be chosen than those in large hospitals. In part (e), what about a sample with patients in every hospital?

12. FTFTT. We must have a cohort or case control study to get enough cases (§3.7, §3.8).

13. FFFTF. In a case control study we start with a group with the disease, the cases, and a group without the disease, the controls (§3.8).

Solution to Exercise 3E

1. Many cases of infection may be unreported, but there is not much that could be done about that. Many organisms produce similar symptoms, hence the need for laboratory confirmation. There are many sources of infection, including direct transmission, hence the exclusion of cases exposed to other water supplies and to infected people.
2. Controls must be matched for age and sex as these may be related to their exposure to risk factors such as handling raw meat. Inclusion of controls who may have had the disease would have weakened any relationships with the cause, and the same exclusion criteria were applied as for the cases, to keep them comparable.
3. Data are obtained by recall. Cases may remember events in relation to the disease more easily that than controls in relation to the same time. Cases may have been think about possible causes of the disease and so be more likely to recall milk attacks. The lack of positive association with any other risk factors suggests that this is not important here.
4. I was convinced. The relationship is very strong and these scavenging birds are known to carry the organism. There was no relationship with any other risk factor. The only problem is that there was little evidence that these birds had actually attacked the milk. Others have suggested that cats may also remove the tops of milk bottles to drink the milk and may be the real culprits (Balfour 1991).
5. Further studies: Testing of attacked milk bottles for campylobacter (have to wait for the next year). Possibly a cohort study, asking people about history of bird attacks and drinking attacked milk, then follow for future campylobacter (and other) infections. Possibly an intervention study. Advise people to protect their milk and observe the subsequent pattern of infection.

Solution to Exercise 4M: multiple choice questions 14 to 19

14. TFFTF. §4.1. Parity is quantitative and discrete, height and blood pressure are continuous.
15. TTFTF. §4.1. Age last birthday is discrete, exact age includes years and fraction of a year.
16. FFTFT. §4.4, §4.6. It could have more than one mode, we cannot say. Standard deviation is less than variance if the variance is greater than one (§4.7–8).
17. TTTFT. §4.2–4. Mean and variance only tell us the location and spread of the distribution (§4.6–7).
18. TFTFT. §4.5–7. Median $= 2$, the observations must be ordered before the central one is found, mode $= 2$, range $= 7 - 1 = 6$, variance $= 22/4 = 5.5$.

2	2 9
3	3 3 3 4 4 4 6 6 6 6 7 7 8 8 8 9
4	0 0 0 1 1 1 2 3 4 4 4 5 6 7 7 7 8 9 9
5	0 1
6	0

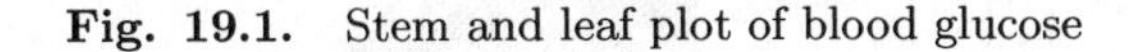

Fig. 19.1. Stem and leaf plot of blood glucose

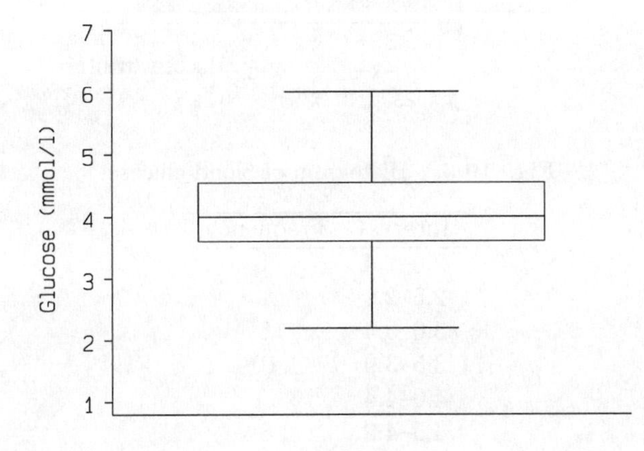

Fig. 19.2. Box and whisker plot of blood glucose

19. FFFFT. §4.6–8. There would be more observations below the mean than above, because the median would be less than the mean. Most observations will be within one standard deviation of the mean whatever the shape. The standard deviation measures how widely the blood pressure is spread between people, not for a single person, which would be needed to estimate accuracy. See also §15.2.

Solution to Exercise 4E

1. The stem and leaf plot is shown in Figure 19.1:
2. Minimum = 2.2, maximum = 6.0. The median is the average of the 20th and 21st ordered observations, since the number of observations is even. These are both 4.0, so the median is 4.0. The first quartile is between the 10th and 11th, which are both 3.6. The third quartile is between the 30th and 31st observations, which are 4.5 and 4.6. We have $q = 0.75$, $i = 0.75 \times 41 = 30.75$, and the quartile is given by $4.5 + (4.6 - 4.5) \times 0.75 = 4.575$ (§4.5). The box and whisker plot is shown in Figure 19.2.
3. The frequency distribution is derived easily from the stem and leaf plot:

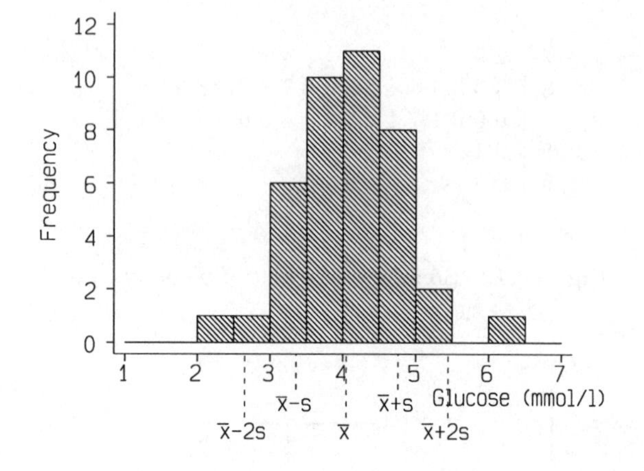

Fig. 19.3. Histogram of blood glucose

Interval	Frequency
2.0–2.4	1
2.5–2.9	1
3.0–3.4	6
3.5–3.9	10
4.0–4.4	11
4.5–4.9	8
5.0–5.4	2
5.5–5.9	0
6.0–6.4	1
Total	40

4. The histogram is shown in Figure 19.3. The distribution is symmetrical.
5. The mean is given by

$$\sum x_i = 16.2, \quad \bar{x} = \frac{\sum x_i}{n} = \frac{16.2}{4} = 4.05$$

The deviations and their squares are as follows:

	x_i	$x_i - \bar{x}$	$(x_i - \bar{x})^2$
	4.7	0.65	0.422 5
	4.2	0.15	0.022 5
	3.9	−0.15	0.022 5
	3.4	−0.65	0.422 5
Total	16.2	0.00	0.890 0

There are $n - 1 = 4 - 1 = 3$ degrees of freedom. The variance is given by

$$s^2 = \frac{\text{sum of squares about the mean}}{\text{degrees of freedom}} = \frac{0.89}{3} = 0.296\,67$$

$$s = \sqrt{s^2} = \sqrt{0.296\,67} = 0.544\,67$$

6. As before, the sum is $\sum x_i = 16.2$, The sum of squares about the mean is then given by $\sum x_i^2 = 66.5$ and

$$\sum x_i^2 - \frac{(\sum x_i)^2}{n} = 66.5 - \frac{16.2^2}{4} = 0.89$$

This is the same as found in 5 above, so, as before,

$$s^2 = \frac{0.89}{3} = 0.296\,67, \quad s = 0.544\,67$$

7. For the mean we have $\sum x_i = 162.2$,

$$\bar{x} = \frac{\sum x_i}{n} = \frac{162.2}{40} = 4.055$$

The sum of squares about the mean is given by:

$$\sum x_i^2 - \frac{(\sum x_i)^2}{n} = 676.74 - \frac{162.2^2}{40} = 19.019$$

There are $n - 1 = 40 - 1 = 39$ degrees of freedom. The variance is given by

$$s^2 = \frac{\text{sum of squares about the mean}}{\text{degrees of freedom}} = \frac{19.109}{39} = 0.487\,667$$

For the standard deviation, $s = \sqrt{s^2} = \sqrt{0.487\,667} = 0.698$.
9. For the limits, $\bar{x} - 2s = 4.055 - 2 \times 0.698 = 2.659$, $\bar{x} - s = 4.055 - 0.698 = 3.357$, $\bar{x} = 4.055$, $\bar{x} + s = 4.055 + 0.698 = 4.753$, and $\bar{x} + 2s = 4.055 + 2 \times 0.698 = 5.451$. Figure 19.3 shows the mean and standard deviation marked on the histogram. The majority of points fall within one standard deviation of the mean and nearly all within two standard deviations of the mean. Because the distribution is symmetrical, it extends just beyond the $\bar{x} \pm 2s$ points on either side.

Solution to Exercise 5M: multiple choice questions 20 to 24

20. FTTTT. §5.1, §5.2. Without a control group we have no idea how many would get better anyway (§2.1). 66.67% is 2/3. We may only have 3 patients.
21. TFFTT. §5.2. To three significant figures, it should be 1730. We round up because of the 9. To six decimal places it is 1729.543710.
22. FTTFT. This is a bar chart showing the relationship between two variables (§5.5). See Figure 19.4. Calendar time has no true zero to show.
23. TTFFT. §5.9, §5A. There is no logarithm of zero.
24. FFTTT. §5.5–7. A histogram (§4.3) and a pie chart (§5.4) each show the distribution of a single variable.

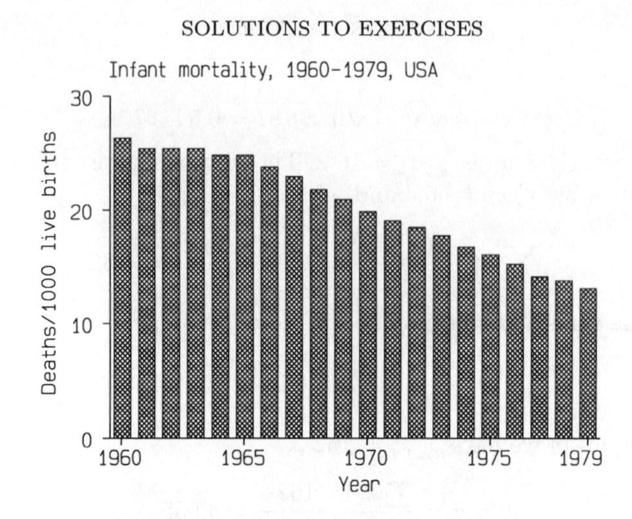

Fig. 19.4. A dubious graph revised

Table 19.2. Calculations for a pie chart for the Tooting Bec data

Category	Frequency	Relative frequency	Angle
schizophrenia	474	0.323 11	116
affective illness	277	0.188 82	68
organic brain syndrome	405	0.276 07	99
subnormality	58	0.039 54	14
alcoholism	57	0.038 85	14
other	196	0.133 61	48
Total	1 467	1.000 00	359

Solution to Exercise 5E

1. This is the frequency distribution of a qualitative variable, so a pie chart can be used to display it. The calculations are set out in Table 19.2. Notice that we have lost one degree through rounding errors. We could work to fractions of a degree, but the eye is unlikely to spot the difference. The pie chart is shown in Figure 19.5.
2. See Figure 19.6.
3. There are several possibilities. In the original paper, Doll and Hill used a separate bar chart for each disease, similar to in Figure 19.7.
4. This is a frequency distribution of a quantitative variable, so a histogram is appropriate. See Figure 19.8.
5. Line graphs can be used here, as we have simple time series (Figure 19.9). For an explanation of the difference between years, see §13E.

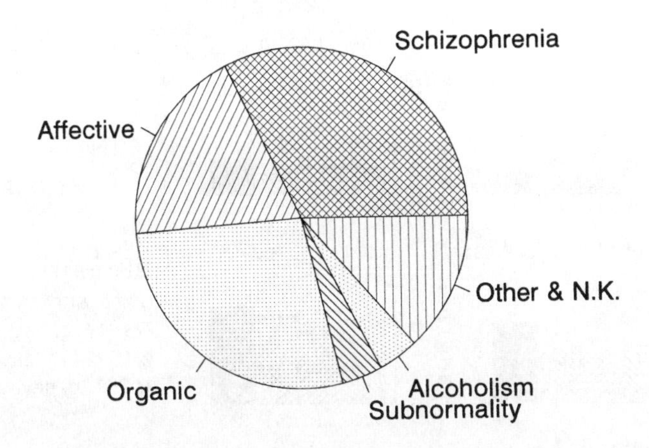

Fig. 19.5. Pie chart showing the distribution of patients in Tooting Bec Hospital by diagnostic group

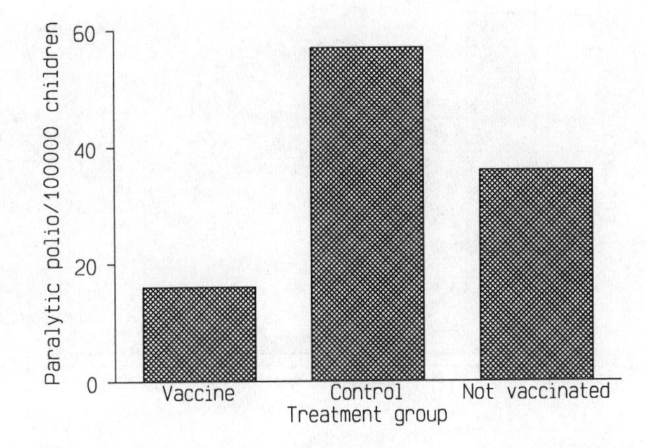

Fig. 19.6. Bar chart showing the results of the Salk vaccine trial

Solution to Exercise 6M: multiple choice questions 25 to 31

25. TTFFF. §6.2. If they are mutually exclusive they cannot both happen. There is no reason why they should be equiprobable or exhaustive, the only events which can happen (§6.3).

26. TFTFT. For both, the probabilities are multiplied, $0.2 \times 0.05 = 0.01$ (§6.2). Clearly the probability of both must be less than that for each one. The probability of both is 0.01, so the probability of X alone is $0.20 - 0.01 = 0.19$ and the probability of Y alone is $0.05 - 0.01 = 0.04$. The probability of having X or Y is the probability of X alone + probability of Y

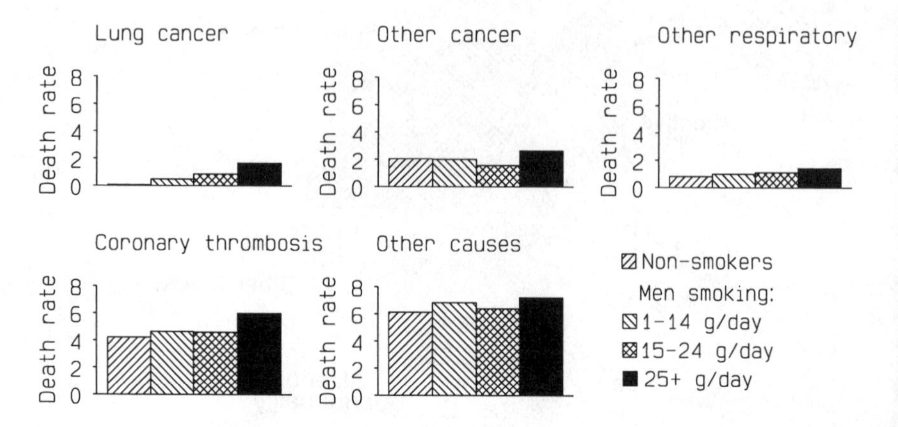

Fig. 19.7. Mortality in British doctors by smoking habits, after Doll and Hill (1956)

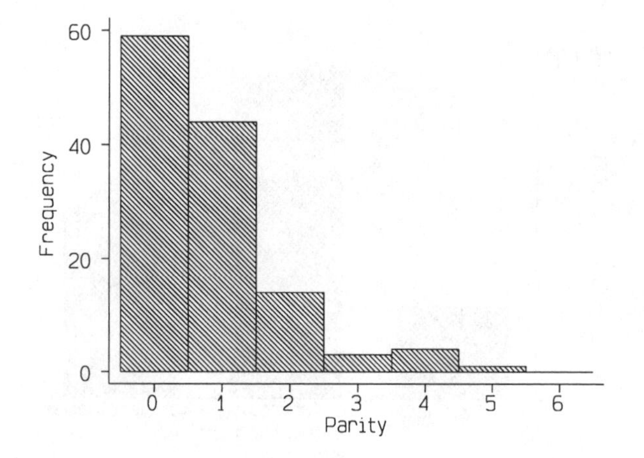

Fig. 19.8. Histogram showing parity of women attending antenatal clinics at St. George's Hospital

alone + probability of X and Y together, because these are three mutually exclusive events. Having X and having Y are not mutually exclusive as she can have both. Having X tells us nothing about whether she has Y. If she has X the probability of having Y is still 0.05, because X and Y are independent.

27. TFTFF. §6.4. Weight is continuous. Patients respond or not with equal probability, being selected at random from a population where the probability of responding varies. The number of red cells might follow a Poisson distribution (§6.7); there is no set of independent trials. The *number* of

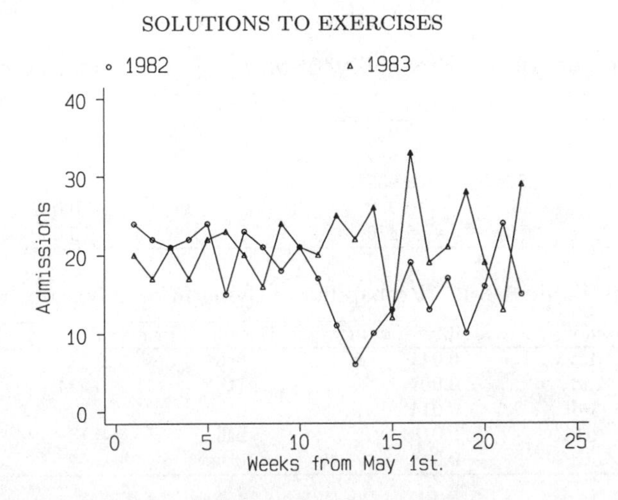

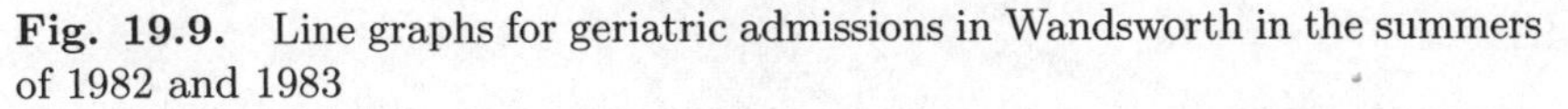

Fig. 19.9. Line graphs for geriatric admissions in Wandsworth in the summers of 1982 and 1983

hypertensives follows a Binomial distribution, not the proportion.
28. TTTTF. The probability of clinical disease is $0.5 \times 0.5 = 0.25$. The probability of carrier status = probability that father passes the gene and mother doesn't + probability that mother passes the gene and father doesn't = $0.5 \times 0.5 + 0.5 \times 0.5 = 0.5$. Probability of not inheriting the gene = $0.5 \times 0.5 = 0.25$. Probability of not having clinical disease = $1 - 0.25 = 0.75$. Successive chidren are independent, so the probabilities for the second child are unaffected by the first (§6.2).
29. FTTFT. §6.3–4. The expected number is one (§6.6). The spins are independent (§6.2). At least one tail means one tail (PROB = 0.5) or two tails (PROB = 0.25). These are mutually exclusive, so the probability of at least one tail is $0.5 + 0.25 = 0.75$.
30. FTTFT. §6.6. $\text{E}(X = 2) = \mu + 2$, $\text{VAR}(2X) = 4\sigma^2$.
31. TTTFF. §6.6. The variance of a difference is the sum of the variances. Variances cannot be negative. $\text{VAR}(-X) = (-1)^2 \times \text{VAR}(X) = \text{VAR}(X)$.

Solution to Exercise 6E

1. Probability of survival to age 10. This illustrates the frequency definition of probability. 959 out of 1 000 survive, so the probability is 959/1 000 = 0.959.
2. Survival and death are mutually exclusive, exhaustive events, so PROB(survives) + PROB(dies) = 1. Hence PROB(dies) = $1 - 0.959 = 0.041$.
3. These are the number surviving divided by 1 000 (Table 19.3). The events are not mutually exclusive, e.g. a man cannot survive to age 20 if he

Table 19.3. Probability of surviving to different ages

Survive to age	Probability	Survive to age	Probability
10	0.959	60	0.758
20	0.952	70	0.524
30	0.938	80	0.211
40	0.920	90	0.022
50	0.876	100	0.000

Table 19.4. Probability of dying in each decade

Decade	Probability of dying	Decade	Probability of dying
1st	0.041	6th	0.118
2nd	0.007	7th	0.234
3rd	0.014	8th	0.313
4th	0.018	9th	0.189
5th	0.044	10th	0.022

does not survive to age 10. This does not form a probability distribution.
4. The probability is found by

$$\begin{aligned} \text{PROB(aged 60 survives to 70)} &= \frac{\text{number alive at 70}}{\text{number alive at 60}} \\ &= 524/758 \\ &= 0.691 \end{aligned}$$

5. Independent events. PROB(survival 60 to 70) $= 0.691$,
PROB(both survive) $= 0.691 \times 0.691 = 0.477$.
6. The proportion surviving on average is the probability of survival $=$ 0.691. So a proportion of 0.691 of the 100 survive. We expect $0.691 \times 100 =$ 69.1 to survive.
7. The probability is found by

$$\begin{aligned} \text{PROB(death in 2nd)} &= \text{PROB(survives to 2nd)} \\ &\quad - \text{PROB(survives to 3rd)} \\ &= 0.959 - 0.952 \\ &= 0.007 \end{aligned}$$

8. As in 7, we find probabilities of dying for each decade (Table 19.4). This is a set of mutually exclusive events and they are exhaustive—there is no other decade in which death can take place. The sum of the probabilities is therefore 1.0. The distribution is shown in Figure 19.10.
9. We find the expected values or mean of a probability distribution by summing each value times its probability (§6.4), to give life expectancy at birth $=$ 66.6 years (Table 19.5).

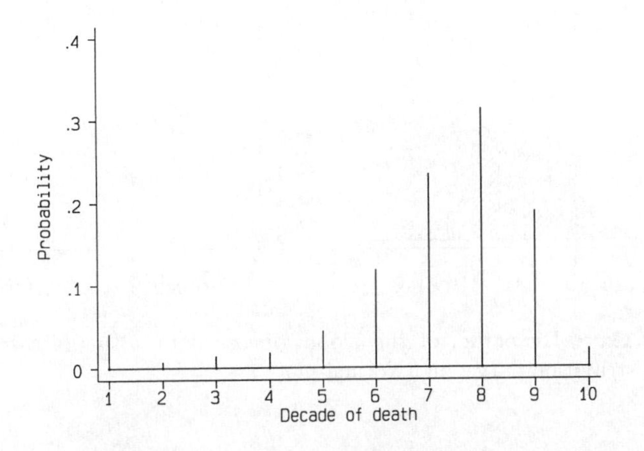

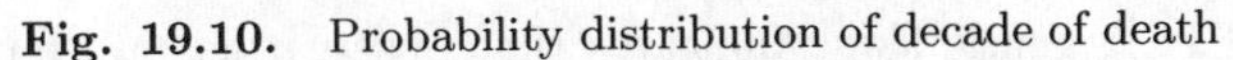
Fig. 19.10. Probability distribution of decade of death

Table 19.5. Calculation of expectation of life

5×0.041	=	0.205
15×0.007	=	0.105
25×0.014	=	0.350
35×0.018	=	0.630
45×0.044	=	1.980
55×0.118	=	6.490
65×0.234	=	15.210
75×0.313	=	23.475
85×0.189	=	16.065
95×0.022	=	2.090
Total		66.600

Solution to Exercise 7M: multiple choice questions 32 to 37

32. TTTFT. §7.2–4.
33. FFFTT. Symmetrical, $\mu = 0, \sigma = 1$ (§7.3, §4.6).
34. TTFFF. §7.2. Median = mean. The Normal distribution has nothing to do with normal physiology. 2.5% will be less than 260, 2.5% will be greater than 340 litres/min.
35. FTTFF. §4.6, §7.3. The sample size should not affect the mean. The relative sizes of mean, median and standard deviation depend on the shape of the frequency distribution.
36. TFTTF. §7.2, §7.3. Adding, subtracting or multiplying by a constant, or adding or subtracting an independent Normal variable gives a Normal distribution. X^2 follows a very skew Chi-squared distribution with one degree of freedom and X/Y follows a t distribution with one degree of freedom (§7A).

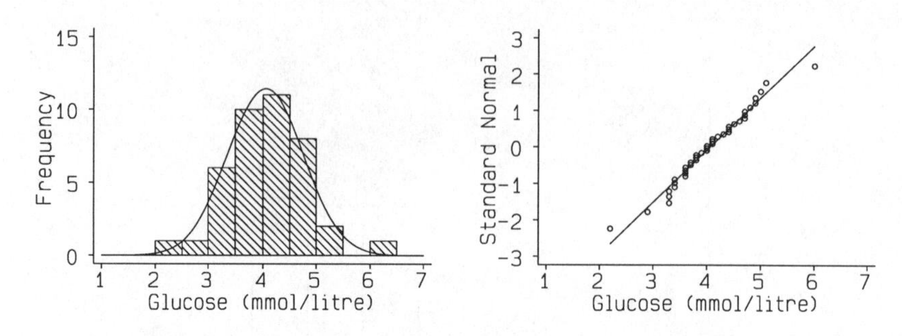

Fig. 19.11. Histogram of the blood glucose data with the corresponding Normal distribution curve, and Normal plot

37. TTTTT. A gentle slope indicates that observations are far apart, a steep slope that there are many observations close together. Hence gentle-steep-gentle ('S' shaped) indicates long tails (§7.5).

Solution to Exercise 7E

1. The box and whisker plot shows a very slight degree of skewness, the lower whisker being shorter than the upper and the lower half of the box smaller than the upper. From the histogram it appears that the tails are a little longer than the Normal curve of Figure 7.10 would suggest. Figure 19.11 shows the Normal distribution with the same mean and variance superimposed on the histogram, which also indicates this.
2. We have $n = 40$. For $i = 1$ to 40 we want to calculate $(i - 0.5)/n = (2i - 1)/2n$. This gives us a probability. We use Table 7.1 to find the value of the Normal distribution corresponding to this probability. For example, for $i = 1$ we have

$$\frac{2i - 1}{2n} = \frac{2 - 1}{2 \times 40} = \frac{1}{80} = 0.0125$$

From Table 7.1 we cannot find the value of x corresponding to $\Phi(x) = 0.0125$ directly, but we see that $x = -2.3$ corresponds to $\Phi(x) = 0.011$ and $x = -2.2$ to $\Phi(x) = 0.014$. $\Phi(x) = 0.0125$ is midway between these probabilities so we can estimate the value of x as midway between -2.3 and -2.2, giving -2.25. This corresponds to the lowest blood glucose, 2.2. For $i = 2$ we have $\Phi(x) = 0.0375$. Referring to the table we have $x = -1.8$, $\Phi(x) = 0.036$ and $x = -1.7$, $\Phi(x) = 0.045$. For $\Phi(x) = 0.0375$ we must have x just above -1.8, about -1.78. The corresponding blood glucose is 2.9. We do not have to be very accurate because we are only using this plot for a rough guide. We get a set of probabilities as follows:

i	$(2i-1)/2n = \Phi(x)$	x	blood glucose
1	$1/80 = 0.0125$	-2.25	2.2
2	$3/80 = 0.0375$	-1.78	2.9
3	$5/80 = 0.0625$	-1.53	3.3
4	$7/80 = 0.0875$	-1.36	3.3

and so on. Because of the symmetry of the Normal distribution, from $i = 21$ onwards the values of x are those corresponding to $40 - i + 1$, but with a positive sign. The Normal plot is shown in Figure 19.11.
3. The points do not lie on a straight line. There are pronounced bends near each end. These bends reflect rather long tails of the distribution of blood glucose. If the line showed a steady curve, rising less steeply as the blood glucose value increased, this would show simple skewness which can often be corrected by a log transformation. This would not work here; the bend at the lower end would be made worse.

The deviation from a straight line is not very great, compared, say, to Figure 7.20. As we see in Chapter 10, such small deviations from the Normal do not usually matter.

Solution to Exercise 8M: multiple choice questions 38 to 43

38. FFTFF. §8.2. Variability of observations is measured by the standard deviation, s. Standard error of the mean $= \sqrt{s^2/n}$.
39. FTFTF. §8.3. The sample mean is always in the middle of the limits.
40. FTFFT. $\text{SE}(\bar{x}) = s/\sqrt{n}$, d.f. $= n - 1$.
41. TTTFF. §8.1–2, §6.4) Variance is $p(1-p)/n = 0.1 \times 0.9/100 = 0.000\,9$. The *number* in the sample with the condition follows a Binomial distribution, not the proportion.
42. FFTTT. It depends on the variability of FEV1 and the number in the sample (§8.2). The sample should be random (§3.3–4).
43. FFTTF. §8.3–4. It is unlikely that we would get these data if the population rate were 10%, but not impossible.

Solution to Exercise 8E

1. The interval will be 1.96 standard deviations less than and greater than the mean. The lower limit is $0.810 - 1.96 \times 0.057 = 0.698$ mmol/litre. The upper limit is $0.810 + 1.96 \times 0.057 = 0.922$ mmol/litre.
2. For the diabetics, the mean is 0.719 and the standard deviation 0.068, so the lower limit of 0.698 will be $(0.698 - 0.719)/0.068 = -0.309$ standard deviations from the mean. From Table 7.1, the probability of being below this is 0.38, so the probability of being above is $1 - 0.38 = 0.62$. Thus the probability that an insulin-dependent diabetic would be within the reference interval would be 0.62 or 62%. This is the proportion we require.

3. The standard error of a mean is estimated by $s/\sqrt{n}$. For the diabetics: $s = 0.068, n = 227, \mathrm{SE} = 0.068/\sqrt{227} = 0.004\,51$ mmol/litre. For the controls: $s = 0.057, n = 140, \mathrm{SE} = 0.057/\sqrt{140} = 0.004\,82$ mmol/litre.
4. The 95% confidence interval is the mean $\pm$ 1.96 standard errors. For the controls, $0.810 - 1.96 \times 0.004\,82$ to $0.810 + 1.96 \times 0.004\,82$ gives us 0.801 to 0.819 mmol/litre. This is much narrower than the interval of part 1. This is because the confidence interval tells us how far the *sample mean* might be from the population mean. The 95% reference interval tells us how far an *individual observation* might be from the population mean.
5. The groups are independent, so the standard error of the difference between means is given by:

$$\begin{aligned} \mathrm{SE}(\bar{x}_1 - \bar{x}_2) &= \sqrt{se_1^2 + se_2^2} \\ &= \sqrt{0.00\,451^2 + 0.004\,82^2} \\ &= 0.006\,60 \end{aligned}$$

6. The difference between the means is $0.719 - 0.810 = -0.091$ mmol/litre. The 95% confidence interval is thus $-0.091 - 1.96 \times 0.006\,60$ to $-0.091 + 1.96 \times 0.006\,60$, giving -0.104 to -0.078. Hence the mean plasma magnesium level for insulin dependent diabetics is between 0.078 and 0.104 mmol/litre below that of non-diabetics.
7. Although the difference is significant, this would not be a good test because the majority of diabetics are within the 95% reference interval.

Solution to Exercise 9M: multiple choice questions 44 to 49

44. FTFFF. There is evidence for a relationship (§9.6), which is not necessarily causal. There may be other differences related to coffee drinking, such as smoking (§3.8).
45. FFFFT. The null hypothesis is that the *population* means are equal (§9.7). Significance is a property of the sample, not the population. $\mathrm{SE}(\bar{x}_1 - \bar{x}_2) = \sqrt{\mathrm{SE}(\bar{x}_1)^2 - \mathrm{SE}(\bar{x}_2)^2}$ (§8.5).
46. TTFTT. §9.2. It is quite possible for either to be higher and deviations in either direction are important (§9.5). $n = 16$ because the subject giving the same reading on both gives no information about the difference and is excluded from the test. The order should be random, as in a cross-over trial (§2.6).
47. FFFFT. The trial is small and the difference may be due to chance, but there may also be a large treatment effect. We must do a bigger trial to increase the power (§9.9). Adding cases would completely invalidate the test. If the null hypothesis is true, the test will give a 'significant' result one in 20 times. If we keep adding cases and doing many tests we have

a very high chance of getting a 'significant' result on one of them, even though there is no treatment effect (§9.10).
48. TFTTF. Large sample methods depend on estimates of variance obtained from the data. This estimate gets closer to the population value as the sample size increases (§9.7, §9.8). The chance of an error of the first kind is the significance level set in advance, say 5%. The larger the sample the more likely we are to detect a difference should one exist (§9.9). The null hypothesis depends on the phenomena we are investigating, not on the sample size.
49. FTFFT. We cannot conclude causation in an observational study (§3.6–8), but we can conclude that there is evidence of a difference (§9.6). 0.001 is the probability of getting so large a difference if the null hypothesis were true (§9.3).

Solution to Exercise 9E

1. Both control groups are drawn from populations which were easy to get to, one being hospital patients without gastro-intestinal symptoms, the other being fracture patients and their relatives. Both are matched for age and sex; Mayberry *et al.* also matched for social class and marital status. Apart from the matching factors, we have no way of knowing whether cases and controls are comparable, or any way of knowing whether controls are representative of the general population. This is usual in case control studies and is a major problem with this design.
2. There are two obvious sources of bias: interviews were not blind and information is being recalled by the subject. The latter is particularly a problem for data about the past. In James' study subjects were asked what they used to eat several years in the past. For the cases this was before a definite event, onset of Crohn's disease, for the controls it was not, the time being time of onset of the disease in the matched case.
3. The question in James' study was 'what did you to eat in the past?', that in Mayberry *et al.* was 'what do you eat now?'
4. Of the 100 patients with Crohn's disease, 29 were current eaters of cornflakes. Of 29 cases who knew of the cornflakes association, 12 were ex-eaters of cornflakes, and among the other 71 cases 21 were ex-eaters of cornflakes, giving a total of 33 past but not present eaters of cornflakes. Combining these with the 29 current consumers, we get 62 cases who had at some time been regular eaters of cornflakes. If we carry out the same calculation for the controls, we obtain $3 + 10 = 13$ past eaters and with 22 current eaters this gives 35 sometime regular cornflakes eaters. Cases were more likely than controls to have eaten cornflakes regularly at some time, the proportion of cases reporting having eaten cornflakes being almost twice as great as for controls. Compare this to James' data, where 17/68

= 25% of controls and 23/34 = 68% of cases, 2.7 times as many, had eaten cornflakes regularly. The results are similar.
5. The relationship between Crohn's disease and reported consumption of cornflakes had a much smaller probability for the significance test and hence stronger evidence that a relationship existed. Also, only one case had never eaten cornflakes (it was also the most popular cereal among controls).
6. Of the Crohn's cases, 67.6% (i.e. 23/34) reported having eaten cornflakes regularly compared to 25.0% of controls. Thus cases were 67.6/25.0 = 2.7 times as likely as controls to report having eaten cornflakes. The corresponding ratios for the other cereals are: wheat, 2.7; porridge, 1.5; rice, 1.6; bran, 6.1; muesli, 2.7. Cornflakes does not stand out when we look at the data in this way. The small probability simply arises because it is the most popular cereal. The P value is a property of the sample, not of the population.
7. We can conclude that there is no evidence that eating cornflakes is more closely related to Crohn's disease than is consumption of other cereals. The tendency for Crohn's case to report excessive eating of breakfast foods before onset of the disease may be the result of greater variation in diet than in controls, as they try different foods in response to their symptoms. They may also be more likely to recall what they used to eat, being more aware of the effects of diet because of their disease.

Solution to Exercise 10M: multiple choice questions 50 to 56

50. FFTFT. §10.2. It is equivalent to the Normal distribution method (§8.3).
51. FTFTF. §10.3. Whether the (population) means are equal is what we are trying to find out. The large sample case is like the Normal test of (§9.7), except for the common variance estimate. It is valid for any sample size.
52. FTTFF. The assumption of Normality would not be met for a small sample t test (§10.3) without transformation (§10.4), but for a large sample the distribution followed by the data would not matter (§9.7). The sign test is for paired data. We have measurements, not qualitative data.
53. FTTFF. §10.5. The more different the sample sizes are, the worse is the approximation to the t distribution. When both samples are large, this becomes a large sample Normal distribution test (§9.7). Grouping of data is not a serious problem.
54. TFFTT. A P value conveys more information than a statement that the difference is significant or not significant. A confidence interval would be even better. What is important is how well the diagnostic test discriminates, i.e. by how much the distributions overlap, not any difference in mean. Semen count cannot follow a Normal distribution because two

Table 19.6. Differences and means for static compliance

Patient	Constant	Decelerating	Difference	Mean
1	65.4	72.9	−7.5	69.15
2	73.7	94.4	−20.7	84.05
3	37.4	43.3	−5.9	40.35
4	26.3	29.0	−2.7	27.65
5	65.0	66.4	−1.4	65.70
6	35.2	36.4	−1.2	35.80
7	24.7	27.7	−3.0	26.20
8	23.0	27.5	−4.5	25.25
9	133.2	178.2	−45.0	155.70
10	38.4	39.3	−0.9	38.85
11	29.2	31.8	−2.6	30.50
12	28.3	26.9	1.4	27.60
13	46.6	45.0	1.6	45.80
14	61.5	58.2	3.3	59.85
15	25.7	25.7	0.0	25.70
16	48.7	42.3	6.4	45.50

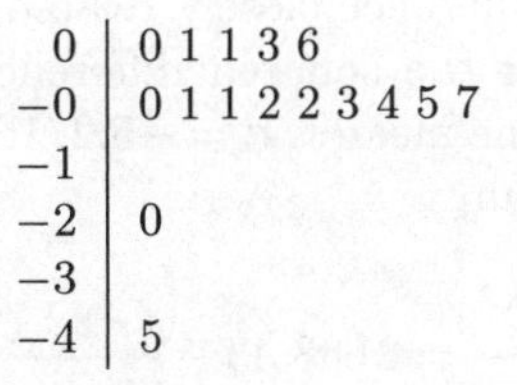

Fig. 19.12. Stem and leaf plot for compliance

standard deviations exceeds the mean and some observations would be negative (§7.4). Approximately equal numbers make the t test very robust but skewness reduces the power (§10.5).
55. FTTFT. §7A. For a Normal distribution $\bar{x}$ and s^2 are independent. s^2 will follow this distribution multiplied by $\sigma^2/(n-1)$, where σ^2 is the population variance. $\bar{x}/\sqrt{s^2/n}$ follows a t distribution only if the mean of the population distribution is zero (§10.1).
56. FTTFT. §10.9. Sums of squares and degrees of freedom add up, mean squares do not. Three groups gives two degrees of freedom. We can have any sizes of groups.

Solution to Exercise 10E

1. The differences for compliance are shown in Table 19.6. The stem and leaf plot is shown in Figure 19.12.
2. The plot of difference against mean is Figure 19.13. The distribution is

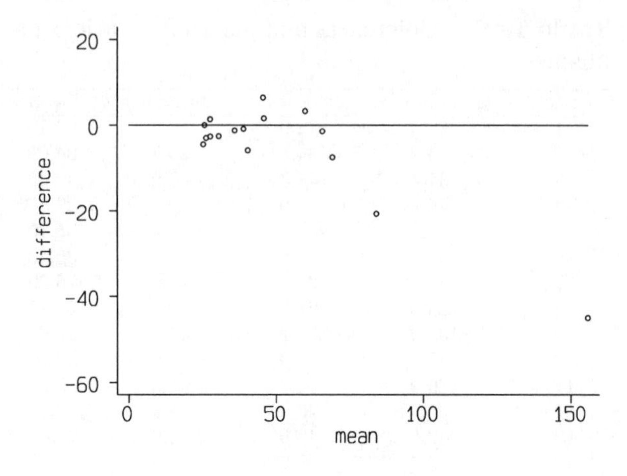

Fig. 19.13. Difference vs mean for compliance

highly skewed and the difference closely related to the mean.
3. The sum and sum of the squared differences are $\sum d_i = -82.7$ and $\sum d_i^2 = 2\,648.43$, hence the mean is $\bar{d} = -5.2/16 = -5.168\,75$. For the sum of squares about the mean

$$\sum d_i^2 - \frac{(\sum d_i)^2}{n} = 2\,648.43 - \frac{(-82.7)^2}{16} = 2\,220.974\,38$$

and hence the variance is $s^2 = 2\,220.974\,38/15 = 148.064\,96$. The standard deviation is $s = \sqrt{148.064\,96} = 12.168$. The standard error of the mean difference is

$$\sqrt{\frac{s^2}{n}} = \sqrt{\frac{148.064\,96}{16}} = 3.042\,05$$

4. We have 15 degrees of freedom and from Table 7.1 the 5% point of the t distribution is 2.13. The 95% confidence interval is $-5.168\,75 - 2.13 \times 3.042\,05$ to $-5.168\,75 + 2.13 \times 3.042\,05$, giving -11.6 to $+1.3$.
5. Table 19.7 shows the log transformed data, using logs to base 10, with their differences and sums. The stem and leaf plot is shown in Figure 19.14. This is a little unwieldy and we can condense it by pairing the first significant digits. The difference against the mean is shown in Figure 19.15. The differences are still related to the mean but not nearly so strongly as in Figure 19.13. The distribution is more symmetrical and the use of the t distribution seems much more reasonable than for the untransformed data. The sum and sum of the squared differences are $\sum d_i = -0.459$ and $\sum d_i^2 = 0.050\,687$, hence the mean is $\bar{d} = -0.459/16 = -0.0286\,88$. The sum of squares about the mean is $0.050\,687 - (-0.459)^2/16 = 0.037\,519$ and hence the variance is $s^2 = 0.037\,519/15 = 0.002\,501\,3$. The standard

Stem	Leaf
0.06	2
0.05	
0.04	
0.03	
0.02	2 4
0.01	5
0.00	0
−0.00	9
−0.01	0 4
−0.02	
−0.03	7
−0.04	2 7 9
−0.05	
−0.06	3
−0.07	7
−0.08	
−0.09	
−0.10	8
−0.11	
−0.12	6

Condensed version:

Stem	Leaf
0.06	2
0.04	
0.02	2 4
0.00	0 5
−0.00	9 0 4
−0.02	7
−0.04	2 7 9
−0.06	3 7
−0.08	
−0.10	8
−0.12	6

Fig. 19.14. Stem and leaf plots for log compliance

deviation is $s = \sqrt{0.002\,501\,3} = 0.050\,013$. The standard error of the mean difference is

$$\sqrt{\frac{s^2}{n}} = \sqrt{\frac{0.002\,501\,3}{16}} = 0.012\,503$$

6. The 95% confidence interval is $-0.028\,688 - 2.13 \times 0.012\,503$ to $-0.028\,688 + 2.13 \times 0.012\,503$ which gives $-0.055\,312$ to $-0.002\,057$. This has not been rounded, because we need to transform them first. If we transform these limits back by taking the antilogs we get 0.880 to 0.995. This means that the compliance with a decelerating waveform is between 0.880 and 0.995 times that with a constant waveform. There is some evidence that waveform has an effect, whereas with the untransformed data the confidence interval for the difference included zero. Because of the skewness of the raw data the confidence interval was too wide.

7. We can conclude that there is some evidence of a reduction in mean compliance, which could be up to 12% (from $(1 - 0.880) \times 100$), but could also be negligibly small.

Table 19.7. Difference and mean for log transformed compliance (to base 10)

Patient	Constant	Decelerating	Difference	Mean
1	1.816	1.863	−0.047	1.839 5
2	1.867	1.975	−0.108	1.921 0
3	1.573	1.636	−0.063	1.604 5
4	1.420	1.462	−0.042	1.441 0
5	1.813	1.822	−0.009	1.817 5
6	1.547	1.561	−0.014	1.554 0
7	1.393	1.442	−0.049	1.417 5
8	1.362	1.439	−0.077	1.400 5
9	2.125	2.251	−0.126	2.188 0
10	1.584	1.594	−0.010	1.589 0
11	1.465	1.502	−0.037	1.483 5
12	1.452	1.430	0.022	1.441 0
13	1.668	1.653	0.015	1.660 5
14	1.789	1.765	0.024	1.777 0
15	1.410	1.410	0.000	1.410 0
16	1.688	1.626	0.062	1.657 0

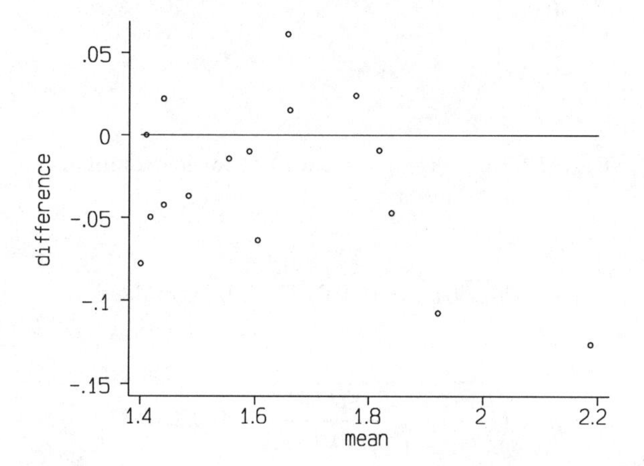

Fig. 19.15. Difference vs mean for log compliance

Solution to Exercise 11M: multiple choice questions 57 to 61

57. FFTTF. Outcome and predictor variables are perfectly related but do not lie on a straight line, so $r < 1$ (§11.9).

58. FTFFF. Knowledge of the predictor tells us something about the outcome variable (§6.2). This is not a straight line relationship. For part of the scale the outcome variable decreases as the predictor increases, then the outcome variable increases again. The correlation coefficient will be close to zero (§11.9). A logarithmic transformation would work if the outcome increased more and more rapidly as the predictor increased (§5.9).

59. FFFTT. A regression line usually has non-zero intercept and slope, which have dimensions (§11.3). Exchanging X and Y changes the line (§11.4)
60. TTFFF. §11.9–10. There is no distinction between predictor and outcome. r should not be confused with the regression coefficient (§11.3).
61. FTTFF. The predictor variable has no error in the regression model (§11.3). Transformations are only used if needed to meet the assumptions (§11.8). There is a scatter about the line (§11.3).

Solution to Exercise 11E

1. The slope is found by

$$b = \frac{\text{sum of products}}{\text{sum of squares}}$$

For females,

$$b_f = \frac{4\,206.9}{1\,444.6} = 2.912\,2$$

For males,

$$b_m = \frac{9\,045.4}{2\,267.5} = 3.989\,2$$

2. For the standard error, we first need the variances about the line:

$$s^2 = \frac{1}{n-2}\left(\sum(y_i - \bar{y})^2 - b^2\sum(x_i - \bar{x})^2\right)$$

then the standard error is

$$\text{SE}(b) = \sqrt{\frac{s^2}{\sum(x_i - \bar{x})^2}}$$

For females:

$$s_f^2 = \frac{1}{43-2}(101\,107.6 - 2.912\,2^2 \times 1\,444.6) = 2\,167.2$$

$$\text{SE}(b_f) = \sqrt{\frac{2\,167.2}{1\,444.6}} = 1.224\,8$$

For males:

$$s_m^2 = \frac{1}{58-2}(226\,873.5 - 3.989\,2^2 \times 2\,267.5) = 3\,406.9$$

$$\text{SE}(b_m) = \sqrt{\frac{3\,406.9}{2\,267.5}} = 1.225\,8$$

3. The standard error of the difference between two independent variables is the square root of the sum of their standard errors squared (§6.6, §8.5):

$$\text{SE}(b_f - b_m) = \sqrt{(\text{SE}(b_f))^2 + (\text{SE}(b_m))^2}$$

$$\begin{aligned} &= \sqrt{1.224\,8^2 + 1.225\,8^2} \\ &= 1.732\,8 \end{aligned}$$

The sample is reasonably large, almost attaining 50 in each group, so this standard error should be fairly well estimated and we can use a large sample Normal approximation. The 95% confidence interval is thus 1.96 standard errors on either side of the estimate. The observed difference is $b_f - b_m = 2.912\,2 - 3.989\,2 = -1.077\,0$. The 95% confidence interval is thus $-1.077\,0 - 1.96 \times 1.732\,8 = -4.5$ to $-1.077\,0 + 1.96 \times 1.732\,8 = 2.3$. If the samples were small, we could do this using the t distribution, but we would need to estimate a common variance. It would be better to use multiple regression, testing the height×sex interaction (§17.3).

4. For the test of significance the test statistic is observed difference over standard error:

$$\frac{b_f - b_m}{\text{SE}(b_f - b_m)} = \frac{-1.077\,0}{1.732\,8} = -0.62$$

If the null hypothesis were true, this would be an observation from a Standard Normal distribution. From Table 7.2, $P > 0.5$.

Solution to Exercise 12M: multiple choice questions 62 to 66

62. TFTFF. §10.3, §12.2. The sign and Wilcoxon tests are for paired data (§9.2, §12.3). Rank correlation looks for the existence of relationships between two ordinal variables, not a comparison between two groups (§12.4, §12.5).

63. TTFFT. §9.2, §12.2, §10.3, §12.5. The Wilcoxon test is for interval data (§12.3).

64. FTFTT. §12.5. There is no predictor variable in correlation. Log transformation would not affect the rank order of the observations.

65. FTFFT. If Normal assumptions are met the methods using them are better (§12.7). Estimation of confidence intervals using rank methods is difficult. Rank methods require the assumption that the scale is ordinal, i.e. that the data can be ranked.

66. TFTTF. We need a paired test: t, sign or Wilcoxon (§10.2, §9.2, §12.3).

Solution to Exercise 12E

1. The differences are shown in Table 19.6. We have 4 positive, 11 negative and 1 zero. Under the null hypothesis of no difference, the number of positives is from the Binomial distribution with $p = 0.5$, $n = 15$. We have $n = 15$ because the single zero contributes no information about the direction of the difference. For PROB($r \leq 4$) we have

$$
\begin{array}{lclcl}
\text{PROB}(r=4) & = & \frac{15!}{4!\times 11!}\times(0.5)^{15} & = & 0.041\,66 \\
\text{PROB}(r=3) & = & \frac{15!}{3!\times 12!}\times(0.5)^{15} & = & 0.013\,89 \\
\text{PROB}(r=2) & = & \frac{15!}{2!\times 13!}\times(0.5)^{15} & = & 0.003\,20 \\
\text{PROB}(r=1) & = & \frac{15!}{1!\times 14!}\times(0.5)^{15} & = & 0.000\,46 \\
\text{PROB}(r=0) & = & \frac{15!}{0!\times 15!}\times(0.5)^{15} & = & 0.000\,03 \\
\hline
\text{PROB}(r\leq 4) & & & = & 0.059\,24
\end{array}
$$

If we double this for a two sided test we get 0.118 48, again not significant.

2. Using the Wilcoxon matched pairs test we get

Diff.	−0.9	−1.2	−1.4	1.4	1.6	−2.6	−2.7	−3.0
Rank	1	2	3.5	3.5	5	6	7	8
Diff.	3.3	−4.5	−5.9	6.4	−7.5	−20.7	−45.0	
Rank	9	10	11	12	13	14	15	

As for the sign test, the zero is omitted. Sum of ranks for positive differences is $T = 3.5 + 5 + 9 + 12 = 29.5$. From Table 12.5 the 5% point for $n = 15$ is 25, which T exceeds, so the difference is not significant at the 5% level. The three tests give similar answers.

3. Using the log transformed differences in Table 19.7, we still have 4 positives, 11 negatives and 1 zero, with a sign test probability of 0.11848. The transformation does not alter the direction of the changes and so does not affect the sign test.

4. For the Wilcoxon matched pairs test on the log compliance:

Diff.	−0.009	−0.010	−0.014	0.015	0.022	0.024
Rank	1	2	3	4	5	6
Diff.	−0.037	−0.042	−0.047	−0.049	0.062	−0.063
Rank	7	8	9	10	11	12
Diff.	−0.077	−0.108	−0.126			
Rank	13	14	15			

Hence $T = 4 + 5 + 6 + 11 = 26$. This is just above the 5% point of 25 and is different from that in the untransformed data. This is because the transformation has altered the relative size of the differences. This test assumes interval data. By changing to a log scale we have moved to a scale where the differences are more comparable, because the change does depend on the magnitude of the original value. This does not happen with the other rank tests, the Mann Whitney U test and rank correlation coefficients, which involve no differencing.

5. Although there is a possibility of a reduction in compliance it does not reach the conventional level of significance.

6. The conclusions are broadly similar, but the effect on compliance is more strongly suggested by the t method. Provided the data can be transformed to approximate Normality the t distribution analysis is more powerful, and as it also gives confidence intervals more easily, I would prefer it.

Solution to Exercise 13M: multiple choice questions 67 to 74

67. FTFTF. §13.1, §13.3. $(5 - 1) \times (3 - 1) = 8$ d.f., $80\% \times 15 = 12$ cells must have expected frequencies > 5. It is O.K. for an *observed* frequency to be zero.
68. TTFTF. §13.1, §13.9. The two tests are independent. There are $(2 - 1) \times (2 - 1) = 1$ d.f. With such large numbers Yates' correction does not make much difference. Without it we get $\chi^2 = 124.5$, with it we get $\chi^2 = 119.4$ (§13.5.).
69. TTFTT. Chi-squared for trend and τ_b will both test the null hypothesis of no trend in the table, but an ordinary chi-squared test will not (§13.8). The odds ratio (OR) is an estimate of the relative risk for a case-control study (§13.7).
70. TTTTT. §13.4–5. The factorials of large numbers can be difficult to calculate.
71. TFFFF. §13.3. 80% of 4 is greater than 3, so all expected frequencies must exceed 5. The sample size can be as small as 20, if all row and column totals are 10.
72. TTFFF. The test compares proportions in matched samples (§13.9). For a relationship, we use the chi-squared test (§13.1). PEFR is a continuous variable, we use the paired t method (§10.2). For two independent samples we use the chi-squared test (§13.1).
73. TFTFF. For a two by two table with small expected frequencies we can use Fisher's exact test or Yates' correction (§13.4–5). McNemar's test is inappropriate because the groups are not matched (§13.9).
74. TTTTF. §13.7.

Solution to Exercise 13E

1. The heatwave appears to begin in week 10 and continue to include week 17. This period was much hotter than the corresponding period of 1982.
2. There were 178 admissions during the heatwave in 1983 and 110 in the corresponding weeks of 1982. We could test the null hypothesis that these came from distributions with the same admission rate and we would get a significant difference. This would not be convincing, however. It could be due to other factors, such as the closure of another hospital with resulting changes in catchment area.
3. The cross-tabulation is shown in Table 19.8.
4. The null hypothesis is that there is no association between year and period, in other words that the distribution of admissions between the periods will be the same for each year. The expected values are shown in Table 19.9.
5. The chi-squared statistic is given by:

Table 19.8. Cross-tabulation of time period by year for geriatric admissions

Year	Period			Total
	before heatwave	during heatwave	after heatwave	
1982	190	110	82	382
1983	180	178	110	468
Total	370	288	192	850

Table 19.9. Expected frequencies for Table 19.8

Year	Period			Total
	before heatwave	during heatwave	after heatwave	
1982	166.3	129.4	86.3	382.0
1983	203.7	158.6	105.7	468.0
Total	370.0	288.0	192.0	850.0

$$\begin{aligned}\sum \frac{(O-E)^2}{E} &= \frac{(190-166.3)^2}{166.3} + \frac{(110-129.4)^2}{129.4} + \frac{(82-86.3)^2}{86.3} \\ &\quad + \frac{(180-203.7)^2}{203.7} + \frac{(178-158.6)^2}{158.6} + \frac{(110-105.7)^2}{105.7} \\ &= 11.806\end{aligned}$$

There are 2 rows and 3 columns, giving us $(2-1)\times(3-1) = 2$ degrees of freedom. Thus we have chi-squared $= 11.8$ with 2 degrees of freedom. From Table 14.3 we see that this has probability of less than 0.01. The data are not consistent with the null hypothesis. The evidence supports the view that admissions rose by more than could be ascribed to chance during the 1983 heatwave. We cannot be certain that this was due to the heatwave and not some other factor which happened to operate at the same time.
6. We could see whether the same effect occurred in other districts between 1982 and 1983. We could also look at older records to see whether there was a similar increase in admissions, say for the heatwaves of 1975 and 1976.

Solution to Exercise 14M: multiple choice questions 75 to 80

75. TFFTT. §14.2.
76. TFTTT. A t test could not be used because the data do not follow a Normal distribution (10.3). The expected frequencies will be too small for a chi-squared test (§13.3), but a trend test would be O.K. (§13.8). A goodness of fit test could be used (§13.10).
77. FTTFT. A small-sample, paired method is needed (§14.4).
78. FTTTT. §14.5.

79. FFFFT. Regression, correlation and paired t methods need continuous data (§11.3, §11.9, §10.2). Kendall's τ can be used for *ordered* categories.
80. TFTFF. §14.2.

Solution to Exercise 14E

1. Overall preference: we have one sample of patients, of whom 12 preferred A, 14 preferred B and 4 did not express a preference. We can use a Binomial or sign test (§9.2), only considering those who expressed a preference. Those for A are positives, those for B are negatives. We get two-sided P = 0.85, not significant.

Preference and order: we have the relationship between two variables, preference and order, both nominal. We set up a two way table and do a chi-squared test. For the 3 by 2 table we have two expected frequencies less than five, so we must edit the table. There are no obvious combinations, but we can delete those who expressed no preference, leaving a 2 by 2 table, $\chi^2 = 1.3$, 1 d.f., P > 0.05.

2. The data are paired so we use a paired t test (§10.2). The assumption of a Normal distribution for the differences should be met as PEFR itself follows a Normal distribution fairly well. We get $t = 6.45/5.05 = 1.3$, d.f. = 31, which is not significant. Using t=2.04 (Table 10.1) we get a 95% confidence interval of −3.85 to 16.75 litres/min.

3. We must use the total number of patients we randomized to treatments, in an intention to treat analysis (§2.5). Thus we have 1 721 active treatment patients including 15 deaths, and 1 706 placebo patients with 35 deaths. A chi-squared test gives us $\chi^2 = 8.3$, d.f. = 1, P < 0.01. A comparison of two proportions gives a difference of −0.011 8 with 95% confidence interval −0.019 8 to −0.003 8 (§8.6) and test of significance using the Standard Normal distribution gives a value of 2.88, P < 0.01, (§9.8).

4. Both variables have very non-Normal distributions. Nitrite is highly skew and pH is bimodal. It might be possible to transform the nitrites to a Normal distribution but the transformation would not be a simple one. The zero prevents a simple logarithmic transformation, for example. Because of this, regression and correlation are not appropriate and rank correlation can be used. Spearman's $\rho = 0.58$ and Kendall's $\tau = 0.40$, both giving a probability of 0.004.

5. We have two large samples and can do the Normal comparison of two means (§8.5). The standard error of the difference is 0.0178 s and the observed difference is 0.02 s, giving a 95% confidence interval of −0.015 to 0.055 for the excess mean transit time in the controls. If we had all the data, for each case we could calculate the mean MTT for the two controls matched to each case, find the difference between case MTT and control mean MTT, and use the one sample method of §8.3.

6. The unequal steps in the visual acuity scale suggest that it is best treated as an ordinal scale, so the sign test is appropriate. Pre minus post, there are 10 positive differences, no negative differences and 7 zeros. Thus we refer 0 to the Binomial distribution with $p = 0.5$ and $n = 10$. The probability is given by

$$\frac{10!}{10! \times 0!} \times 0.5^0 \times 0.5^{10} = 0.00098$$

For a two sided test we double this to give P=0.002. The contrast sensitivity test is a measurement, and hence an interval scale. We could carry out the paired t test or the Wilcoxon signed rank test on the differences. The distribution of the differences is grouped, as the scale is clearly discrete, but not skewed, so either would be valid. For the paired t test, the mean difference (pre−post) is −0.335, standard deviation = 0.180, standard error of mean = $0.180/\sqrt{17} = 0.044$ and the t statistic for testing the null hypothesis that the population mean is zero is $t = 0.335/0.044 = 7.61$, with 16 degrees of freedom, P < 0.001. For the Wilcoxon signed-rank test, all the differences are negative, so $T = 0$, which is also highly significant. For the relationship between visual acuity and the contrast sensitivity test, visual acuity is ordinal so we must use rank correlation. Spearman's $\rho = -0.49$, P = 0.05, Kendall's $\tau = -0.40$, P = 0.06.
7. We want to test for the relationship between two variables, which are both presented as categorical. We use a chi-squared test for a contingency table, $\chi^2 = 38.1$, d.f. = 6, P < 0.001. One possibility is that some other variable, such as the mother's smoking or poverty, is related to both maternal age and asthma. Another is that there is a cohort effect. All the age 14–19 mothers were born during the second world war, and some common historical experience may have produced the asthma in their children.

Solution to Exercise 15M: multiple choice questions 81 to 86

81. TFTTF. §15.2. Unless the measurement process changes the subject, we would expect the difference in mean to be zero.
82. TFTFF. §15.4. We need the sensitivity as well as specificity. There are other things, dependent on the population studied, which may be important too, like the positive predictive value.
83. FTTTF. §15.4. Specificity, not sensitivity, measures how well people without the disease are eliminated.
84. TTFFF. §15.5. The 95% reference interval should not depend on the sample size.
85. FFFFT. §15.5. We expect 5% of 'normal' men to be outside these limits. The patient may have a disease which does not produce an abnormal haematocrit. This reference interval is for men, not women who may have

a different distribution of haematocrit. It is dangerous to extrapolate the reference interval to a different population. In fact, for women the reference interval is 35.8 to 45.4, putting a woman with a haematocrit of 48 outside the reference interval. A haematocrit outside the 95% reference interval suggests that the man may be ill, although it does not prove it.
86. TFTTT. §15.6. As time increases, rates are based on fewer potential survivors. Withdrawals during the first interval contribute half an interval at risk. If survival rates change those subjects starting later in calendar time, and so more likely to be withdrawn, will have a different survival to those starting earlier. The first part of the curve will represent a different population to the second. The longest survivor may still be alive and so become a withdrawal.

Solution to Exercise 15E

1. The blood donors were used because it was easy to get the blood. This would produce a sample deficient in older people, so it was supplemented by people attending day centres. This would ensure that these were reasonably active, healthy people for their age. Given the problem of getting blood and the limited resources available, this seems a fairly satisfactory sample for the purpose. The alternative would be to take a random sample from the local population and try to persuade them to give the blood. There might have been so many refusals that volunteer bias would make the sample unrepresentative anyway. The sample is also biassed geographically, being drawn from one part of London. In the context of the study, where we wanted to compare diabetics with normals, this did not matter so much, as both groups came from the same place. For a reference interval which would apply nationally, if there were a geographical factor the interval would be biassed in other places. To look at this we would have to repeat the study in several places, compare the resulting reference intervals and pool as appropriate.
2. We want normal, healthy people for the sample, so we want to exclude people with obvious pathology and especially those with disease known to affect the quantity being measured. However, if we excluded all elderly people currently receiving drug therapy we would find it very difficult to a sufficiently large sample. It is indeed 'normal' for the elderly to be taking analgesics and hypnotics, so these were permitted.
3. From the shape of the histogram and the Normal plot, the distribution of plasma magnesium does indeed appear Normal.
4. The reference interval, outside which about 5% of normal values are expected to lie, is $\bar{x} - 2s$ to $\bar{x} + 2s$, or $0.810 - 2 \times 0.057$ to $0.810 + 2 \times 0.057$, which is 0.696 to 0.924, or 0.70 to 0.92 mmol/litre.
5. As the sample is large and the data Normally distributed the standard

error of the limits is approximately

$$\sqrt{\frac{3s^2}{n}} = \sqrt{\frac{3 \times 0.057^2}{140}} = 0.008\,343\,9$$

For the 95% confidence interval we take 1.96 standard errors on either side of the limit, $1.96 \times 0.008\,343\,9 = 0.016$. The 95% confidence interval for the lower reference limit is $0.696 - 0.016$ to $0.696 + 0.016 = 0.680$ to 0.712 or 0.68 to 0.71 mmol/litre. The confidence interval for the upper limit is $0.924 - 0.016$ to $0.696 + 0.016 = 0.908$ to 0.940 or 0.91 to 0.94 mmol/litre. The reference interval is well estimated as far as sampling errors are concerned.
6. Plasma magnesium did indeed increase with age. The variability did not. This would mean that for older people the lower limit would be too low and the upper limit too high, as the few above this would all be elderly. We could simply estimate the reference interval separately at different ages. We could do this using separate means but a common estimate of variance, obtained by one-way analysis of variance (§10.9). Or we could use the regression of magnesium on age to get a formula which would predict the reference interval for any age. The method chosen would depend on the nature of the relationship.

Solution to Exercise 16M: multiple choice questions 87 to 92

87. FTFFF. §16.1. It is for a specific age group, not age adjusted. It measures the number of deaths per person at risk, not the total number. It tells us nothing about age structure.
88. FTTTT. §16.4. The life table is calculated from age specific death rates. Expectation of life is the expected value of the distribution of age at death if these mortality rates apply (§6E). It usually increases with age.
89. TFTTF. The SMR (§16.3) for women who had just had a baby is lower than 100 (all women) and 105 (stillbirth women). The confidence intervals do not overlap so there is good evidence for a difference. Women who had had a stillbirth may be less or more likely than all women to commit suicide, we cannot tell. We cannot conclude that giving birth prevents suicide—it may be that optimists conceive, for example.
90. TFFFF. §16.3. Age effects have been adjusted for. It may also be that heavy drinkers become publicans. It is difficult to infer causation from observational data. Men at high risk of cirrhosis of the liver, i.e. heavy drinkers, may not become window cleaners, or window cleaners who drink may change their occupation, which requires good balance. Window cleaners have low risk. The 'average' ratio is 100, not 1.0.

Table 19.10. Age specific mortality rates for volatile substance abuse, Great Britain, and calculation of SMR for Scotland

Age group	Great Britain A.S.M.R.s per million per year	per thousand per 13 years	Scotland population (thousands)	Scotland expected deaths
0–9	0.00	0.000 00	653	0.000 00
10–14	0.79	0.010 30	425	4.377 50
15–19	2.58	0.033 58	447	15.010 26
20–24	0.87	0.011 37	394	4.479 78
25–29	0.32	0.004 15	342	1.419 30
30–39	0.08	0.001 08	659	0.711 72
40–49	0.03	0.000 33	574	0.189 42
50–59	0.09	0.001 12	579	0.648 48
60+	0.03	0.000 37	962	0.355 94
Total				27.192 40

91. FFFTF. §16.6. A life table tells us about mortality, not population structure. A bar chart shows the relationship between two variables, not their frequency distribution (§5.5).
92. TFFFT. §16.1, §16.2, §16.5. Expectation of life does not depend on age distribution (§16.4).

Solution to Exercise 16E

1. We obtain the rates for the whole period by dividing the number of deaths in an age group by the population size. Thus for ages 10–14 we have 44/4 271 = 0.010 30 cases per thousand population. This is for a 13 year period so the rate per year is 0.010 30/13 = 0.000 79 per 1 000 per year, or 0.79 per million per year. Table 19.10 shows the rates for each age group. The rates are unusual because they are highest among the adolescent group, where mortality rates for most causes are low. Anderson *et al.* (1985) note that '...our results suggest that among adolescent males abuse of volatile substances currently account for 2% of deaths from all causes ...'. The rates are also unusual because we have not calculated them separately for each sex. This is partly for simplicity and partly because the number of cases in most age groups is small as it is.
2. The expected number of deaths by multiplying the number in the age group in Scotland by the death rate for the period, i.e. per 13 years, for Great Britain. We then add these to get 27.19 deaths expected altogether. We observed 48, so the SMR is 48/27.19 = 1.77, or 177 with Great Britain as 100.
3. We find the standard error of the SMR by $\sqrt{O}/E = \sqrt{48}/27.19 = 0.2548$. The 95% confidence interval is then $1.77 - 1.96 \times 0.2548$ to $1.77 + 1.96 \times 0.2548$, or 1.27 to 2.27. Multiplying by 100 as usual, we get 127 to 227. The observed number is quite large enough for the Normal

approximation to the Poisson distribution to be used.
4. Yes, the confidence interval is well away from zero. Other factors relate to the data collection, which was from newspapers, coroners, death registrations etc. Scotland has different newspapers and other news media and a different legal system to the rest of Great Britain. It may be that the association of deaths with VSA is more likely to be reported there than in England and Wales.

Solution to Exercise 17M: multiple choice questions 93 to 97

93. TFTFT. §17.2. It is the ratio of the regression sum of squares to the total sum of squares.
94. FTFFF. §17.2. There were 37 + 1 = 38 observations. There is a highly significant ethnic group effect. The non-significant sex effect does *not* mean that there is no difference (§9.6). There are three age groups, so two degrees of freedom. If the effect of ethnicity were due entirely to age, it would have disappeared when age was included in the model.
95. TTTTF. §17.8. A four-level factor has three dummy variables (§17.6). If the effect of white cell count were due entirely to smoking, it would have disappeared when smoking was included in the model.
96. TTTFT. §17.4
97. FFFFT. §17.9. Boys have a lower risk of readmission than girls, shown by the negative coefficient, and hence a longer time before being readmitted. Theophiline is related to a lower risk of readmission but we cannot conclude causation. Treatment may depend on the type and severity of asthma.

Solution to Exercise 17E

1. The difference is highly significant (P < 0.001) and is estimated to be between 1.3 and 3.7, i.e. volumes are higher in group 2, the trisomy-16 group.
2. From both the Normal plot and the plot against number of pairs of somites there appears to be one point which may be rather separate from the rest of the data, an outlier. Inspection of the data showed no reason to suppose that the point was an error, so it was retained. Otherwise the fit to the Normal distribution seems quite good. The plot against number of pairs of somites shows that there may be a relationship between mean and variability, but this very small and will not affect the analysis too much. There is also a possible nonlinear relationship, which should be investigated. (The addition of a quadratic term did not improve the fit significantly.)
3. Model difference in sum of squares= $207.139 - 197.708 = 9.431$, residual sum of squares $= 3.384$, F ratio $= 9.431/3.384 = 2.79$ with 1 and 36 degrees of freedom, corresponding to $t = 1.67$, P > 0.1, not significant.

Solution to Exercise 18M: multiple choice questions 98 to 100

98. TTFTT. §9.9. Power is a property of the test, not the sample. It cannot be zero, as even when there is no population difference at all the test may be significant.
99. TTTTF. §18.5. If we keep on adding observations and testing, we are carrying out multiple testing and so invalidate the test (§9.10).
100. TTFFT. §18.1. Power is not involved in estimation.

Solution to Exercise 18E

1. The standard error of the reference limit is approximately $\sqrt{3s^2/n}$ (§15.5), the width of its confidence interval is 4 times this, and the width of the reference interval is $4s$. Hence

$$\begin{aligned} 0.2 &= \frac{4\sqrt{3s^2/n}}{4s} \\ 0.2^2 &= \frac{3}{n} \\ n &= \frac{3}{0.04} \\ &= 75 \end{aligned}$$

2. The accuracy is to two standard errors, and for a proportion this is $2\sqrt{p(1-p)/n}$. The maximum value of this is when $p = 0.5$. Two percentage points is 0.02, so from §18.2

$$\begin{aligned} 0.02 &= 2\sqrt{\frac{0.5 \times (1-0.5)}{n}} \\ 0.02^2 &= 2^2 \times \frac{0.5 \times 0.5}{n} \\ n &= \frac{4 \times 0.25}{0.0004} \\ &= 2500 \end{aligned}$$

3. This is a comparison of two proportions (§18.5). We have $p_1 = 0.15$ and $p_2 = 0.15 \times 0.9 = 0.135$, a reduction of 10%. With a power of 90% and a significance level of 5%, we have

$$\begin{aligned} n &= \frac{10.5\,(p_1(1-p_1) + p_2(1-p_2))}{(p_1 - p_2)^2} \\ &= \frac{10.5 \times (0.15 \times (1-0.15) + 0.135 \times (1-0.135))}{(0.15 - 0.135)^2} \end{aligned}$$

$$= \quad 11\,399.5$$

Hence we need 11 400 in each group, 22 800 patients altogether. With a power of 80% and a significance level of 5%, we have

$$\begin{aligned} n &= \frac{7.9\,(p_1(1-p_1)+p_2(1-p_2))}{(p_1-p_2)^2} \\ &= \frac{7.9\times(0.15\times(1-0.15)+0.135\times(1-0.135))}{(0.15-0.135)^2} \\ &= 8\,576.77 \end{aligned}$$

Hence we need 8 577 in each group, 17 154 patients altogether. Lowering the power reduces the required sample size, but, of course, reduces the chance of detecting a difference if there really is one.

4. This is the comparison of two means (§18.4). We estimate the sample size for a difference of one standard deviation, $\mu_1 - \mu_2 = \sigma$. With a power of 90% and a significance level of 5%, the number in each group is given by

$$\begin{aligned} n &= \frac{10.5\times 2\sigma^2}{(\mu_1-\mu_2)^2} \\ &= \frac{10.5\times 2\sigma^2}{\sigma^2} \\ &= 10.5\times 2 \\ &= 21 \end{aligned}$$

Hence we need 21 in each group. If we have unequal samples and $n_1 = 100$, n_2 is given by

$$\begin{aligned} (\mu_1-\mu_2)^2 &= 10.5\times\sigma^2\left(\frac{1}{n_1}+\frac{1}{n_2}\right) \\ \sigma^2 &= 10.5\times\sigma^2\left(\frac{1}{100}+\frac{1}{n_2}\right) \\ \frac{1}{n_2} &= \frac{1}{10.5}-\frac{1}{100} \\ \frac{1}{n_2} &= 0.095\,238-0.01 \\ n_2 &= \frac{1}{0.085\,238} \\ &= 11.7 \end{aligned}$$

and so we need 12 subjects in the disease group.

References

Altman, D.G. (1982). Statistics and ethics in medical research. In *Statistics in Practice*, (ed. S.M. Gore and D.G. Altman). British Medical Association, London.

Altman, D.G. (1991). *Practical Statistics for Medical Research*, Chapman and Hall, London.

Altman, D.G. and Bland, J.M. (1983). Measurement in medicine: the analysis of method comparison studies.. *The Statistician,* **32**, 307–17.

Anderson, H.R., Bland, J.M., Patel, S., and Peckham, C. (1986). The natural history of asthma in childhood. *Journal of Epidemiology and Community Health,* **40**, 121–9.

Anderson, H.R., MacNair, R.S., and Ramsey, J.D. (1985). Deaths from abuse of substances, a national epidemiological study. *British Medical Journal,* **290**, 304–7.

Appleby, L. (1991). Suicide during pregnancy and in the first postnatal year. *British Medical Journal,* **302**, 137–40.

Armitage, P. (1975). *Sequential Medical Trials*, Blackwell, Oxford.

Armitage, P. and Berry, G. (1987). *Statistical Methods in Medical Research*, Blackwell, Oxford.

Balfour, R.P. (1991). Birds, milk and campylobacter. *Lancet,* **337**, 176.

Ballard, R.A., Ballard, P.C., Creasy, R.K., Padbury, J., Polk, D.H., Bracken, M., Maya, F.R., and Gross, I. (1992). Respiratory disease in very-low-birthweight infants after prenatal thyrotropin releasing hormone and glucocorticoid. *Lancet,* **339**, 510–5.

Banks, M.H., Bewley, B.R., Bland, J.M., Dean, J.R., and Pollard, V.M. (1978). A long term study of smoking by secondary schoolchildren. *Archives of Disease in Childhood,* **53**, 12–19.

Bewley, B.R. and Bland, J.M. (1976). Academic performance and social factors related to cigarette smoking by schoolchildren. *British Journal of Preventive and Social Medicine,* **31**, 18–24.

Bewley, B.R., Bland, J.M., and Harris, R. (1974). Factors associated with the starting of cigarette smoking by primary school children. *British Journal of Preventive and Social Medicine,* **28**, 37–44.

Bewley, T.H., Bland, J.M., Ilo, M., Walch, E., and Willington, G. (1975). Census of mental hospital patients and life expectancy of those unlikely to be discharged.

British Medical Journal, **4**, 671–5.

Bewley, T.H., Bland, J.M., Mechen, D., and Walch, E. (1981). 'New chronic' patients. *British Medical Journal,* **283**, 1161–4.

Bland, J.M and Altman, D.G. (1986). Statistical methods for assessing agreement between two methods of clinical measurement. *Lancet,* **i**, 307–10.

Bland, J.M., Holland, W.W., and Elliott, A. (1974). The development of respiratory symptoms in a cohort of Kent schoolchildren. *Bulletin Physio-Pathologie Respiratoire,* **10**, 699–716.

Bland, J.M., Bewley, B.R., Banks, M.H., and Pollard, V.M. (1975). Schoolchildren's beliefs about smoking and disease. *Health Education Journal,* **34**, 71–8.

Bland, J.M., Mutoka, C., and Hutt, M.S.R. (1977). Kaposi's sarcoma in Tanzania. *East African Journal of Medical Research,* **4**, 47–53.

Bland, J.M., Bewley, B.R., Pollard, V., and Banks, M.H. (1978). Effect of children's and parents' smoking on respiratory symptoms. *Archives of Disease in Childhood,* **53**, 100–5.

Bland, J.M., Bewley, B.R., and Banks, M.H. (1979). Cigarette smoking and children's respiratory symptoms: validity of questionnaire method. *Revue d'Epidemiologie et Santé Publique,* **27**, 69–76.

Breslow, N.E. and Day, N.E. (1987). *Statistical methods in cancer research. Volume II—the design and analysis of cohort studies,* IARC, Lyon.

British Standards Institution (1979). *Precision of test methods. 1: Guide for the determination and reproducibility of a standard test method (BS5497, part 1),* BSI, London.

Brooke, O.G., Anderson, H.R., Bland, J.M., Peacock, J., and Stewart, M. (1989). The influence on birthweight of smoking, alcohol, caffeine, psychosocial and socio-economic factors. *British Medical Journal,* **298**, 795–801.

Bryson, M.C. (1976). The *Literary Digest* poll: making of a statistical myth. *The American Statistician,* **30**, 184–5.

Burr, M.L., St Leger, A.S., and Neale, E. (1976). Anti-mite measures in mite-sensitive adult asthma: a controlled trial. *Lancet,* **i**, 333–5.

Campbell, M.J. and Gardner, M.J. (1989). Calculating confidence intervals for some non-parametric analyses. In *Statistics with confidence,* (ed. Gardner, M.J. and Altman D.G.). British Medical Journal, London.

Carleton, R.A., Sanders, C.A., and Burack, W.R. (1960). Heparin administration after acute myocardial infarction. *New England Journal of Medicine,* **263**, 1002–4.

Christie, D. (1979). Before-and-after comparisons: a cautionary tale. *British Medical Journal,* **2**, 1629–30.

Colton, T. (1974). *Statistics in Medicine,* Little Brown, Boston.

Conover, W.J. (1980). *Practical Nonparametric Statistics,* John Wiley and Sons,

New York.

Curtis, M.J., Bland, J.M., and Ring, P.A. (1992). The Ring total knee replacement—a comparison of survivorship. *Journal of the Royal Society of Medicine,* **85**, 208–10.

Davies, O.L. and Goldsmith, P.L. (1972). *Statistical Methods in Research and Production*, Oliver and Boyd, Edinburgh.

DHSS (1976). *Prevention and Health: Everybody's Business*, HMSO, London.

Doll, R. and Hill, A.B. (1950). Smoking and carcinoma of the lung. *British Medical Journal,* **ii**, 739–48.

Doll, R. and Hill, A.B. (1956). Lung cancer and other causes of death in relation to smoking: a second report on the mortality of British doctors. *British Medical Journal,* **ii**, 1071–81.

Donnan, S.P.B. and Haskey, J. (1977). Alcoholism and cirrhosis of the liver. *Population Trends,* **7**, 18–24.

Easterbrook. P.J., Berlin, J.A., Gopalan, R., and Mathews, D.R. (1991). Publication bias in clinical research. *Lancet,* **337**, 867–72.

Egero, B. and Henin, R.A. (1973). *The Population of Tanzania*, Bureau of Statistics, Dar es Salaam.

Finney, D.J., Latscha, R., Bennett, B.M., and Hsa, P. (1963). *Tables for Testing Significance in a* 2×2 *Contingency Table*, Cambridge University Press, London.

Fish, P.D., Bennett, G.C.J, and Millard, P.H. (1985). Heatwave morbidity and mortality in old age. *Age and Aging,* **14**, 243–5.

Flint, C. and Poulengeris, P. (1986). *The 'Know Your Midwife' Report*, Caroline Flint, London.

Galton, F. (1886). Regression towards mediocrity in hereditary stature. *Journal of the Anthropological Institute,* **15**, 246–63.

Gardner, M.J. and Altman, D.G. (1986). Confidence intervals rather than P values: estimation rather than hypothesis testing. *British Medical Journal,* **292**, 746–50.

Gazet, J-C., Markopoulos, C., Ford, H.T., Coombes, R.C., Bland, M., and Dixon, R.C. (1988). Preliminary communication—Prospective trial of tamoxifen versus surgery in elderly patients with breast cancer. *Lancet,* , 679–80.

Glasziou, P.P. and Mackerras, D.E.M. (1993). Vitamin A supplementation in infectious disease: a meta-analysis. *British Medical Journal,* **306**, 366–70.

Hart, P.D. and Sutherland, I. (1977). BCG and vole bacillus in the prevention of tuberculosis in adolescence and early adult life. *British Medical Journal,* **2**, 293–5.

Healy, M.J.R. (1968). Disciplining medical data. *British Medical Bulletin,* **24**, 210–4.

Hedges, B.M. (1978). Question wording effects: presenting one or both sides of a case. *The Statistician*, **28**, 83–99.

Hickish, T., Colston, K., Bland, J.M., and Maxwell, J.D. (1989). Vitamin D deficiency and muscle strength in male alcoholics. *Clinical Science*, **77**, 171–6.

Hill, A.B. (1962). *Statistical Methods in Clinical and Preventive Medicine*, Churchill Livingstone, Edinburgh.

Hill, A.B. (1977). *A Short Textbook of Medical Statistics*, Hodder and Stoughton, London.

Holland, W.W., Bailey, P., and Bland, J.M. (1978). Long-term consequences of respiratory disease in infancy. *Journal of Epidemiology and Community Health*, **32**, 256–9.

Holten, C. (1951). Anticoagulants in the treatment of coronary thrombosis. *Acta Medica Scandinavica*, **140**, 340–8.

Huff, D. (1954). *How to Lie with Statistics*, Gollancz, London.

Huskisson, E.C. (1974). Simple analgesics for arthritis. *British Medical Journal*, **4**, 196–200.

James, A.H. (1977). Breakfast and Crohn's disease. *British Medical Journal*, **1**, 943–7.

Johnson, F.N. and Johnson, S. (eds) (1977). *Clinical Trials*, Blackwell, Oxford.

Johnston, I.D.A., Anderson, H.R., Lambert, H.P., and Patel, S. (1983). Respiratory morbidity and lung function after whooping cough. *Lancet*, **ii**, 1104–8.

Kaste, M., Kuurne, T., Vilkki, J., Katevuo, K., Sainio, K., and Meurala, H. (1982). Is chronic brain damage in boxing a hazard of the past?. *Lancet*, **ii**, 1186–8.

Kendall, M.G. (1970). *Rank correlation methods*, Charles Griffin, London.

Kendall, M.G. and Babington Smith, B. (1971). *Tables of Random Sampling Numbers*, Cambridge University Press, Cambridge.

Kendall, M.G. and Stuart, A. (1968). *The Advanced Theory of Statistics, 2nd. ed., vol. 3*, Charles Griffin, London.

Kendall, M.G. and Stuart, A. (1969). *The Advanced Theory of Statistics, 3rd. ed., vol. 1*, Charles Griffin, London.

Lancet (1980). BCG: bad news from India. *Lancet*, **i**, 73–4.

Lee, K.L., McNeer, J.F., Starmer, F.C., Harris, P.J., and Rosati, R.A. (1980). Clinical judgements and statistics: lessons form a simulated randomized trial in coronary artery disease. *Circulation*, **61**, 508–15.

Lemeshow, S., Hosmer, D.W., Klar, J., and Lwanga, S.K. (1990). *Adequacy of Sample Size in Health Studies*, John Wiley and Sons, Chichester.

Leonard, J.V, Whitelaw, A.G.L., Wolff, O.H., Lloyd, J.K., and Slack, S. (1977). Diagnosing familial hypercholesterolaemia in childhood by measuring serum cholesterol. *British Medical Journal*, **1**, 1566–8.

Levine, M.I. and Sackett, M.F. (1946). Results of BCG immunization in New York City. *American Review of Tuberculosis,* **53**, 517–32.

Lindley, M.I. and Miller, J.C.P. (1955). *Cambridge Elementary Statistical Tables,* Cambridge University Press, Cambridge.

Lucas, A., Morley, R., Cole, T.J., Lister, G., and Leeson-Payne, C. (1992). Breast milk and subsequent intelligence quotient in children born preterm. *Lancet,* **339**, 510–5.

Luthra, P., Bland, J.M., and Stanton, S.L. (1982). Incidence of pregnancy after laparoscopy and hydrotubation. *British Medical Journal,* **284**, 1013.

Machin, D. and Campbell, M.J. (1987). *Statistical Tables for the Design of Clinical Trials,* Blackwell, Oxford.

Mantel, N. (1966). Evaluation of survival data and two new rank order statistics arising in its consideration.. *Cancer Chemotherapy Reports,* **50**, 163–70.

Mather, H.M., Nisbet, J.A., Burton, G.H., Poston, G.J., Bland, J.M., Bailey, P.A., and Pilkington, T.R.E. (1979). Hypomagnesaemia in diabetes. *Clinica Chemica Acta,* **95**, 235–42.

Matthews, D.E and Farewell, V. (1987). *Using and understanding medical statistics,* Karger, Basel; New York.

Matthews, J.N.S., Altman, D.G., Campbell, M.J., and Royston, P. (1990). Analysis of serial measurements in medical research. *British Medical Journal,* **300**, 230–35.

Maugdal, D.P., Ang, L., Patel, S., Bland, J.M., and Maxwell, J.D. (1985). Nutritional assessment in patients with chronic gastro-intestinal symptoms: comparison of functional and organic disorders. *Human Nutrition: Clinical Nutrition,* **39**, 203–12.

Maxwell, A.E. (1970). Comparing the classification of subjects by two independent judges. *British Journal of Psychiatry,* **116**, 651–5.

Maxwell, J.D., Patel, S.P., Bland, J.M., Lindsell, D.R.M., and Wilson, A.G. (1983). Chest radiography compared to laboratory markers in the detection of alcoholic liver disease. *Journal of the Royal College of Physicians of London,* **17**, 220–3.

Mayberry, J.F., Rhodes, J., and Newcombe, R.G. (1978). Breakfast and dietary aspects fo Crohn's disease. *British Medical Journal,* **2**, 1401.

Mckie, D. (1992). Pollsters turn to secret ballot. *The Guardian,* London, 24 August, p.20.

Meier, P. (1977). The biggest health experiment ever: the 1954 field trial of the Salk poliomyelitis vaccine. In *Statistics: a Guide to the Biological and Health Sciences,* (ed. J.M. Tanur, *et al.*). Holden-Day, San Francisco.

Mitchell, E.A., Bland, J.M., Thompson, J.M.D. (1994). Risk factors for readmission to hospital for asthma. *Thorax,* **49**, 33–36.

Morris, J.A. and Gardner, M.J. (1989). Calculating confidence intervals for relative risks, odds ratios and standardized ratios and rates. In *Statistics with confidence,* (ed. Gardner, M.J. and Altman D.G.). British Medical Journal, London.

MRC (1948). Streptomycin treatment of pulmonary tuberculosis. *British Medical Journal,* **2**, 769–82.

Newcombe, R.G. (1992). Confidence intervals: enlightening or mystifying. *British Medical Journal,* **304**, 381–2.

Newnham, J.P., Evans, S.F., Con, A.M., Stanley, F.J., Landau, L.I. (1993). Effects of frequent ultrasound during pregnancy: a randomized controlled trial. *Lancet,* **342**, 887–91.

Norris, D.E, Skilbeck, C.E., Hayward, A.E., and Torpy, D.M. (1985). *Microcomputers in Clinical Practice,* John Wiley and Sons, Chichester.

OPCS (1991). *Mortality statistics, Series DH2, No 16,* HMSO, London.

OPCS (1992). *Mortality statistics, Series DH1, No 24,* HMSO, London.

OPCS (1992b). *Mortality statistics, Series DH1, No 25,* HMSO, London.

Osborn, J.F. (1979). *Statistical Exercises in Medical Research,* Blackwell, Oxford.

Oldham, H.G., Bevan, M.M., McDermott, M. (1979). Comparison of the new miniature Wright peak flow meter with the standard Wright peak flow meter.. *Thorax,* **34**, 807–8.

Paraskevaides, E.C., Pennington, G.W., Naik, S., and Gibbs, A.A. (1991). Prefreeze/post-freeze semen motility ratio. *Lancet,* **337**, 366–7.

Pearson, E.S. and Hartley, H.O. (1970). *Biometrika Tables for Statisticians, volume 1,* Cambridge University Press, Cambridge.

Pearson, E.S. and Hartley, H.O. (1972). *Biometrika Tables for Statisticians, volume 2,* Cambridge University Press, Cambridge.

Pocock, S.J. (1983). *Clinical Trials: A Practical Approach,* John Wiley and Sons, Chichester.

Pocock, S.J. and Hughes, M.D. (1990). Estimation issues in clinical trials and overviews. *Statistics in Medicine,* **9**, 657–71.

Pritchard, B.N.C, Dickinson, C.J., Alleyne, G.A.O, Hurst, P, Hill, I.D., Rosenheim, M.L., and Laurence, D.R. (1963). Report of a clinical trial from Medical Unit and MRC Statistical Unit, University College Hospital Medical School, London. *British Medical Journal,* **2**, 1226–7.

Radical Statistics Health Group (1976). *Whose Priorities?,* Radical Statistics, London.

Reader, R., *et al.* (1980). The Australian trial in mild hypertension: report by the management committee. *Lancet,* **i**, 1261–7.

Rose, G.A., Holland, W.W., and Crowley, E.A. (1964). A sphygmomanometer for epidemiologists. *Lancet,* **i**, 296–300.

Rodrigues, L. and Kirkwood, B.R. (1990). Case-control designs in the study of common diseases: updates on the demise of the rare disease assumption and the choice of sampling scheme for controls. *International Journal of Epidemiology,* **19**, 205–13.

Rowe, D. (1992). Mother and daughter aren't doing well. *The Guardian,* London, 14 July, p.33.

Royston, P., and Altman, D.G. (1994). Regression using fractional polynomials of continuous covariates: parsimonious parametric modelling. *Applied Statistics,* **43**, 429–467.

Samuels, P., Bussel, J.B., Braitman, L.E., Tomaski, A., Druzin, M.L., Mennuti, M.T., and Cines, D.B. (1990). Estimation of the risk of thrombocytopenia in the offspring of pregnant women with presumed immune thrombocytopenia purpura. *New England Journal of Medicine,* **323**, 229–35.

Schapira, K., McClelland, H.A, Griffiths, N.R., and Newell, D.J. (1970). Study on the effects of tablet colour in the treatment of anxiety states. *British Medical Journal,* **2**, 446–9.

Schmid, H. (1973). Kaposi's sarcoma in Tanzania: a statistical study of 220 cases. *Tropical Geographical Medicine,* **25**, 266–76.

Siegel, S. (1956). *Non-parametric Statistics for the Behavioural Sciences,* McGraw-Hill Kagakusha, Tokyo.

Sibbald, B., Addington Hall, J., Brenneman, D., Freeling, P. (1994). Telephone versus postal surveys of general practitioners. *British Journal of General Practice,* **44**, 297–300.

Snedecor, G.W. and Cochran, W.G. (1980). *Statistical Methods, 7th edn.,* Iowa State University Press, Ames, Iowa.

South-east London Screening Study Group (1977). A controlled trial of multiphasic screening in middle-age: results of the South-east London Screening Study. *International Journal of Epidemiology,* **6**, 357–63.

Southern, J.P., Smith, R.M.M, and Palmer, S.R. (1990). Bird attack on milk bottles: possible mode of transmission of *Campylobacter jejuni* to man. *Lancet,* **336**, 1425–7.

Stuart, A. (1955). A test for homogeneity of the marginal distributions in a two-way classification. *Biometrika,* **42**, 412.

'Student' (1908). The probable error of a mean. *Biometrika,* **6**, 1–24.

'Student' (1931). The Lanarkshire Milk Experiment. *Biometrika,* **23**, 398–406.

Thomas, P.R.S., Queraishy, M.S., Bowyer, R., Scott, R.A.P., Bland, J.M., Dormandy, J.A. (1993). Leucocyte count: a predictor of early femoropopliteal graft failure. *Cardiovascular Surgery,* **1**, 369–72.

Thompson, S.G. (1993). Controversies in meta-analysis: the case of the trials of serum cholesterol reduction. *Statistical methods in medical research,* **2**, 173–92.

Todd, G.F. (1972). *Statistics of Smoking in the United Kingdom, 6th ed.,* Tobacco Research Council, London.

Tukey, J.W. (1977). *Exploratory Data Analysis*, Addison-Wesley, New York.

Turnbull, P.J., Stimson, G.V., and Dolan, K.A. (1992). Prevalence of HIV infection among ex-prisoners. *British Medical Journal,* **304**, 90–1.

Victora, C.G. (1982). Statistical malpractice in drug promotion: a case-study from Brazil. *Social Science and Medicine,* **16**, 707–9.

White, P.T., Pharoah, C.A., Anderson, H.R., and Freeling, P. (1989). Improving the outcome of chronic asthma in general practice: a randomized controlled trial of small group education. *Journal of the Royal College of General Practitioners,* **39**, 182–6.

Whittington, C. (1977). Safety begins at home. *New Scientist,* **76**, 340–2.

Williams, E.I., Greenwell, J., and Groom, L.M. (1992). The care of people over 75 years old after discharge from hospital: an evaluation of timetabled visiting by Health Visitor Assistants. *Journal of Public Health Medicine,* **14**, 138–44.

Wroe, S.J., Sandercock, P., Bamford, J., Dennis, M., Slattery, J., and Warlow, C. (1992). Diurnal variation in incidence of stroke: Oxfordshire community stroke project. *British Medical Journal,* **304**, 155–7.

Index

abridged life table 297–8
absolute difference 268–9
absolute value 234
accepting null hypothesis 136–7
accidents 51–3
addition rule 87
admissions to hospital 85, 252, 351, 366–7
age 51–3, 264, 306–15, 369
age in life table *see* life table
age specific mortality rate 292, 297–7, 300, 304–5, 372
age standardized mortality rate 73, 293–4, 300
age standardized mortality ratio 295–6, 300, 304–5, 372
agreement 269–73
AIDS 57, 76–7, 166–9, 170, 176, 315–7
alcoholics 306–15
alcoholism 275
allocation to treatment 6–13, 15, 21
  alterations to 11–12, 21
  alternate 6–7, 10–11
  alternate dates 11–12
  cheating in 11
  minimization 12–13
  non-random 10–13, 21
  physical randomization 11–12
  random 7–10, 15
  systematic 10–11
  using envelopes 12
  using hospital number 11
alpha error 137
alternate allocation 6–7, 10–11
alternative hypothesis 133, 135–8
ambiguous questions 50
analgesics 15, 18–19
analysis of covariance 317
analysis of variance 170–6, 258–9, 316–17
  assumptions 171, 173–5
  balanced 316
  in estimation of measurement error 267–8
  Friedman 317
  Kruskal-Wallis 212, 259–60
  in meta-analysis 324
  multi-way 316–17
  one-way 170–6
  in regression 309–11
  two-way 316–17
  using ranks 212, 259–60, 317
angina pectoris 15–16, 134–5, 212–14
animal experiments 5, 16–17, 20–1, 33
anticoagulant therapy 11–12, 20, 139
antidiuretic hormone 192
antilogarithm 83
anxiety 18, 205
ARC 57, 170, 176
area under the curve 101–4, 107–9, 167–9
  probability 101–4, 107–9
  serial data 167–9
arithmetic mean 56
arterial oxygen tension 179
arthritis 15, 18
assessment 19–20
association 225–7
asthma 22, 262, 264, 328, 368–9
attack rate 300
attribute 46
average *see* mean
AVP 192
AZT 76–7, 166–9

back-transformation 164, 268
backwards regression 323
bar chart 73–4, 348–50
bar notation 57
Bartlett's test 169, 174–5
base of logarithm 81–3
baseline 77
baseline hazard 321
BASIC 105
Bayes' theorem 287
BCG vaccine 6–7, 10, 17, 33, 80
beta error 137, 333
between groups sum of squares 172
bias
  in allocation 11–12
  in assessment 19–20
  publication 323–4
  in question wording 40–2
  recall 39, 344, 358
  in reporting, 39, 41

response 18–19
in sampling 28, 31, 279
volunteer 31
biceps skinfold 162–4, 207–10, 335
bimodal distribution 53–4
binary variable *see* dichotomous variable
Binomial distribution 89–91, 94–5, 101, 104–7, 110, 125
mean and variance 94
and Normal distribution 90–1, 104–7
probability 89–90
in sign test 134–5, 243
biological variation 266
birds 44–5, 250, 344
birth rate 300, 302
birthweight 147
blind assessment 19–20
blocks 9
blood pressure 19, 27, 186, 265–6
Bonferroni method 146–7
box and whisker plot 56–7, 65–6, 345, 354
boxers 250
boxes 92–3
breast feeding 149
breathlessness 73–4
British Standards Institution 267
bronchitis 127–9, 142, 227, 229

Campylobacter jejuni 43–5, 250, 344
C-T scanner 5–6, 67
caesarian section 25, 343
calculation error 69
calibration 189
cancer 32–9, 41, 67–9
breast 37
lung 32, 34–9, 67–9, 237–8, 296
oesophagus 73–4, 77–8
parathyroid
registry 38
cards 8, 11–12, 49
carry-over effect 15
case control study 36–40, 44–5, 149–51, 236, 320, 344, 357–8
case fatality rate 300
cataracts 263–4, 369
categorical data 46–7, 369
cats 344
cause of death 69–73
cause of disease 33
cell of table 225
censored observations 279–80, 306, 320–2
census 26–7
decennial 26, 291
hospital 27, 47, 85
local 26–7
national 26
years 297, 291
centile 55–6, 114, 276–9
central limit theorem 105–7
chart
bar *see* bar chart
pie *see* pie chart
cheating in allocation 11
Chi-squared distribution 114–17, 227–8
contingency tables 227–8, 245–7
degrees of freedom 115–16, 227
logistic regression 319
and sample variance 116, 130
table 228
chi-squared test 225–31, 234–5, 238–41, 245–7, 256–7, 259–60
contingency table 225–31, 245–7, 256–7, 259–60, 366–9
continuity correction 234, 256–7, 260
degrees of freedom 227, 246–7
goodness of fit 244–5
logrank test 286
sample size 337
trend 238–41, 256–7, 259–60
validity 230–1, 234–5, 241
children *see* school children
choice of statistical method 254–264
cholesterol 53–4, 323
cigarette smoking *see* smoking
cirrhosis 294–6, 304, 315
class interval 48–51
class interval 48–51
class variable 315–16
clinical trials 5–20, 32, 323–6
allocation to treatment 6–15
assessment 19–20
combining results from 323–6
cross-over 15–16
double blind 19–20
ethics 19, 23
intention to treat 14–15, 343, 368
placebo effect 18–20
randomized 7–10
sample size
selection of subjects size 16–18
sequential 23
cluster sampling 30–1
Cochran 230
coefficient of variation 267–8
coefficients in regression 184, 186–8, 308–10, 318–20, 321–2
Cox 321–2
logistic 318–20
multiple 308–10
simple linear 184, 186–8
coeliac disease 162–4, 208, 335
cohort study 35–6, 344
cohort, hypothetical life table 296
coins 7–8, 27, 86–93

colds 68, 241–3
combinations 96
combining data from different studies 323–6
common cold *see* colds
common estimate 323–6
common odds ratio 325
common proportion 141–3
common variance 159–61, 171
comparison
  of means 126–7, 139–41, 159–61, 170–6, 334–7, 341, 356, 375
  of methods of measurement 269–73
  multiple *see* multiple comparisons
  of proportions 127–9, 141–3, 227, 229, 241–3, 256–7, 337–8, 341, 368, 374–5
  of regression lines 203, 363–4
  of two groups 126–9, 139–43, 159–61, 206–12, 254, 255–7, 334–8, 341, 356, 368, 374–5
  of variances 169, 258
  within one group 155–9, 212–15, 241–3, 254, 257–8, 336–7
compliance 179, 224, 358–62, 364–5
computer
  diagnosis 286–8
  random number generation 8–9, 105
  statistical analysis 3, 166, 170, 197, 306, 308, 316–17
conception 138
conditional test 245
confidence interval 123–9
  correlation coefficient 196
  difference between two means 126–7, 132, 159–61, 356
  difference between two proportions 127–8, 243
  difference between two regression coefficients 203, 363–4
  hazard ratio 321–2
  mean 123–5, 132, 155–6, 331–2, 356
  odds ratio 236, 238
  or significance test 139, 141, 222
  percentile 277–9
  predicted value in regression 190
  proportion 125–6, 332
  quantile, 277–9
  ratio of two proportions 128–9
  reference interval 277–9, 290, 371, 374
  regression coefficient 187
  regression estimate 188
  sample size 331–2
  SMR 296, 304, 372
  survival probability 282
  transformed data 164
  using rank order 212, 215
confidence limits 124–5
constraint 116, 246–7
contingency table 225
continuity correction 220–1, 234, 243
  chi-squared test 234
  Kendall's rank correlation coefficient 221
  Mann Whitney U test 230
  McNemar's test 243
continuous variable 46, 48–9, 75, 86–7, 92, 101–4, 320–1
  in diagnostic test 274–5
contrast sensitivity 264, 369
control group
  case control study 37–9, 344, 357
  clinical trial 5–7
controlled trial *see* clinical trial
cornflakes 149–51, 357–8
coronary artery disease 34, 145, 323
coronary thrombosis 11–12, 36
correlation 192–9, 215, 258–9
  assumptions 195–6
  between repeated measurements 336–7
  coefficient 192–4
  confidence interval 196
  Fisher's z transformation 195–6, 339–40
  intra-class 269
  linear relationship 194
  matrix 197
  negative 193
  positive 193
  product moment 193
  r 193–4
  r^2 195
  rank *see* rank correlation
  and regression 195–6, 309
  repeated observations 197–9
  sample size 338–40
  significance test 195–7
  table of 197
  table of sample size 340
  zero 193–4
cough 34–5, 41, 125–9, 140–2, 227, 229, 238–41, 250, 318–20
counselling 41
covariance analysis 317
Cox regression 320–2
Crohn's disease 149–51, 162–4, 208, 335, 357–8
cross-classification 225, 366–7
cross-over trial 15–16, 133, 336
cross-sectional study 34
cross-tabulation 225, 366–7
crude death rate 292
crude mortality rate 292, 300

cumulative frequency 47, 49–50, 55
cumulative survival probability 281, 297
cushion volume 329–30, 373
cut-off point 274–5, 280

death 26, 69–73, 95, 99–100, 279
death certificate 26, 291
death rate *see* mortality rate
decennial census 26, 291
decimal places 69, 266
decimal point 69
decimal system 68–9
decision tree 286–8
degrees of freedom 59, 116
  analysis of variance 171–2
  Chi-squared distribution 116
  chi-squared test 227
  F distribution 117
  F test 169
  goodness of fit test 245–6
  logistic regression 319
  logrank test 280
  regression 187, 309, 311
  sample size calculations 331
  t distribution 116, 153–4
  t method 153–4, 159
  variance estimate 59–63, 66, 94, 346–7
delivery 25, 225, 343
demography 296
denominator 67–8
dependent variable 182
depressive symptoms 18
Derbyshire 125
detection, below limit of 280
deviance 318–9
deviation
  from assumptions 157–9, 161, 165–6, 173–5, 191–2
  from mean 58–9, 346
  from regression line 182
diabetes 131–2, 355–6
diagnosis 47, 85, 273–6, 315
diagnostic test 132, 273–6, 355–6
diagrams 72–81, 85
  bar *see* bar chart
  pie *see* pie chart
  scatter *see* scatter diagram
diarrhoea 170, 316–17
dice 8–9, 86–7, 89–90, 119
dichotomous variable 255–60, 306, 315, 317–20, 322, 324
difference against mean plot 158, 179, 270–2, 359–60, 362
differences 134–5, 155–9, 179, 212–14, 270–2, 336, 359–62, 364–5
differences between two groups 126–8, 132, 139–143 159–61, 206–12, 254–7, 334–8, 356
digit preference 266
direct standardization 292–4
discharge 47–8
discrete data 46, 48
discriminant analysis 287
distribution
  Binomial *see* Binomial distribution
  Chi-squared *see* Chi-squared distribution
  cumulative frequency *see* cumulative frequency distribution
  F *see* F distribution
  frequency *see* frequency distribution
  Normal *see* Normal distribution
  Poisson *see* Poisson distribution
  probability *see* probability distribution
  Rectangular *see* Rectangular distribution
  t *see* t distribution
  Uniform *see* Uniform distribution
distribution-free methods 205, 222
diurnal variation 244–5
doctors 14, 36, 67, 85, 294–6, 350
Doppler ultrasound 147
double blind 19–20
double placebo 18
drug 68
dummy variables 315–16, 318–20, 324–5
Duncan's multiple range test 175

e 82–3
election 28, 31, 41
electoral roll 30, 32
embryos 329, 373
enumeration district 26
envelopes 12
enzyme concentration 341, 375
epidemiological studies 33–9, 323
equality, line of 270–1
error
  alpha 137
  beta 137, 333
  calculation 69
  first kind 137
  measurement 266–9
  second kind 137, 333
  term in regression model 182, 187
  type I 137
  type II 137, 333
estimate 59, 119–32, 323
estimation 119–32, 331–2
ethical approval 32
ethics 4, 19, 23

ex-prisoners 126
expectation 91–2
  of a distribution 91–2
  of Binomial distribution 94
  of Chi-squared distribution 116
  of life 100, 297–9, 300, 302–3, 352–3
  of sum of squares 63–4, 96–8, 116
expected frequency 225–6, 245–6
expected number of deaths 294–5
expected value *see* expectation, expected frequency
experimental unit 22–3
experiments 5
  animal 5, 16–17, 20–1, 33
  clinical *see* clinical trials
  design of 5, 25
  factorial 23
  laboratory 5, 16, 20–1
expert system 286–8
external assessor 20

F distribution 115, 117, 330
F test 169, 172, 309, 311, 316, 330, 373
factor 315–16
factorial 89, 96, 172
factorial experiment 23
false negative 274–5
false positive 274–5
family of distributions 89
Farr 1
fat absorption 76–7, 167
fatality rate 300
fertility 138, 302–3
fertility rate 300
FEV1 48–53, 55–6, 60–1, 122–5, 180–1, 183–6, 276–7, 309, 331–2
fever tree 33–4
Fisher 2
Fisher's exact test 231–5, 247–8, 256–7, 259–60
Fisher's z transformation 195–6 339–40
five figure summary 56
five year survival rate 281
follow-up, lost to 280
forced expiratory volume *see* FEV1
forward regression 322
fourths 55
fractured rib 275
frequency 47, 67–8, 225
  cumulative 47–50
  density 51–3, 101–3
  distribution 47–54, 65–6, 101–3, 345–6, 348
  expected 225–6
  per unit 51–3
  polygon 50
  and probability 86, 101–3
  proportion 67
  relative 47–9, 51–3, 101–3
  tally system 49, 53
  in tables 72, 225

G.P. 41
Gabriel's test 176
gallstones 282–6, 321–2
Galton 181–2
gastric pH 263, 368
Gaussian distribution *see* Normal distribution
gee whiz graph 78–9
geometric mean 164–5
geriatric admissions 85, 252, 351, 366–7
gestational age 192
glucose 65–6, 118, 345–7, 354–5
glue sniffing *see* volatile substance abuse
goodness of fit test 244–5
Gossett *see* Student
gradient 180–1
graft failure 327
graphs 72–81, 85
group comparison *see* comparisons
grouping of data 165

hazard ratio 321–2
health 40
health centre 216
healthy population 276, 289, 320
heart transplants 261
heatwave 85, 252, 351, 366–7
height 75, 77, 86–7, 92–4, 111, 155, 180–1, 183–96, 203–4, 306–15, 363–4
heteroscedasticity 173
Hill 2
histogram 50–4, 66, 72, 101–3, 300–1, 346, 348, 350, 354
historical controls 6
HIV 57, 126, 170, 176
holes 93–4
homogeneity of odds ratios 325
homogeneity of variance *see* uniform variance
homoscedasticity 173
hospital admissions 85, 252, 351, 366–7
hospital census 27, 47, 85
hospital controls 38
house dust mite 262, 368
housing tenure 225, 315
Huff 79–80
hypercholesterolaemia 53–4
hypertension 90, 262–3, 368
hypothesis, alternative *see* alternative hypothesis
hypothesis, null *see* null hypothesis

ICD 69–71
ileostomy 262, 368
incidence rate 300
independent events 87, 352
independent groups 126–9, 139–43, 159–61, 170–6, 206–12
independent random variables 93–4
independent trials 89
independent variable in regression 182
India 17, 33
indirect standardization 292–6
infant mortality rate 300
instrumental delivery 25, 343
intention to treat 14–15, 343
inter-pupil distance 327
inter-quartile range 58
interaction 308, 311–12, 324–6, 330, 373
intercept 180–1
International Classification of Disease 69–71
interval estimate 123
interval scale 205, 212, 255–260, 369
interval, class 48
intra-class correlation coefficient

Kaplan-Meier survival curve 282
Kaposi's sarcoma 68 215–17
Kendall's rank correlation coefficient 217–21, 241, 258–60, 368–9
    continuity correction 221
    in contingency tables 241
    τ 217
    table 220
    tau 217
    ties 218
Kendall's test for two groups 211
Kent 241–3
Know Your Midwife trial 25, 342–3
knowledge based system 287–8
Korotkov sounds 265
Kruskal-Wallis test 212, 259–60

laboratory experiment 5, 16, 20–1
lactulose 170, 173–6
Lanarkshire milk experiment 12
laparoscopy 138
large sample 124, 126–7, 139–43, 166, 256–8, 331–2
least squares 182–5, 199–200, 308
left censored data 280
Levene test 169
life expectancy 100, 297–300, 302–3, 352–3
life table 99–100, 280–1, 292, 296–9
limits of agreement 271–2
line graph 76–8, 348, 351
line of equality 270–1
linear constraint 116, 246–7
linear regression *see* regression, multiple regression
linear relationship 180–204, 238–41
linear trend in contingency table 238–41
Literary Digest 31
lithotrypsy 282
log *see* logarithm, logarithmic
log hazard 321
log odds 235, 248–9, 317–20
log odds ratio 236, 238, 248–9, 320
logarithm 81–3, 128–9
logarithm of proportion 248
logarithm of ratio 128–9
logarithmic scale 80–1
logarithmic transformation 112, 161–5, 174
    and coefficient of variation 268
    and confidence interval 164
    geometric mean 164–5
    to equal variance 162, 164, 192, 268
    to Normal distribution 112, 161–4, 179, 355, 360–2, 368
    base of 81–3
    variance of 129, 248
logistic regression 287, 317–20, 322, 325–6
logit transformation 235, 248–9, 317–20
Lognormal distribution 82, 112
logrank test 282–6, 322
longitudinal study 35–6
loss to follow-up 280
Louis 1
lung cancer 34–5, 67–9, 95, 237–8, 296
lung function *see* FEV1, PEFR, mean transit time, vital capacity

magnesium 131–2, 289–90, 355–6
malaria 33–4
Mann-Whitney U test 161, 206–12, 220–1, 256, 259
    continuity correction 220–1
    Normal approximation 210–11, 220–1
    table 208
    tables of 211–12
    ties 208–9, 211
    and two sample t method 206, 209–11
mannitol 57, 170, 172–3, 316–17
Mantel's method for survival data 286
Mantel-Haenszel
    method for combining two by two tables 325
    method for trend 241
marginal totals 225–6
matched samples 241–3, 155–9, 212–14, 257–8, 336–7, 359–63, 364–5
matching 38–9, 44–5

maternal age 264, 369
maternal mortality rate 300
maternity care 25
mathematics 2–3
matrix 308
maximum 56, 65, 167, 345
maximum voluntary contraction 306–15
McNemar's test 241–3, 257–8
mean transit time 263, 368
mean 56–8, 65–6
    arithmetic 56
    comparison of two 126–7, 139–41, 159–61, 334–7, 356
    confidence interval for 123–5, 132, 155–9, 356
    geometric 164–5
    of population 123–4, 331–2
    of probability distribution 91–4, 102–4
    of a sample 56–8, 65–6, 346–7
    sample size 331–2, 334–7
    sampling distribution of 119–20, 124–5
    standard error of 123–5, 132, 152, 356
measurement 265–6
measurement error 266–9
measurement methods 269–73
median 54–8, 214, 345
Medical Research Council 9
meta-analysis 323–6
methods of measurement 269–73
mice 21–2, 329, 373
midwives 25, 342–3
mild hypertension 262–368
milk 12, 43–5, 250, 344
mini Wright peak flow meter *see* peak flow meter
minimization 12–13
minimum 56, 65, 345
misleading graphs 77–80
missing denominator 68
missing zero 77–8
mites 262, 368
mode 53–4
modulus 234
mortality 15, 36, 69–73, 85, 341, 350, 374–5
mortality rate 36, 291–4
    age specific 292, 296–7
    age standardized 293–4
    crude 292
    infant 300
    neonatal 300
    perinatal 300
mosquitos 33–4
MTT *see* mean transit time
multifactorial methods 306–30
multiple comparisons 175–6
multiple regression 306–16, 329–30
    analysis of variance for 309–11
    and analysis of variance 316–17
    assumptions 308, 314–15
    class variable 315–16
    coefficients 308, 310
    computer programs 306, 308, 316
    correlated predictor variables 310–11
    degrees of freedom 309, 311
    dichotomous predictor 315
    dummy variables 315–16
    F test 309, 311, 316
    factor 315–16
    interaction 308, 311–12, 329–30, 373
    least squares 308
    linear 308, 313
    in meta-analysis 324
    non-linear 308, 313–15, 373
    Normal assumption 314
    outcome variable 306
    polynomial 313–15
    predictor variable 306, 310–11, 315–16
    quadratic term 313, 315, 373
    qualitative predictors 315–16
    R^2 309–10
    reference class 316
    residual variance 309
    residuals 314–15, 329–30, 373
    significance tests 309–11
    standard errors 310
    stepwise 322–3
    sum of squares 309, 311, 373
    t tests 309–11, 316
    transformations 314
    uniform variance 314
    variance ratio 309
    variation explained 309–10
multiple significance tests 144–7, 167
multiplicative rule 87, 89
multiway analysis of variance 316–17
muscle strength 306–15
mutually exclusive events 87, 89, 351–2
myocardial infarction 341, 374–5

Napier 82
natural history 32
natural logarithm 82
natural scale 80–1
negative predictive value 275
neonatal mortality rate 300
New York 6–7, 10
Newman-Keuls test 175
nitrite 263, 368
nominal scale 205, 255–60
non-parametric methods 205, 222

non-significant 145
none detectable 280
Normal curve 104, 107–8
normal delivery 25, 343
Normal distribution 90, 104–18
  in confidence intervals 123–5, 256–7, 259, 368
  in correlation 195–7
  derived distributions 114–17
  independence of sample mean and variance 116
  as limit 104–7
  and normal range 276–7
  of observations 111–14, 152, 205, 255–60, 354–5
  and reference interval 276–7, 289, 370
  in regression 182, 314
  in significance tests 139–43, 256–7, 259, 368
  standard error of sample standard deviation 130
  in t method 152–4
  tables 108–9
Normal plot 112–14, 118, 157–8, 354–5
Normal probability paper 112
normal range *see* reference interval
null hypothesis 133, 135–8

observational studies 5, 26–45
observed and expected frequencies 225–6
occupation 95
odds 235, 317–20
odds ratio 236–8, 248–9, 256–7, 320, 325–6
oesophogeal cancer 73–4, 77–8
Office of Population Censuses and Surveys 291
on treatment analysis 15
one sided percentage point 108–9
one sided test 137–8, 233
one tailed test 137–8, 233
opinion poll 31, 41, 314, 374
ordered nominal scale 255–60
ordinal scale 205, 214, 255–60, 369
outcome variable 182, 185, 306, 317
outliers 57, 191, 373
overview 323–6

P value 1, 136–7
$p_a(O_2)$ 179
pain 15–16
paired data 134–5, 155–9, 166, 212–15, 241–3, 257–8, 336–7, 359–62, 364–5, 368
  McNemar's test *see* McNemar's test
  sample size 336–7
  sign test *see* sign test
  t method *see* t methods
  Wilcoxon *see* test Wilcoxon test
parameter 89
parametric methods 205–222
parathyroid cancer 280–2
parents 318–20
parity 48, 85, 244
passive smoking 34–5
peak expiratory flow rate *see* PEFR
peak flow meter 156, 158, 267, 270–3
peak value 167
Pearson's correlation coefficient *see* correlation coefficient
PEFR 127, 140–1, 144, 156, 158–61, 186, 203–4, 262, 267–9, 270–3, 363–4, 368
percentage 67, 72
percentage point 108–9, 341, 374
percentile 55, 276–9
perinatal mortality rate 300
permutation 96
pH 263, 368
phlegm 141, 144
phosphomycin 68
physical mixing 11–12
pictogram 78
pie chart 72–3, 79, 348–9
pie diagram *see* pie chart
pilot study 331, 335, 337
Pitman's test 258
placebo 18–20
point estimate 123
Poisson distribution 94–5, 105–7, 162, 244, 245–6, 248, 295–6
poliomyelitis 13–15, 20, 67, 85, 349
polygon *see* frequency polygon
polynomial regression 313–15
population 27, 33, 36, 39, 86, 331–2
  census 26–7, 291
  estimate 291
  mean 123–4, 331–2
  national 26, 291
  projection 299
  pyramid 300–3
  restricted 33
  standard deviation 121–2
  statistical usage 27
  variance 121–2
positive predictive value 275
power 143–5, 332–5, 340
precision 265–6
predictor variable 182, 185, 306, 310–11, 315–16, 320–2, 323
pregnancy 25, 48, 342–3
presenting data 67–85
presenting tables 71–2
prevalence 35, 89, 275–6, 300

prevention 33
probability 86–118
  addition rule 87
  density function 103–4
  distribution 87–8, 91–4, 101–3, 352–3
  of dying 99–100, 296–7, 351–2
  multiplication rule 87
  paper 112
  in significance tests 133–6
  of survival 99–100, 351–2
  that null hypothesis is true 136
product moment correlation coefficient *see* correlation coefficient
pronethalol 15–16, 133–6, 212–14
proportion 67–8, 72, 317–20
  confidence interval for 125–6, 332
  denominator 68
  difference between two 127–8, 141–3, 241–3, 337–8
  as outcome variable 317–20
  ratio of two 128–9
  sample size 332, 337–8
  standard error 125–6, 332
  in tables 72
  of variability explained 187, 195
proportional frequency 47
proportional hazards model 320–2
prospective study 35
protocol 265
pseudo-random 8
publication bias 323–4
pulmonary tuberculosis *see* tuberculosis
pyramid 300–3

q-q plot *see* quantile-quantile plot
quadratic term 313, 315, 373
qualitative data 46, 255–60, 315–16
quantile 54–6, 113–14, 276–9
quantile-quantile plot 113–14
  confidence interval 278–9
quantitative data 46, 48
quartile 55–6, 65, 345
quasi-random sampling 31
questionnaires 36, 40–2
quota sampling 28

r 193–4
r^2 195
r_S 215
R^2 309–10
radiological appearance 20
random allocation 7–10, 15, 17, 20–1, 25
random blood glucose 65–6
random numbers 8, 10, 29, 30
random sampling 9, 28–30, 38, 89
random variable 86–118
  addition of a constant 92–3
  difference between two 93–4
  expected value of 91–4
  mean of 91–4
  multiplied by a constant 93
  sum of two 93
  variance of 91–4
randomization 7–10, 15, 17, 20–1, 25
randomizing devices 8, 86, 89
range 58, 276
  inter-quartile 58
  normal *see* reference interval
  reference *see* reference interval
rank 206
rank correlation 215–21, 258–60, 368–9
  choice of 220, 258–60
  Kendall's 217–21, 258–60
  Spearman's 215–17, 258–60
rank order 206
rank sum test 206–214
  one sample *see* Wilcoxon
  two sample *see* Mann Whitney
rate 67–8, 72
  age specific mortality 292, 296–7, 300
  age standardized mortality 293–4, 300
  attack 300
  birth 300, 302
  case fatality 300
  crude mortality 292, 300
  denominator 67–8
  fertility 300
  five year survival 281
  incidence 300
  infant mortality 300
  maternal mortality 300
  mortality 291–4
  multiplier 67, 292
  neonatal mortality 300
  perinatal mortality 300
  prevalence 300
  response 31
  stillbirth 300
  survival 281
ratio
  odds *see* odds ratio
  of proportions 128–9
  scale 255
rats 21
raw data 164
recall bias 39, 344, 358
reciprocal transformation 162–4
Rectangular distribution 105–6
reference class 316, 319
reference interval 32, 131–2, 276–9, 289–90, 355–6
  confidence interval 277–9, 289–90, 355–6

by direct estimation 278–9
sample size 341, 374
using Normal distribution 276–7, 289–90, 355–6
using transformation 277
refusing treatment 13–15
register of deaths 26
regression 180–92, 199–204, 258–9, 306
analysis of variance for 309
assumptions 182
backward 322–2
coefficient 184, 186–8
comparing two lines 203, 363–4
confidence interval 187
in contingency table 233–41
and correlation 195–6
Cox 320–2
dependent variable 182
deviations from 182
deviations from assumptions 191–2
equation 184
error term 182, 187
estimate 188–9
explanatory variable 182
forward 322–3
gradient 180–1
independent variable 182
intercept 180–1
least squares 182–5, 199–200
line 182
linear 184
logistic 317–20
multiple *see* multiple regression
outcome variable 182, 185
outliers 191
perpendicular distance from line 183
polynomial *see* polynomial regression
prediction 188–90
predictor variable 182, 185
proportional hazards 320–2
residual sum of squares 187
residual variance 187
residuals 190–1
significance test 187–8
simple linear 184
slope 180–1
standard error 186–8, 190, 210–13
stepwise 322–3
sum of products 184
sum of squares about 187, 309
sum of squares due to 187
towards the mean 181–2, 186
variability explained 187, 195
variance about line 187, 200–1
X on Y 185–6
rejecting null hypothesis 136–7
relationship between variables 34, 73–7, 180–204, 215–21, 225–35, 238–41, 254, 258–60, 306–30
relative frequency 47, 51–3, 101–3
relative risk 129, 236–8, 320
reliability 269
repeatability 32, 266–9
repeated observations 166–9, 197–9
representative sample 27–8, 31, 34
reproducibility
residual mean square 172, 267
residual standard deviation 187, 267
residual sum of squares 172, 309, 311
residual variance 171, 309
residuals 162–4, 173–4
about regression line 190–1, 314–15
plots of 162–4, 173–4, 191, 314–15, 329–30, 373
within groups 162, 173–4
respiratory disease 31, 36
respiratory symptoms 31, 34–5, 41, 125–9, 140–2, 227, 229, 238–43, 250
response bias 18–19
response rate 31
retrospective study 39
risk 129
risk factor 38, 323–4, 344
RND(X) 105
robustness to deviations from assumptions 165–6

s^2 59
saline 13–14
Salk vaccine 13–15, 17, 19–20, 67, 349
sample 86
large 124, 126–7, 166, 256–8, 331–2
mean *see* mean
size *see* size of sample
small 126, 152–66, 222, 256–8
variance *see* variance
sampling 27–35
in clinical studies 32–3, 289, 370
cluster 30
distribution 119–20, 124–5
in epidemiological studies 33–5
experiment 62–3, 119–22
frame 29
multi-stage 30
quasi-random 30–1
quota 28
random 28–30
simple random 29–30
stratified 30
systematic 30–1
scanner 5–6
scatter diagram 75–6, 181–1
scattergram *see* scatter diagram

school children 12–14, 17, 22, 30–1, 34–5, 41, 125–9, 140–2, 238–44, 250, 318–20
schools 22, 30–1, 34
screening 15, 22, 80, 262–3, 273–6
selection of subjects 16–18, 31, 37–9
in case control studies 37–9
in clinical trials 16–18
self 31
self selection 31
semen analysis 178
semi-parametric 322
sensitivity 273–6
sequential trials 23
serial measurements 166–9
sex 70–1
sign test 134–5, 157, 205, 212, 224, 243, 257–8, 364–5, 368–9
signed rank test *see* Wilcoxon
significance and importance 138–9
significance and publication 323–4
significance level 137, 143
significance tests 133–51
multiple 144–7
and sample size 332
in subsets 145–6
significant difference 136
significant digits *see* significant figures
significant figures 68–71, 265–6
size of sample 31, 143, 331–41
accuracy of estimation 340
correlation coefficient 338–40
and estimation 331–2
paired samples 336–7
reference interval 341, 374
and significance tests 332–4
single mean 331–2
single proportion 332, 374
two means 334–7, 375
two proportions 337–8, 374–5
skew distribution 54, 57, 66, 114, 162, 165–6, 355
skinfold thickness 162–4, 207–10, 335
slope 180–1
small samples 152–66, 222, 257–8
smoking 22, 31, 33–9, 41, 67, 73–4, 237–41, 318–20, 350
SMR 295–6, 304–5, 372
Snow 1
somites 329–30, 373
South East London Screening Study 15
Spearman's rank correlation coefficient 215–17, 257–8, 368–9
table 217
ties 217
specificity 273–6
square root transformation 162–4, 174
squares, sum of *see* sum of squares
standard age specific mortality rates 294–5
standard deviation 58, 60–1, 65–6
degrees of freedom for 61–3, 66, 116
of differences 152–5
of population 120–1
of probability distribution 92–4, 103–4
of sample 60–1, 65–6, 116, 347
of sampling distribution 120–1
and standard error 123
standard error of 130
within subjects 267–8
standard error 120–3
correlation coefficient 195–6, 339
difference between two means 126–7, 132, 334, 337, 356
difference between two proportions 127–8, 141–3, 243, 337
difference between two regression coefficients 203
different in significance test and confidence interval 143
log hazard ratio 321
log odds ratio 236, 248–9
logistic regression coefficient 320
mean 120–3, 132, 331–2, 356
normal range
percentile
predicted value in regression 190
proportion 125–6, 332, 374
quantile 277
ratio of two proportions 128
reference interval 277, 370–1, 374
regression coefficient 186–8, 310, 316, 320–1
regression estimate 188–9
SMR 295–6, 372
standard deviation 130
survival rate 281, 337
Standard Normal deviate 112–14, 220–1
Standard Normal distribution 107–16, 139, 152–4, 333–4
standard population 293
standardized mortality rate 73, 293–4
standardized mortality ratio 295–6, 304–5
statistic 1, 46
test 136, 333
vital 300
statistical significance *see* significance test
stem and leaf plot 51–3, 55, 65, 179, 345, 359–61
step function 50, 281–2
step-down 322–3
step-up 322–3
stepwise regression 322–3
stillbirth rate 300

stratification 30
strength 306–15
strength of evidence 133, 136–7, 358
streptomycin 10–11, 17, 19–20, 80, 230–1
stroke 5–6, 244–5
Stuart test 243, 257–8
Student 12, 155
Student's t distribution *see* t distribution
Studentized range 175
subsets 145–6
success 89–90
suicide 303
sum of products about mean 184, 192–3
sum of squares 58–9, 63–6, 116, 171–2, 309, 311
  about mean 58–9, 63–6, 116, 346–7
  about regression 187, 309, 311
  due to regression 187, 309, 311
  expected value of 63–4, 96–8, 116
summary statistics 324
summation 57
survey 28, 89
survival 10, 279–86, 320–2
  analysis 279–86, 320–2
  curve 281–2, 284, 320
  probability 281–2, 285–6
  rate 281
  time 162, 279–86
symmetrical distribution 53–4, 57

t distribution 115–16, 152–5
  degrees of freedom 116, 153–4, 156–7
  and Normal distribution 116, 152–4
  shape of 153
  table 154
t methods 114, 152–66
  assumptions 157–9, 161, 179, 359–62
  confidence intervals 156–7, 161
  deviations from assumptions 157–9, 161, 165–6
  difference between means in matched sample 155–9, 179, 257.360–1
  difference between means in two samples 159–61, 256, 259
  one sample 155–9, 179, 257, 360–1, 365
  paired 155–9, 166, 179, 212, 214, 257, 360–1, 365, 368–9
  regression coefficient 187–8, 309–11, 316
  single mean 155–9, 176
  two sample 159–61, 256, 259, 315
  unpaired *same as* two sample 159–61
table of probability distribution
  Chi-squared 228
  correlation coefficient 197
  Kendall's τ 220
  Mann Whitney U 208
  Normal 108–9
  Spearman's ρ 217
  t 154
  Wilcoxon matched pairs 213
table sample size for correlation coefficient 339
tables of random numbers 8, 10, 29–30
tables, presentation of 71–2
tables, two way 225–41
tails of distributions 54, 355–6
tally system 49, 53
Tanzania 68, 215–17
TB *see* tuberculosis
telephone survey 41
temperature 10, 69, 85, 205, 253, 328
test statistic 136, 333
test, diagnostic 132, 273–6, 355–6
three dimensional effect in graphs 79
ties in rank tests 208–9, 211–14,
ties in sign test 134
time 166, 320
time series 76–7, 166–9, 348, 351
time to peak 167
time, survival *see* survival time
total sum of squares 171
transformations 161–6
  and confidence intervals 164
  Fisher's z 195–6, 339–40
  logarithmic 112, 114, 174, 179, 359–62, 365
  logit 235, 248–9, 368
  to Normal distribution 112, 114, 161–4, 166, 174
  reciprocal 162–4
  and significant figures 266
  square root 162–4, 174
  to uniform variance 162–4, 166, 174, 192, 268
treated group 5–7
treatment 5–7, 323–4
trend in contingency tables 238–41
  chi-squared test 238–41
  Kendall's τ_b 241
  Mantel-Haenzsel 241
triglyceride 55–7, 61, 111–12, 277–9
trisomy-16 329–30, 373
true difference 143
true negative 275
true positive 275
tuberculosis 6–7, 10, 17, 80–1
Tukey 51, 56
Tukey's Honestly Significant Difference 175
tumour growth 21
two sample t test *see* t methods
two sample trial 16
two-sided percentage point 108–9

two-sided test 137–8
two-tailed test 137–8
type I error 137
type II error 137, 333

Uniform distribution 105–6, 245
uniform variance 159, 162, 164–6, 173–5, 182, 314
unimodal distribution 53
unit of analysis 22
urinary infection 68
urinary nitrite 263, 368

vaccine 6–7, 10, 13–15, 17, 19, 57
validity of chi-squared test 230–1, 234–5, 241
variability 58–63, 266
variability explained by regression 187, 195
variable 46
    categorical 46
    continuous 46, 48
    dependent 182
    dichotomous 255–60
    discrete 46, 48
    explanatory 182
    independent 182
    nominal 205, 255–60
    ordinal 205, 214, 255–60
    outcome 182, 185, 306, 317
    predictor 182, 185
    qualitative 46
    quantitative 46
    random *see* random variable
variance 58–63, 66
    about regression line 187, 200–1
    analysis of *see* analysis of variance
    common 159–60, 171
    comparison in paired data 258
    comparison of several 169
    comparison of two 169, 258
    degrees of freedom for 59, 61–3, 123, 345–6
    estimate 58–63, 123
    of logarithm 129, 248
    population 123
    of probability distribution 91–4, 103–4
    of random variable 91–4
    ratio 117, 309
    residual 187, 200–1, 314
    sample 58–63, 66, 94, 96–8, 116, 345–6
    uniform 159, 162, 164–6, 173–5, 182, 314
    within subjects 267–8
variation, coefficient of 267–8
visual acuity 263–4, 369
vital capacity 75, 77
vital statistics 300
vitamin A 325–6
vitamin D 113–14
volatile substance abuse 305–5, 372–3
volunteer bias 6, 13–14, 31
volunteers 5, 16–17
VSA *see* volatile substance abuse

Wandsworth Health District 85, 252, 351
weight gain 20–1
wheeze 264
whooping cough 263, 368
Wilcoxon test 212–14, 257–8, 365, 369
    matched pairs 212–14, 257–8, 365, 369
    one sample 212–14, 257–8, 365, 369
    signed rank 212–14, 257–8, 365, 369
    table 213
    ties 212–14
    two sample 211
withdrawn from follow-up 280
within group residuals *see* residuals
within groups sum of squares 171
within groups variance 171
within subjects variation 266–8
Woolf's test 325
Wright peak flow meter *see* peak flow meter

$\bar{x}$ 57
X-ray 19–20, 80, 275

Yates' correction 234–5, 243, 256–7, 260

z test 139–43, 256–7, 259
z transformation 195–6, 339–40
zero, missing 77–9
zidovudine 76–7, 166–9

% symbol 72
! (factorial) 96
| (absolute value) 234
α 137
β 137
χ^2 115–16
μ 92
ϕ 107
Φ 108
ρ 215–17
Σ 57
σ^2 92
τ 217